THE HOME-SCALE FOREST GARDEN

How to Plan, Plant, and Tend a Resilient Edible Landscape

DANI BAKER

Chelsea Green Publishing
White River Junction, Vermont
London, UK

Unless otherwise noted, all photographs by Dani Baker.

Project Manager: Alexander Bullett
Editor: Fern Marshall Bradley
Assistant Editor: Natalie Wallace
Copy Editor: Laura Jorstad
Proofreader: Angela Boyle
Indexer: Shana Milkie
Designer: Melissa Jacobson

Printed in the United States of America.
First printing May 2022.
10 9 8 7 6 5 4 3 2 23 24 25 26 27

Our Commitment to Green Publishing
Chelsea Green sees publishing as a tool for cultural change and ecological stewardship. We strive to align our book manufacturing practices with our editorial mission and to reduce the impact of our business enterprise in the environment. We print our books using vegetable-based inks whenever possible. This book may cost slightly more because it was printed on paper from responsibly managed forests, and we hope you'll agree that it's worth it. *The Home-Scale Forest Garden* was printed on paper supplied by Versa that is certified by the Forest Stewardship Council.®

Library of Congress Cataloging-in-Publication Data
Names: Baker, Dani, 1949– author.
Title: The home-scale forest garden: how to plan, plant, and tend a resilient edible landscape / Dani Baker.
Other titles: How to plan, plant, and tend a resilient edible landscape
Description: First edition. | White River Junction, Vermont: Chelsea Green Publishing, 2022.
| Includes bibliographical references and index.
Identifiers: LCCN 2022003452 | ISBN 9781645020981 (paperback)
Subjects: LCSH: Woodland gardening. | Organic gardening.
Classification: LCC SB439.6 .B35 2022 | DDC 635/.0484—dc23/eng/20220214
LC record available at https://lccn.loc.gov/2022003452

Chelsea Green Publishing
White River Junction, Vermont, USA
London, UK
www.chelseagreen.com

PRAISE FOR *THE HOME-SCALE FOREST GARDEN*

"*The Home-Scale Forest Garden* is a thoroughly enjoyable read, with lots of good photos and illustrations. I particularly enjoyed reading about forest gardening in a colder winter climate and found Dani's strategies for dealing with wet flooded ground, very heavy soil, deer attacks, and many other challenges both fascinating and inspirational. This book should be of great use to anybody making a forest garden on any scale."

—MARTIN CRAWFORD,
author of *Creating a Forest Garden*; founder, Agroforestry Research Trust

"Dani Baker enriches cold-climate forest gardening with candid details of successes and (importantly) failures in her decade-old forest garden. *The Home-Scale Forest Garden* serves as a guide to anyone who wishes to plant one, and includes valuable experience with challenges including some very wet soils. Featuring over 200 beautiful color photographs from the garden."

—ERIC TOENSMEIER,
author of *Perennial Vegetables*; coauthor of *Edible Forest Gardens*

"This book brings to life the visual beauty and the diverse productivity a perennial landscape can offer to anyone looking to get started or improve on what they've created. Forest gardening can often feel intimidating, but Dani Baker's unique experience coupled with her friendly tone and plenty of details on the best plants, placement, and companions provides anyone interested in building a food forest with a plethora of material to help them succeed."

—STEVE GABRIEL,
coauthor of *Farming the Woods*; author of *Silvopasture*; extension specialist, Cornell Small Farms Program

"Dani Baker generously shares the gems of her forest garden journey, offering insights of her successes, mishaps, and inspirations that culminate in the delicious fruits of her labor. Her vision, supported by a depth of teachers and study, guides readers in learning the art of edible abundance. Whether your goals include greater harvests, landscape beauty, or diversity of species, I highly recommend *The Home-Scale Forest Garden* as a resource for the land steward in us all."

—KATRINA BLAIR,
author of *The Wild Wisdom of Weeds*

"We need many more examples of forest gardens, food forests, and permaculture orchards, and *The Home-Scale Forest Garden* will help inspire and direct you on the adventure of planning, planting, and tending your garden abundance. Dani Baker does a wonderful job introducing the uninitiated to permaculture principles in a clearly understandable way. Once you experience *abundance* from your forest garden, your life changes. A forest garden, as Dani explains so simply and beautifully, will get you to that *abundance*."

—STEFAN SOBKOWIAK,
permaculture educator and YouTuber; owner, Miracle Farms: The Permaculture Orchard

"Working with the natural world in our gardens has never been more important with the challenges we now face through climate change. Forest gardening offers hope, inspiration, and solutions for the future, and Dani Baker shows gardeners how to embrace this important method on a smaller scale. Build resilience and create a low maintenance edible haven with this accessible guide."

—KIM STODDART, editor, *The Organic Way* magazine; coauthor of *The Climate Change Garden*

*To Steve Gabriel, Martin Crawford, David Jacke,
Sepp Holzer, Lee Reich, Eric Toensmeier,
Jonathan Bates, Toby Hemenway, and
Stefan Sobkowiak, whose works inspired me
to create the Enchanted Edible Forest.*

CONTENTS

Part 1: Planning and Planting Your Garden

Part 2: Plants for Every Level

Part 3: Creating Plant Groupings

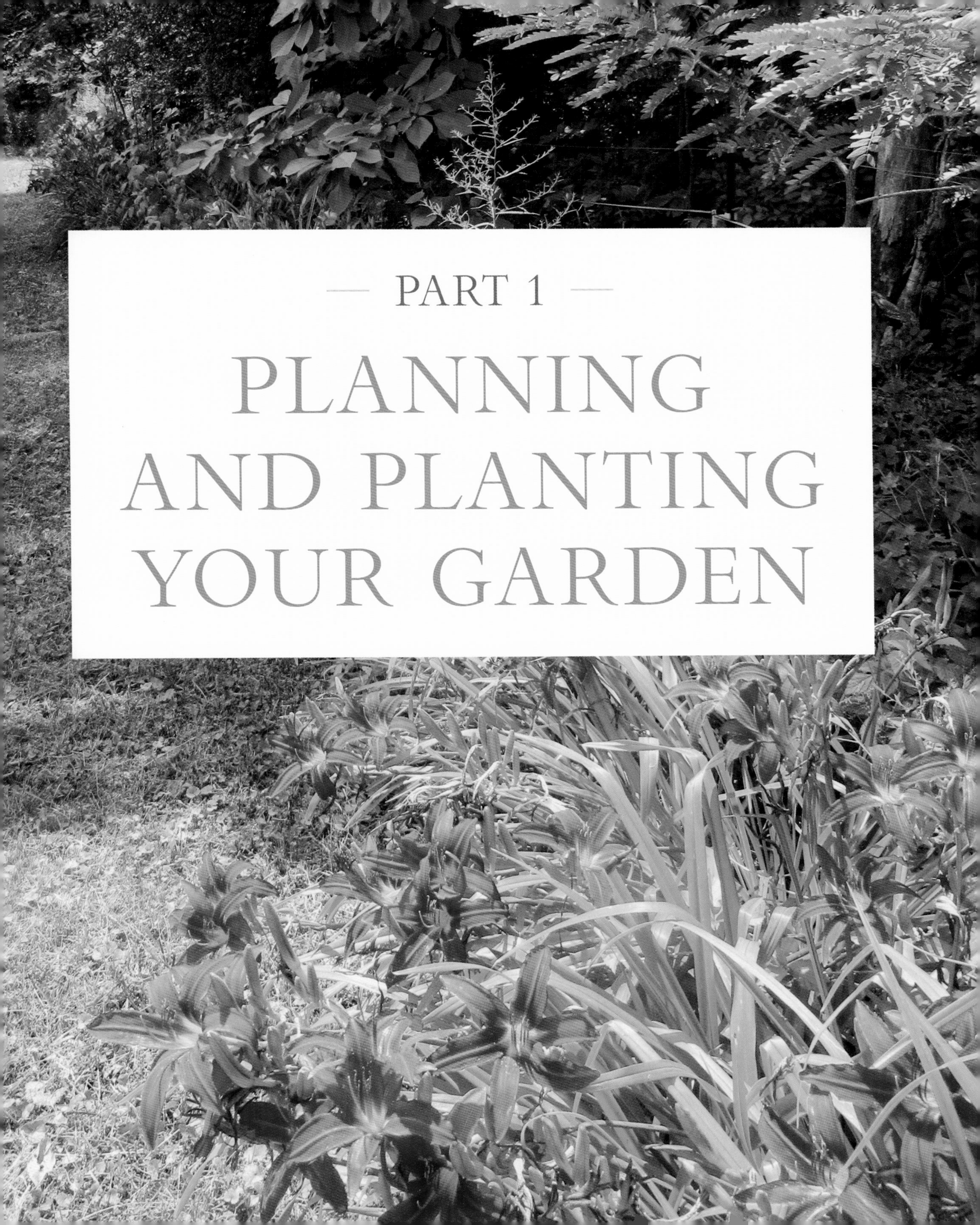

— PART 1 —

PLANNING AND PLANTING YOUR GARDEN

Figure 1.1. Summer's lush foliage beckons the visitor to enter the Enchanted Edible Forest.

CHAPTER 1

STARTING A FOREST GARDEN ADVENTURE

What is a forest garden, and why would you consider developing one instead of planting a traditional vegetable garden, a row of berry bushes, or an orchard? How does a forest garden differ from these other forms of gardening? The simplest explanation is that a forest garden is modeled after nature. In nature, no one weeds, waters, adds nutrients, sprays for pests and disease, or applies mulch. Nature takes care of all of these plant needs with no human intervention. How does she do it?

First, in nature, the ground is completely covered naturally with some form of organic matter. Think of a mountain meadow replete with wildflowers or the cushy floor of a dense woods blanketed with a deep layer of fallen leaves or pine needles. In both of these environments, unless there is a disturbance like a fire or a storm that uproots a tree, there are no patches of exposed soil.

When there is a disturbance that bares the ground, nature provides a series of healing remedies, a natural succession of plants that, with time, prepare the soil for more permanent plants to become established. The succession begins with annual "weeds." Then deep-rooted, broad-leaved plants sprout and grow, and their long roots pull up essential nutrients from the subsoil. As these first plants die back in the colder months, nutrient-rich organic matter is built up on and in the soil. In addition, plants that have a special relationship with the bacteria that transform or fix atmospheric nitrogen into natural forms of nitrogen fertilizer move in, and they make this important nutrient available to all the surrounding plants. Once the soil conditions are suitable, a variety of woody shrubs and then taller trees also take up residence in their preferred niches. Beneficial insects, birds, and animals find homes in the leaf litter on the ground, in piles of fallen branches, and amid the leafy canopy. These creatures provide natural pest control for the plants as well as additional fertilizer.

Since the ground is covered with a thick layer of plants or natural mulch, it stays cool and moisture is preserved. When it rains, the water readily sinks into the mulch and soil beneath. As the overhead canopy of tree and shrub foliage expands, more dew is collected on cool nights, both absorbed by the leaves and filtered to the ground. The increasing shade also contributes to keeping the ground cooler and conserving moisture. Each year as autumn leaves fall, the natural mulch is replenished and nutrients are recycled.

Thus, in a forest garden, nature serves all the roles or functions filled by humans in a conventional system of agriculture. Why wouldn't you want the same attributes in your garden, especially since nature is doing most of the work? Plus, most of the

Figure 1.2. As autumn approaches, the garden captivates with copious blooms and a touch of fall color.

plants in a forest garden are perennial, meaning they are planted once and live for many years. The annual labor that growing vegetables entails—planting seeds or sets, cultivating, mulching, watering—is vastly reduced.

Important, too, is the fact that a forest garden can serve as a storehouse of carbon. Thus, in a small way it acts as a buffer to global warming caused by the rising levels of atmospheric carbon dioxide and other greenhouse gases. As the years go by and the ground in a forest garden remains undisturbed while the canopy matures, more and more carbon is sequestered in the trunks, branches, roots, and leaves of the growing plants as well as in the earth in the form of organic matter that has been transformed into humus, a stable form of soil carbon. The ability of humus to attract and hold moisture and nutrients increases the resiliency of the garden in the face of extreme and extended weather events such as torrential rains and more frequent and sustained droughts. The diversity of edible plants in a forest garden, from berry bushes and vines to nut trees to herbs, edible flowers, and perennial vegetables, also provides a hedge against the vagaries of weather. Even when some plants have meager yields due to unfavorable conditions, others may well produce bumper crops.

A forest garden does more than just maintain positive ecological conditions—over time, it improves them. As more carbon is sequestered and more humus is created, the garden becomes *regenerative,* bettering its ecological state rather than merely sustaining it.

The process I've described does not unfold in a single year or even a few years. Once planted, a forest garden evolves with time in ever-changing and often unpredictable ways. But you don't have to wait an eternity to reap results. You will begin to harvest some herbs, perennial vegetables, and berries from the first year. Three to five years after

Figure 1.3. The winter landscapes' silence and stillness evoke a sense of peace and serenity.

an initial planting, a forest garden begins to pop, taking on a life of its own and increasingly meeting its own needs while providing an increasing abundance of food for the gardener.

Moreover, something fantastical happens when you create a planting that emulates nature. Imagine the way you feel wandering a woodland path—an ethereal ambience pervades this kind of landscape. Visitors to my garden often spontaneously describe it as "magical," and more often than not they don't want to leave. I often lose track of time myself as I wander in my forest garden, returning to my house to find two hours have gone by! You, too, will experience this phenomenon when you embark on your own forest garden adventure.

Your forest garden needn't be a huge undertaking. You can start with a limited area, such as a foundation bed along one side of your house where you install edible bushes and ground covers instead of plants that are merely ornamental. Plant one fruit tree surrounded with other edible plants in your front or backyard; or develop an edible hedge to screen your house from your neighbor's. The options are limited only by your imagination. Throughout this book, you'll find descriptions of plants and plant groupings from small- to large-scale to stimulate ideas for your own plot.

My Edible Forest Garden Story

I want to share with you how I came to plant my Enchanted Edible Forest. When I was approaching retirement from a career as a psychologist in the early 2000s, I was concerned about how I would fill all the idle time without the structure of a formal workday. If I owned some land, I thought, I could dabble in landscaping, maybe keep a couple of horses.

My partner, David Belding, and I talked it over, and we decided to go for it. We live in a rural part of northern New York State, and at that time sizable tracts of land were both available and affordable. (Nowadays it's a different story.) I called a real estate agent who'd helped me buy my first house in the area several years before and told her I wanted to buy some land. She asked how large a parcel I was considering. "A hundred acres seems like a nice round number," I replied. The property we eventually found was just that big: 102 acres (41 ha) on Wellesley Island in the Thousand Islands region of the St. Lawrence River, a stone's throw from Canada.

I fell in love with the land the minute I saw it. From my youth I have been very attached to the landscape of the foothills of the Adirondack Mountains, where my family spent their summers. The first time we visited the Wellesley Island property, I ran up one of the rocky ridges. On top, the bare rock was partially covered by moss and lichens and surrounded by woods with lots of white pines, exactly like the terrain I remembered from my childhood. In fact, the Thousand Islands region has the same geological underpinning as the Adirondacks, with similar fauna and flora. We made an offer on the property on the spot.

Wellesley Island comprises a little over 13 square miles (34 sq k) of rocky ridges, mixed hardwood forests, and wetlands, interspersed with some patches of higher ground suitable for pasture or growing crops. Our land mirrors that of the island as a whole. Because of the rocky underlay, much of the island has thin soils. Where the soil is deeper, it is primarily heavy clay.

Figure 1.4. Spring's vibrant hues kindle feelings of hope and joy.

Our property was a former dairy farm, established in the early 1800s. There were so many dairy farms on the island that in its heyday, Wellesley Island sported its own cheese factory. In the middle of the 20th century, most of the dairies went out of business. The previous owner of our property, who had died at some time before we purchased the land, had continued cutting hay as well as raising vegetables to sell to campers from the state park down the road.

We spent the first year exploring our new land. I had never lived on a property more than ½ acre (0.2 ha) in size. It took me months to wrap my head around one so large. In fact, I sometimes lost my way wandering in the woods, or among the fields and wetlands. As we explored the property, Dave and I discovered remnants of the farm's past: an old fence post here, a discarded farm implement there.

We did not buy the land with the intention of farming it. That fall, however, we attended a course offered by our local cooperative extension called Building Your Small Farm Dream. Dave had had a childhood dream of operating an organic farm. In his imagination it would be 200 acres (81 ha) and he would be self-sufficient: He likes to say he would go to town only once a year whether he needed to or not. I, on the other hand, grew up tending our family's ½-acre lot after we moved to a brand-new home built on a former cornfield. At first, the lot was an expanse of sod with no landscaping. I was given the tasks of digging holes for trees, creating beds for annual flowers and vegetables, and then planting and caring for all the gardens. They say if you do as an adult what you enjoyed doing when you were 12 years old, you will be happy. And I enjoyed all aspects of gardening at that age. Later, as an adult, I planted a small vegetable garden whenever I had some land available to work with. The Small Farm Dream course guided Dave and me to explore our personal interests and skills, as well as the resources available on the property. As we discovered the interests we shared and the abundant resources that our property offered, we were so inspired that we decided to try our hand at farming.

The following spring, we prepared a small plot, planted a vegetable garden, became certified as an organic farm, and sold almost everything we grew. The year after that, we doubled the size of the vegetable garden and added our first livestock—two pigs, Oscar and Mayer. Meanwhile, my fantasy of owning two horses evaporated when, despite over a year's worth of riding lessons, I could not overcome my fear of falling off. At the same time, I realized that farming was the answer to my concern about having too much idle time. When you are a farmer, you don't have any!

As the years went by, we increased the vegetable operation to ¾ acre (0.3 ha) and acquired more livestock: chickens, ducks, goats, beef cows, and more pigs. I am the vegetable and fruit gardener on the farm, and Dave is the animal person. Dave likes to tell people, "If it grows in the dirt, Dani cares for it; if it poops on the dirt, David takes care of it."

In 2012, the seventh year of our farm's evolution, I must have had the seven-year itch. I attended a workshop on permaculture at our local cooperative extension office. The instructor was Dr. Steve Gabriel from Cornell University. I had never heard the term *permaculture* before, but the ideas that Steve presented—harnessing the full power of nature to provide sustenance, manage pests and disease, and create self-sufficiency and resilience in a garden—were like a revelation. They made so much sense that I decided there and then to plant my own edible forest garden. I came home from the workshop and told Dave of my goal, and he was very supportive. I explained that first of all, I was going to need a fence to demarcate the garden's boundaries and—equally if not more important—to keep out the deer and any hungry goats or other farm animals. Dave scouted out a site. He was not willing to give up any of the pastureland because it is a scarce resource on our farm. He settled on a 100-by-200-foot (30 × 61 m) area (roughly ½ acre) about 50 feet (15 m) from the road that runs along the northeast side of our land. It was a site we weren't using for any particular purpose, and it was covered with scrubby brush and a few trees. Dave graciously built the fence for me—a heavy-duty, pounded-post, high-tensile electric wire fence.

Contending with clearing the brush was quite a project, and we made use of some of our farm animals (as I describe in chapter 3) as well as some heavy equipment. Once the brush was cleared, I could observe the land and learn about the variety of habitats that the land afforded. Hopefully you will be able to start with a clear site, but just as in nature, any site, no matter how challenging, can become fruitful with thoughtful planning.

You may be thinking that ½ acre is much bigger than what you envision for your edible garden, and that's okay. Even if you are contemplating growing a few herbs or berries beside your foundation or a fruit tree or two in your yard, this book will guide you to create a planting that is not only fruitful but self-sustaining and aesthetically pleasing. If your plot is small, the advice I share about grouping plants and using all the vertical space available will inspire you to design a plan that provides more food than you thought possible in your limited space. As figure 2.3 on page 14 demonstrates, a forest garden can be as small as 25 square feet (2.3 sq m)! Whatever the scale of your forest garden project, you'll find that the lessons I share in this book will be useful as you undertake planning and planting.

I wanted to create a large forest garden because my vision included hosting parties, events, and possibly even weddings there. That meant designing a gathering area on a scale that could accommodate such events. As a result, the garden design included

The Goat Wedding

Once Dave had built the fence demarcating the plot that was to become our edible forest, we decided to have a little fun before we started the project in earnest. One of our farm customers wanted her female goat bred (stud service is occasionally part of our goat farming operation), and so we decided to have a "Goat Wedding." Our pastor agreed to officiate the ceremony. It was early July, the perfect time for an outdoor wedding and for starting a gardening adventure.

The wedding guests were Dave and myself, plus our four very hardworking volunteers who were participating in the Worldwide Opportunities on Organic Farms (WWOOF) program that summer. (The WWOOF program allows people interested in experiencing life on an organic farm to donate their labor in exchange for room and board.) I don't think our volunteers could have predicted that they would be attending a ceremony like this one, though!

One of the volunteers adorned the horns of the soon-to-be-wedded goats with garlands of gloriosa daisies. Our pastor blessed the bride and groom, and prayed that they would be fruitful. Then the newlyweds promenaded off into the edible forest for their honeymoon. The local TV news station even featured the story on the evening broadcast with the headline "Goats Get Married, No Kidding."

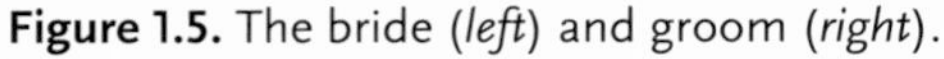

Figure 1.5. The bride (*left*) and groom (*right*).

hardscaping that involved use of heavy machinery. I learned some valuable lessons from that experience, which I share later in the book.

This brings me to the importance of planning when it comes to a perennial landscape. Yes, an annual vegetable garden involves planning, too: You need to prepare the ground, order your seeds, and decide what you will plant where. But come the end of the growing season, you clean up the garden, and you can start a fresh plan the following year. In a forest garden, your perennial plants may live for years, even decades. So your choices have long-lasting consequences. Plus, the start-up costs may be much greater, considering the cost of woody shrubs and trees compared with vegetable seeds or a six-pack of tomato transplants, along with the cost of landscape features that may require contracted labor and equipment. In order to maximize your chances of success, careful observation of your land is an essential first step. This will include

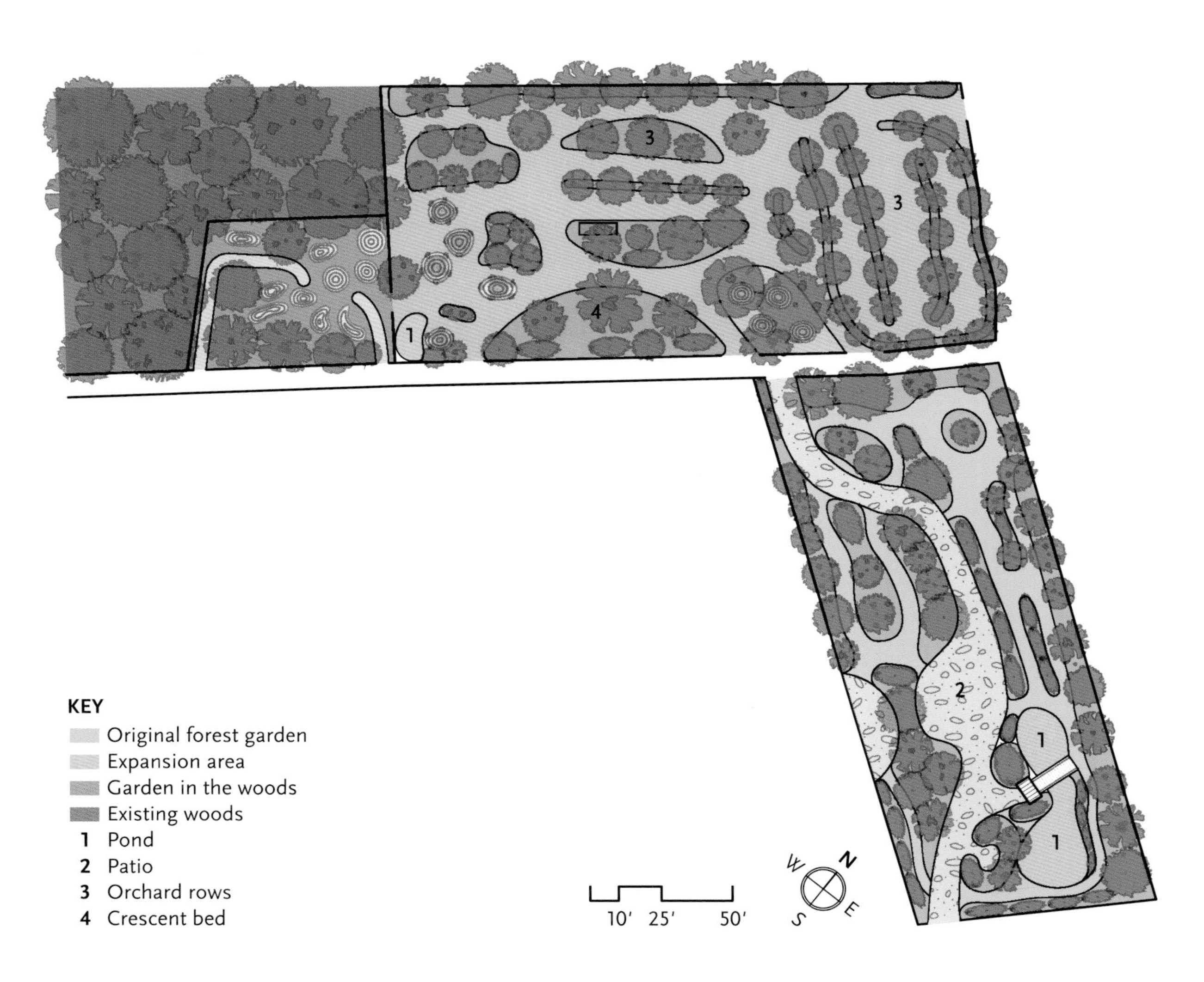

Figure 1.6. This overhead view of my entire forest garden in its 10th year of development shows the overall layout of the original ½ acre (0.2 ha), the ½-acre expansion, and the garden in the woods. Illustration by Turner Andrasz.

determining the nature of your local climate and your soil. You'll want to consider the type and placement of any hardscaping or infrastructure you want to include in the garden, as well as how you will maintain access to all parts of your garden as plants grow to maturity. Another part of planning is learning about the nature of the plants you want to include so you can place them in the locations that best suit their needs and at the appropriate spacing so they will have room to stretch their limbs to full height and width over time. In chapter 3, I delve into these aspects of planning in more detail, as well as explaining how to capture your vision for your garden on a map of your plot drawn to scale.

With so many considerations that benefit from careful planning, it is important not to rush. After I attended the workshop on permaculture, I wanted to start planting right away, but I soon realized how much planning needed to take place first. For me, observing the land and planning the landscape and initial planting took almost a whole year.

In the process of planning and planting the Enchanted Edible Forest, I immersed myself in literature describing permaculture principles as well as a diverse range of edible plants. (I've included many of the books I consulted in the selected bibliography on page 313.) In my experience, a single reading of any useful text is insufficient. I reread many of the guides in that list on an annual basis as I continue to develop my garden. Each reading reveals gems of information that I was not ready to absorb during previous years. For example, it wasn't until the sixth year of my garden that Martin Crawford's discussion of cover crops in his remarkable book *Creating a Forest Garden*, particularly those that were good for partially shaded pathways, became relevant to me. I subsequently planted some low-growing ryegrass along with Dutch white clover to provide ground cover in such a pathway.

Figure 1.7. A hügelkultur mound built in existing woods.

Three years after planting that ½-acre garden, I decided to expand. The second phase of the garden covers an adjacent ½ acre, which I designed to be constructed primarily with hand labor. I wanted it to be a model for those who want to do it themselves without a large capital investment in hardscape materials or hired labor.

At that time, I was beginning to give presentations about my edible forest garden to gardening groups and at conferences, and many people who attended expected to learn how to grow edibles in a wooded area where there is considerable shade. In response to this expectation, I also decided to expand my garden into the adjacent woods, a dense stand of ash trees with a scattered understory of brush and some herbaceous plants.

There is standing water beneath the trees during most of the spring and fall and even during a wet summer. It was not feasible to plant directly into the ground due to the high water table and root competition from the existing brush and trees. As a solution, I decided to build a series of special raised planting areas called *hügelkultur mounds* under the trees (I explain lots more about hügelkultur in chapter 4). This third part of my edible forest, the

What about Tomatoes?

You may be wondering: What about annual vegetables like tomatoes? Is there a place for them in an edible forest? I still grow annual vegetables, but in market garden beds, which are closer to my house than my forest garden. This is not to say that I have never grown annual vegetables in my forest garden. For example, during the initial establishment phase, annual vegetables served well as ground cover until perennial plants became established. And I installed annual edibles on hügelkultur mounds during their first year to ensure that living roots were stimulating the soil life in the mounds while I made decisions about which perennials I would eventually plant in them.

One unexpected result of becoming a forest gardener is that it helped me become a better vegetable gardener. I realized that some of the permaculture principles that make a forest garden sustainable and efficient can also help reduce the labor input needed for annual vegetables. For example, letting native nitrogen-fixing legumes such as clovers grow among annual vegetables minimizes the need to add nitrogen-rich amendments. I've tucked in suggestions about how to apply permaculture practices in a vegetable garden throughout this book, too—because planting a forest garden doesn't mean you have to give up growing tomatoes!

Many people set up a dedicated plot in their forest gardens where they grow annual vegetables, or they integrate some annual vegetables into their perennial plantings. I have not done this because annual vegetables like salad greens and some culinary herbs often require daily attention, and for me, the closer to the house the vegetable crops are, the better. In fact, there is a permaculture principle prescribing that the plots visited most often be located closest to the house while the least-visited areas, like wild woods, be located the farthest. The forest garden falls somewhere in the middle of this continuum: While it may not require daily attention, it cannot be ignored for days on end, either. Since my forest garden is at some distance from the house, it does not lend itself to intensive visitation. If your forest garden will be close to your home, it may well make sense to integrate an annual vegetable plot within it.

Another reason to separate annual vegetables is that in a forest garden, the goal is to leave the soil undisturbed to the greatest extent possible. In an annual vegetable plot, many tasks such as digging furrows for seeds, planting sets, and cultivating to reduce weed competition violate this precept.

As a forest garden matures, the amount of direct sunlight reaching the ground surface diminishes, but many annual vegetables do best in full sun. Leafy greens do benefit from some shade in hot weather, and it is possible to grow self-seeding annuals like lettuce and radishes in a forest garden. If your plans include a patch that will continue to receive at least eight hours of direct sun as your edible forest matures, or if your design includes one or more sunny hügelkultur mounds dedicated to growing annual vegetables, you may well be able to have your forest garden and eat tomatoes, too.

garden in the woods, is still a project in process as I write this manuscript.

Developing a forest garden turned out to be an amazing adventure. When I first began it, I was unacquainted with most species of edible herbaceous and ground cover plants. I had not seen or heard of many of the small fruits and berries I decided to grow. And my research led me to some familiar (and some completely unfamiliar) fruit and nut trees to include in the design. It was exciting to learn about all the edible plants I could potentially grow in my northern climate. For example, before I began planning the garden, I had never heard of honeyberries, but they have become one of my favorite fruits. Envisioning the garden space, figuring out how to arrange plants in groupings, and caring for the first plantings were all labors of love. Then observing the garden grow, bloom, flourish, and change in unanticipated ways felt immeasurably rewarding.

About This Book

I was motivated to write this book by numerous friends and visitors to the Enchanted Edible Forest who remarked on their wish to have a handbook for starting their own edible forest gardens. Most had tended gardens of annual vegetables that had to be weeded, fertilized, and watered year in and year out. The idea of a perennial edible garden that would grow to be self-sustainable was unique and appealing. As my edible forest has taken shape over the years, so did this book.

In the pages that follow, I provide detailed, practical information based on my experience to help you plan, develop, and reap the bounty from your perennial edible planting—large or small. Part 1 describes the overall process of planning and developing an edible forest garden, and is packed with suggestions about what to consider as you plan and establish your garden, including how to incorporate the permaculture principles that will make it resilient and self-sustaining and help to reduce the need for human labor to maintain the garden over time.

Part 2 includes detailed descriptions of plants for every level of a forest garden, from tall trees to ground covers. Learning about the growth habit, preferred habitat, uses, and aesthetic appeal of a broad range of attractive edible plants will help you decide what to include in every nook of your garden. I've included my personal experience with the plants, including tips to help you succeed and some precautionary tales of missteps to avoid.

In part 3, I focus on the art of combining plants in multilayered groupings. This is one of the most creative, interesting, and engaging aspects of forest gardening. I include descriptions, photos, illustrations, and planting plans for groupings of specific plants in different habitats as possible templates that you can work with and adapt to fit the features of your land. I also offer examples of how others have applied permaculture principles in their gardens to inspire you further to embark on your own edible adventure.

As I write this chapter, the initial ½ acre of the Enchanted Edible Forest has been growing for almost 10 years. In the spring, the garden bursts with blooms whose fragrances permeate the air. The harvest of herbs, berries, fruits, and vegetables continues from April through November. In the fall, the garden is replete with vibrant color from the ground cover to the treetops. In winter its silent, snow-covered presence embodies the promise of next year's abundance. As I recount this adventure with you, its successes and missteps, I hope you will become inspired to embark on your own adventure in edible landscaping, regardless of the size of your plot or the quality of your soil. If I could do it without much prior experience, you can do it, too.

CHAPTER 2

LETTING NATURE DO THE WORK

Bill Mollison, the originator of the term *permaculture*, defines the word in his classic text *Permaculture: A Designer's Manual*. He writes, "Permaculture (permanent agriculture) is the conscious design and maintenance of agriculturally productive ecosystems which have the diversity, stability, and resilience of natural ecosystems."

Because the principles of permaculture are derived from nature, the end result is resilient and regenerative, just as a forest, meadow, or other natural ecosystem is. Permaculture principles have a wide range of applications well beyond an edible forest setting. In fact, permaculturists rely on the basic principles to guide many kinds of decisions—from where they choose to build a toolshed or create a pathway to what types of heating system they install in their house. The concepts of permaculture originated in Australia, but now there are farmers, homesteaders, and homeowners all over the world who are applying its very practical ideas. Although I don't consider myself a permaculturist, I continue to learn about its principles and applications by reading books, attending workshops, and experimentally applying them in my own garden.

I choose to apply the principles I find most relevant to my edible forest, and these six principles are the ones I cover in this chapter: maximizing diversity, maximizing solar absorption, maximizing water conservation, designing for sustainability, building in redundancy, and minimizing human labor into the future. There are lots of ways to maximize water conservation, from growing ground covers to planting in layers for casting cooling shade. Designing for sustainability includes choosing plants to help deter and control pests rather than purchasing chemicals. And redundancy simply means making something foolproof by having backup systems. For example, in my garden there are several sources of water, including a pond, an artesian well, and water held in the plentiful organic matter in raised planting beds.

You'll also find that the first five principles all help with the sixth of reducing human labor going forward.

Maximize Diversity

In nature, there are few monocultures. Most natural landscapes feature a wide variety of plants, and this diversity offers many benefits. It creates niches where beneficial insects such as ladybugs and dragonflies and animals such as birds and toads can make their homes and find sustenance. In the face of unfavorable weather, biodiversity ensures that some species will likely survive or even thrive under the challenging circumstances. Likewise, as the climate continues to change due to global warming, a diverse flora will more likely include some species that can adapt effectively to the changes.

A varied planting makes it harder for pests or diseases to find their object. A monoculture, such as an orchard planted exclusively with apple trees, is like an open invitation to insects to enjoy a sumptuous buffet of their favorite food or to disease organisms to spread unfettered. In contrast, a lone apple tree surrounded by a variety of other trees as well as shrubs, flowers, and ground covers presents many obstacles and distractions to apple pests, making it difficult for them to find and lay eggs on young apple fruits. Disease may spread more slowly and be less devastating. In sum, landscape diversity ensures health and resilience.

I've implemented the principle of maximizing diversity in my garden in several ways, including alternating species, planting mixed hedges and ground covers, and designing to maximize the edge effect.

I alternate species throughout my garden. As much as possible, I avoid placing two plants of a particular species side by side. For example, in a bed containing fruit trees, I included an apple, a cherry, a mulberry, a peach, and a persimmon, all surrounded by an array of other kinds of plants. There are exceptions to this rule. I do plant berry bushes such as honeyberries in a group for ease of U-pick harvesting. And plum trees are an interesting example of a benefit of planting several of the same species together, as I describe in "The Plum Patch" on page 129.

In the United States, most traditional hedges or windbreaks are monocultures, such as a rose hedge or a cedar windbreak. Ground cover plantings tend to be single species, too—a big swath of pachysandra or periwinkle, for example. Whenever I design a hedge or windbreak, though, I make it as varied as possible.

I modeled my hedges in the Enchanted Edible Forest after the ancient hedgerows I observed in the rural parts of England where, I was told, you could guess the relative age of a hedgerow by the extent of diversity found in it. Over the long span of years, birds nesting in the dense greenery of a hedge excrete viable seeds of a wide range of plants. Some of those seeds germinate. It's one of nature's best planting methods, and it effectively increases a hedge's diversity over time. The hedgerows in England contain tall trees, bushes, brambles, and many other species. Following that model, I designed my hedges and windbreaks to include a diversity of trees, bushes, brambles, and ground covers. Whenever I want to add a tree to my garden but can't find a spot for it, my go-to solution is to plant it in a hedge.

In some situations, a uniform ground cover is inevitable because one species is so dominant it displaces others in the area, or because only one type of plant is suited to a challenging habitat. But apart from such circumstances, diversity is also desirable in a planting of ground covers. In most locations, having two or more ground covers interplanted or bunched in groups by species contributes to the overall diversity of the garden. (See part 3 for descriptions of some of my diverse ground cover plantings.)

The overall goal in designing a forest garden is to follow nature's model. In nature the greatest diversity is often found at an edge where two habitats meet, such as where a meadow transitions into the woods. A riverbank or lakeshore is also a natural edge. In an edible forest garden, the intent is to mimic the diversity of plants found in a *forest edge*, possibly the most diverse habitat of all in a temperate climate. At the edge of a natural woodland, there is abundant light, which encourages plant growth in many layers, from the treetops all the way down to the ground surface. This is quite different from the environment farther into the woods, where very little sun reaches the ground level and only shade-loving plants can survive. It's possible to plan a forest garden to allow plenty of patches where ample light reaches the ground so that a wide variety of plants can grow even though the trees and shrubs will cast some shade.

Figure 2.1. Maximizing diversity is desirable at all levels. Here a diverse ground cover in spring includes strawberries, wild blue lupine, perennial blue flax, and coneflower.

Figure 2.2. A wide, grassy path allows early-morning light to bathe all the branches of the cherry trees on the right.

Because incorporating edge into a design enhances diversity, I like to build lots of curves into my beds instead of straight lines. It's simple math—if a straight line is the shortest distance between two points, then a curving line must be a longer edge.

Maximize Solar Absorption

Edges also provide opportunities for light to illuminate the lower levels of plantings. To make sure plenty of light can reach into my forest garden, I designed it with wide, curving access routes around and between collections of plants. I also included patio areas with rounded edges to open up space for light to penetrate. Opening up a planting for more light to penetrate contributes to maximizing solar absorption. This is important for gardens in general because sunlight is the energy source for photosynthesis. Photosynthesis fuels the growth of plant tissues and also the microbiome—the community of fungi, bacteria, and a wide range of other organisms in the soil that all contribute to keeping soil healthy and functioning. And the warmth of the sun is especially important for the ripening of fruit, so the more light that penetrates into all layers of your forest garden, the more evenly and successfully your fruit will ripen.

In addition to creating abundant edges to allow light into the garden, you can maximize solar absorption by utilizing the entire vertical dimension. Typical home landscape plantings don't do this, but your goal should be to capture all of the available sunlight. To accomplish this, design your garden to take advantage of as much vertical space as your plot allows.

In a forest garden, vertical space can be divided into seven layers that overlap in an organic way.

- The **overstory** includes the tallest plants, those that grow to 30 feet (9 m) or taller. Black locust and walnut trees are examples of overstory plants.

- The **understory** contains trees that reach between 10 and 30 feet (3–9 m) tall. This includes most fruit trees and some nuts.
- The **shrub layer** ranges from 3 to 12 feet (1–3.5 m) above the soil surface. In this layer you'll find brambles, berry bushes, and other woody plants.
- The **herbaceous layer**, from 2 to 10 feet (60 cm–3 m) above ground level, contains perennial plants with succulent stems that die down to the ground as the weather cools in the fall. These plants then re-sprout each spring. Examples are most perennial vegetables, flowers, and some herbs.
- The **ground cover layer** is close to the soil surface. These are the clumping, running, suckering, and sometimes vining herbaceous plants that grow up to 2 feet tall and cover the soil. Examples include strawberries, violets, and clovers.
- The **root layer** includes plants that have edible parts growing in the vertical space below the soil surface. Edible licorice and sunchoke are examples.
- The **vine layer** can occupy all levels of vertical space. For instance, grapes and wisteria can climb as high as a tree is tall, if they have strong enough support. Other vines such as perennial sweet pea will self-limit their height to about 10 feet. All vines will trail over the ground if there is nothing nearby to climb up.
- Some permaculturists add an eighth layer, which consists of fungi. The **fungus layer** includes edible mushrooms, such as oyster and shiitake.

Figure 2.3. It's possible to incorporate all seven layers even in a small plot. Here six layers occupy no more than 25 square feet (2.3 sq m). Overstory: honey locust, top center. Understory: apricot, center. Shrub: clove currant, left; red currant, right. Herbaceous: Russian comfrey, foreground. Not visible are pineberries as a ground cover and daylilies (which have edible roots) in the root layer.

Depending on the size of your garden, aim to incorporate as many of these layers as you can. If your space is restricted to a small plot, your edible forest might be made up of shrub, herbaceous, and ground cover layers only, with no overstory or understory trees. In *Paradise Lot*, permaculturists

Eric Toensmeier and Jonathan Bates describe how they omitted the overstory when they planted a permaculture garden on a 1⁄10-acre (405 sq m) city lot. They focused their attention on fruit trees as the tallest layer and then wove in all the other layers beneath.

In some situations, layers can be added later in time as conditions permit. For instance, at first there might be no trees tall enough to serve as trellises for vines, or no shady spots in which to grow mushrooms. Wait until the overstory matures, and then add vines and mushrooms. I planted trees the first year of my garden's development, and then waited three or four years to add shade-adapted vines and ground covers beneath them. In another instance, several years after I planted understory fruit trees, I planted more trees nearby that would eventually surpass the fruit trees in height to occupy the overstory layer above them.

In addition to utilizing as many vertical layers as possible, you can maximize absorption of solar radiation by planting the tallest trees on the north side of a garden. This will minimize the amount of shade they cast on other garden plants. Reserve the sunniest spots in your garden for the plants that need the most sun, place those that can tolerate shade in the shadier spots, and incorporate windbreaks to slow or divert the wind and thus preserve heat. (For more about windbreaks, see "Solving Landscape Problems on page 38.)

Maximize Water Conservation

Plants require water in greater quantity than any other nutrient. With the changes in the Earth's climate due to global warming, drought is more frequent or intense, as are heavy downpours of rain that may run off rather than being absorbed by the soil. I have observed both of these phenomena in my region.

The practice of water conservation is becoming more and more important as we face these weather-related challenges. There are several practices that can be incorporated in an edible forest to capture and conserve water: designing foliage layers to maximize the leaf canopy, keeping all ground surface covered, matching plants to the appropriate habitat, capturing runoff with ponds and swales, and building hügelkultur mounds.

MAXIMIZE LEAF CANOPY

Here again, making use of all vertical space is important. Doing so maximizes water collection and retention in several ways. First, the more foliage is in the vertical plane, the more shade is cast below. Shade keeps the ground cool and minimizes evaporation.

Second, when the air is saturated with moisture, as during a foggy day, or when the air cools at night below the dew point, water condenses on the leaves of plants. The leaves can absorb that water directly—or, if the water becomes heavy enough, it drips off onto the ground and is absorbed by the earth. Filling the vertical space with layers of foliage results in more leaves, and thus more surfaces available for water to condense upon. One foggy day, as I neared an established hedgerow along the border of my property, I felt and heard rain falling from the trees above. But it was not raining. Rather, so much moisture from the saturated air had condensed on the tree canopy that a continual shower of water drops was falling like rain.

A third way that filling vertical space with foliage aids in water conservation is by creating a buffer that reduces the effective intensity of a heavy downpour that might otherwise end up as surface runoff. The abundant leaf and stem surfaces intercept many of the pounding raindrops. The rainwater then travels more gently across and down these surfaces to the soil level, and thus can be absorbed. Tree branches and trunks actually channel rainwater to the ground below.

One final benefit of filling vertical space is that the denser the canopy, the greater the quantity of

Figure 2.4. A many-layered canopy is functional and beautiful!

leaves that eventually drop off in autumn. This wealth of fallen leaves forms a natural mulch that keeps the ground warmer in winter and cooler in summer while also holding in moisture.

COVER THE GROUND

In places where the canopy is limited, another water conservation strategy is to cover the ground with plants or mulch. The thicker and denser the ground cover, the better it dampens the force of rain. This improves absorption and lessens the chance of erosion. Like a dense canopy of tree or shrub foliage, ground cover foliage can shade the ground, keeping the soil cool and reducing evaporation. When I first began planting my garden, I covered bare ground with copious quantities of wood chips until ground cover plants became established. When the ground surface is covered with thick mulch or well-established living plants, there is little need for supplemental watering.

MATCH PLANTS TO SITE

Another way to reduce the need to water and instead conserve this invaluable resource is to match plants to sites that naturally supply the amount of soil moisture they require. In my garden there are several variations in habitat when it comes to moisture. In some areas water is so abundant that the water table rises above the soil surface for several months of the year. But there are also several elevated slopes that drain effectively and become dry when rain is scarce. In still other spots the soil tends to stay evenly moist but not sodden. Matching water-loving plants with moist or wet habitats eliminates the need to supply water from another source, as does placing drought-tolerant plants in spots that tend to be dry.

INCORPORATE PONDS, SWALES, AND MOUNDS

When I was initially studying my land, I noticed that rivulets appeared in three spots during snowmelt and heavy downpours. It seemed natural to capture the water by digging ponds in the areas that the rivulets flowed into. These ponds serve to capture and store large quantities of rain and snowmelt. In the early spring when it's too cold to make use of the water line that we installed for watering the garden, I use buckets to scoop up the pond water to water newly installed bare-root plants. The ponds are also home to numerous beneficial creatures such as frogs, which eat mosquitoes and other pests and are a joy to observe.

Figure 2.5. Besides their useful functions, ponds add aesthetic appeal.

A swale is an excavated strip, oriented to run along the contour on a hillside, that captures and stores surface runoff. A swale is generally paired with a raised parallel mound on its downhill side.

The water collected by the swale slowly seeps into the mound, where plants can then access it. I designed and dug a swale in my garden beside a path, just upslope of a 50-foot-long (15 m) planting mound. I designed this combination so that I could point out the swale as an example of water conservation during garden tours. As it turned out, though, this was my swale folly.

One of my goals for the garden was to make it wheelchair-accessible. After completing the initial landscaping, I invited an accessibility expert to tour my garden and give me feedback regarding this goal. When we walked on the path that included the swale, he immediately pointed out that the slope of the excavation would cause a wheelchair to tip over. As a result, I decided to eliminate that swale. It was an error that cost me some extra time and labor to undo.

I still recommend swales very highly, however. If your landscape can accommodate one or more of these water-capturing features, you may wish to include them in your design. (See "Creating Swales" on page 60 for more details on constructing a swale.)

Hügelkultur mounds are planting beds built up from a base layer of wood that serves as a kind of storage tank to provide water to plants on the mound during dry weather. Conserving water is just one of the many functions of a hügelkultur mound, and I describe these special structures in detail in "Building a Hügelkultur Mound" on page 54.

Design for Sustainability

A self-sustaining garden plan incorporates elements that will supply all of the plants' needs going forward in order to reduce, or ideally eliminate, the need for externally derived inputs such as fertilizer. It's possible to design a garden to include "built-in" sources of nitrogen and other nutrients essential for plant growth, to attract insects and animals that

Figure 2.6. During dry spells, plants on this hügelkultur mound (under construction) will be able to reach water absorbed and stored by the decaying wood in the mound's base layer.

help control and deter plant pests, and to reduce the prevalence of plant disease. These self-sustaining features are primarily provided by plants incorporated into the plan. There are four essential categories of useful plants that help make a forest garden self-sufficient: nitrogen-fixers, nutrient accumulators, beneficial attractors, and aromatic pest confusers. By reducing the need for external inputs such as fertilizers, pesticides, or fungicides, these useful plants also reduce the need for human labor and financial resources to procure and apply such inputs over time. I discuss these four categories of plants and provide examples in "Grouping Plants for Sustainability" beginning on page 259.

Build in Redundancy

Permaculturists pay a lot of attention to the concept of function, and I've learned that thinking through the functions that are important in my forest garden is very helpful. For example, one function is building soil health. Another is preventing pest problems. A rule of thumb is that each important function should be provided in at least three ways—this is what it means to build in redundancy. A succinct way of expressing the need for redundancy is "two is one and one is none." In other words, if an important function is provided in fewer than three ways, the risk of losing it is high. When there are three or more independent sources for a function, there is more assurance that the function will persist.

One practical example of this principle is my strategy for providing multiple independent sources of water for my garden. I dug ponds to capture runoff. I piped in water from our artesian well. I built hügelkultur mounds. And I chose plants that would provide many foliage layers to capture and hold water. Even if the well pump fails and the ponds dry up during a drought, I still have the mounds and the foliage to sequester water. There

Figure 2.7. Ponds bring a feeling of peace and serenity to a landscape, enhancing the experience of visiting the garden.

are other ways to provide water, too. Most homes have an outdoor spigot to which a hose or drip irrigation system can be attached. You can set up a rain barrel to capture rainwater that runs off the roof of your house or a garden shed. And even in a small yard, it's possible to create a mini-pond and build a hügelkultur mound.

There's a corollary to the rule to provide important functions in three ways, which is that every element included in the garden should have at least three useful attributes. When an element serves multiple functions, it acts as a better contributor to the overall goal of redundancy. For example, ponds fulfill several functions in addition to providing a water source. Because they absorb and store the sun's heat, they create a microclimate that extends the growing season in the fall for the surrounding plants. Ponds provide a habitat for frogs, dragonflies, ducks, and other beneficial animals and insects. Useful native plants can germinate and grow on the banks of a pond. Ponds can also be a medium in which to cultivate food. For example, you may be able to harvest native crayfish from a pond or stock it with fish of your choosing. Of course, not every landscape or forest garden will include a pond, but a ground-level or elevated birdbath can serve some of the same functions.

Another important function is nitrogen fixation, which is carried out by bacteria that live in the nodules of legume plants (such as beans); the process converts nitrogen into forms that plant roots can absorb. This is a critical function because, after water, nitrogen is the nutrient required by plants in the greatest quantity. Therefore, the function should also be redundant. In almost every area of my garden, I have incorporated at least three different species of plants that fix nitrogen. For example, I incorporated licorice, clovers, and a redbud tree on one of my hügelkultur mounds. I also allowed hairy vetch to grow from the natural seed bed. (The ground contains many seeds that have been deposited there over time by native and naturalized plants. These seeds will spontaneously germinate and grow when the right conditions are present.) If one or two of these nitrogen-fixing sources fail to thrive, I still have at least one left to help make the nutrient available to surrounding plants.

These nitrogen-fixing plants also meet the standard of having at least three useful attributes. Besides providing nitrogen in available form, one or more of the plants supply edible or useful parts, add aesthetic appeal, grow flowers that attract beneficial insects, and provide ground cover, shade, and habitat for other beneficial creatures.

If you cannot come up with three positive functions of a plant or other element in the garden, don't incorporate it. Positive functions include food, fiber, wood, healing properties, nitrogen fixation, pollen, nutrient accumulation, attractant or habitat for beneficials, and pest deterrence. And to my mind occupying any one of the seven vertical layers counts as a function because the plant is filling an important niche. Aesthetic value counts as well.

Wisteria is an example of a multifunctional plant that I've chosen to include in my edible forest garden even though all parts of the plant are poisonous and do not produce food. I decided to plant wisteria because it has at least four positive attributes. It fixes nitrogen, has beautiful flowers, attracts pollinators, and occupies the vine level. On the other hand, I chose *not* to include another nitrogen fixer, the Kentucky coffeetree (*Gymnocladus dioica*), because I couldn't come up with more than two positive attributes.

Minimize Human Labor Going Forward

People often assume that there will be little labor involved in establishing a perennial planting like a forest garden, but this is a misinterpretation of the permaculture principle of minimizing human labor. If I had known in advance how much time and effort would be required to plan and establish

Figure 2.8. This dense ground cover preserves moisture, outcompetes weeds, attracts beneficials, and contributes nutrients and organic matter, thus minimizing human labor to fulfill these functions.

my edible forest garden, I might well not have undertaken the project! Even now that the garden is well established, there are numerous maintenance chores to do and management decisions to be made that require substantial time or manual labor. (See appendix 3, "Stewardship through the Seasons," for a list of these activities season to season.) To underline this point: I recently spoke with a young man from the Philippines who described how many bountiful forest gardens in his homeland have declined and lost their productivity because they have been neglected.

Nevertheless, there is a difference in the labor requirements for a vegetable garden and a perennial planting such as an edible forest garden of equivalent size. In an annual vegetable operation, the labor of bed preparation, planting, watering, cultivating, weeding, mulching, adding amendments, controlling pests and disease, and harvesting does not diminish from year to year. But with a forest garden, the amount of labor required diminishes substantially as time goes on. I observed this in my garden from the fifth year on. Since the planting is perennial and permanent, I no longer have to do bed preparation or cultivation. As the ground becomes covered with desired plants, the need to add mulch, to weed, to water, and to add more plants diminishes. (Although I continue to add new plants to my garden, the amount of planting I do each year is now significantly reduced.) As the canopy expands each year, more leaves drop in fall, providing a natural mulch. Because the garden includes many plants that provide nutrients and help to prevent or minimize disease and pest problems, the labor and expense of adding amendments to fulfill these functions is minor or nil. The only significant *increase* in labor over the years has been harvesting the garden's ever-expanding bounty of food.

The more features that can be built into a garden to enhance its health and growth, the less human labor is required going forward. As you plan and develop your planting, consider how you can incorporate components that provide nutrients, attract beneficial insects, birds, and other animals, and exclude destructive pests. This will be an evolutionary process. When I began my forest garden, I included various techniques aimed at minimizing human effort, and I added others as I grew in knowledge and experience.

LET GRAVITY HELP

It is always advantageous to harness the power of gravity to help you do your work. In my garden's early development, I had a back hoe dump a large pile of partially composted organic matter on the crest of a small hill, the point of highest elevation in my garden. After that, whenever I needed a load of this material to build a mound or mulch a new planting, I was moving it downhill, aided by gravity. Keep this in mind when you are designing a site for a water source such as a dug pond, too. If the pond is at the highest point in the garden, gravity can move the water through a hose—no pump required. Or if you are transporting it by hand in buckets as I do, you get to carry them downhill instead of up.

HAVE ANIMALS DO THE WORK

If you happen to raise farm animals, they can help you maintain your edible planting. Chickens are excellent at weed and pest control. Even a backyard flock of half a dozen hens will eat weed seeds, ticks, other insect pests, and infested fruit drop. I have not incorporated chickens in my forest garden to date because there has been so much wood chip mulch throughout, and chickens tend to scratch up such a mulch, leaving bare ground exposed. I plan to incorporate chickens in my plum patch in the future, however, to help control the plum curculio beetles that infest the immature plum fruit and cause it to drop to the ground. When chickens consume these drops, they also consume the curculio larvae inside, thus breaking the pest's life cycle. And here's a tip if you don't have chickens: Place a tarp under your plum tree to catch and remove the drops before the larvae have a chance to enter the soil, which is where they pupate and then later emerge as new adult curculios.

Pigs can also be helpful cleaning up fruit drops that can harbor pests and disease. It is handy to run them through an area where tree fruits such as apples have fallen to the ground. Be careful to limit their stay, as they will begin to turn up the soil with their snouts in search of additional food if left too long.

Ducks are avid consumers of slugs, and they are handy near mushroom production logs, because slugs can decimate a mushroom crop. Ducks also like to eat mushrooms, however, so take care to keep them at a distance from the logs when the mushrooms are growing. Frogs and toads can also dispense with slugs for you.

If you don't have domestic animals, many native birds and animals serve beneficial functions. Snakes eat slugs and rodents, birds eat insect pests, and moles eat grubs. For more about how to attract beneficials, see "Structures for Protection and Sustainability" on page 61.

Ducks on Patrol

Ducks can be useful for slug control and cleaning up leafy garden waste. One wet spring I noticed a destructive slug infestation in an asparagus patch I reserved for my personal use. The slugs were climbing up the immature stalks at night, stripping the outer skins and leaving behind an unappetizing trail of slime. I knew ducks enjoyed eating slugs, so decided to assign them to slug patrol. I enclosed the asparagus patch with portable ElectroNet fencing and put the ducks inside. Soon after, I noticed that some new asparagus shoots were broken where they emerged from the soil. I attributed this to trampling by the ducks' broad webbed feet. I figured that was okay as long as the ducks were dispatching the slugs. When a whole week went by during which there were no asparagus shoots to harvest, I felt something wasn't right. I was accustomed to harvesting a dozen or more full-grown shoots each morning. Now there were none but a few broken stalks. After some consideration, I concluded that not only were the ducks eating slugs, but they were also eating my asparagus! I removed the ducks right away, and sure enough, my usual harvest resumed two days later with minimal damage from slugs. In the end, I concluded that the loss of a week's harvest was worth the sacrifice.

Figure 2.9. Ducks on pest patrol. Caution: Keep them away from tempting greens like asparagus!

MATCH PLANTS WITH PREFERRED HABITATS

Human labor is reduced when you take care to locate plants in their ideal habitats. If a plant's location is a good match to its needs for water, shelter, shade, and sun, then nothing further needs to be done for the plant to thrive. If these needs are not met, then human intervention (and consequent labor) is necessary. As you plan your garden landscape, the better you match your plants with their preferred habitats, the better they will fare without any additional effort or expense on your part. I have learned this lesson through experience, as I describe in "My Tamarack Travesty" on page 115.

EMPLOY EXISTING ELEMENTS

Every time you use an existing element for a specified purpose rather than constructing something new, you reduce the input of human labor. For example, rather than building a garden bench or chair from scratch, turn a tree stump already on your garden site into a place to sit. When you want to trellis a vine, you can undertake a labor-intensive project such as building a strong fence or arbor, or you can use an existing tree for a trellis. When I expanded my garden, I left some native trees standing with this purpose in mind. I planted grape, arctic kiwi, wisteria, and akebia vines around the trees and guided their growth onto the tree trunk and limbs.

ENSURE AMPLE ACCESS

As I mentioned earlier, one of my goals for my forest garden was to make it wheelchair-accessible, but that's just one aspect of planning routes for people and, possibly, machinery to travel into and through your garden. As part of the garden design, it is critical to consider building in ample access.

Figure 2.10. A grapevine is trained along the limb of a native tree that serves as a trellis.

For small spaces, this step might not be as critical as it was for my acre (0.4 ha) garden, but don't skip over this point, because establishing access after planting can be challenging.

Dave was adamant about access to the garden in case we needed to do work using a big piece of power equipment. Our tractor required a path at least 10 feet (3 m) wide, for example, so I designed a wide route from the main gate leading to the large gathering area near the center of the garden and on out to the far gate. When I expanded the garden, I included 10-foot-wide access routes there as well, along with narrower paths for golf-cart access.

Providing wide paths can turn out to be invaluable in unanticipated ways. At one point while I was working on the ½-acre (0.2 ha) expansion of the Enchanted Edible Forest, a neighbor offered to give me the muck soil he was digging out of a site where he was deepening a pond. I knew that this soil would be valuable to raise the surface level in an area where the water table was consistently too high to allow healthy plant growth, and so I was grateful for the wide access route Dave had insisted on. My neighbor was able to drive his dump truck along that wide path right to the location where I wanted the muck dirt deposited. Without the built-in access, a huge amount of manual labor would have been required. We would have been stuck hauling in the muck one wheelbarrow at a time.

Even for a small garden, you may want to ensure that a garden cart or golf cart can be maneuvered in and out to deliver plants and mulch materials and to transport out harvests and waste material. Measure the width of equipment you want to use in your garden, and be sure the main access routes can accommodate it. Keep in mind that paths need to be wider than the width of the intended vehicle to provide room for rounding curves and entering intersections. For example, a path for a 4-foot-wide (1.25 m) garden cart or a golf cart should be at least 6 feet (2 m) wide. Also consider that as your garden matures and plants spread more widely, they may begin to obstruct movement along an insufficiently wide path. For narrower footpaths, it is a good idea to observe where people (or pets) tend to walk rather than arbitrarily designing a footpath.

Dog Knows Best

One day I was planting bushes in a bed with a complex system of curved mounds, accompanied by my dog Splash Gordon. I was careful to enter and exit through narrow dips among the mounds that I had planned out as access paths. Splash had another idea—each time he came and went, he cut across one particular mound in a specific spot.

At first, I tried to correct his trajectory, but he insisted on the same route. When I stepped back to consider the situation, I realized he was showing me that another route was much more natural than the one I intended. Humbled, I modified the mounds (and later my graph paper plan) to reflect his superior choice of access route. While it is important to not obstruct access to any part of the garden, it is not necessary to plan in all the routes in advance—just the wide, primary access ones.

Now that you have become acquainted with the permaculture principles that let "nature do the work," it's time to begin planning your own edible forest garden. Chapter 3 will help you specify goals for your garden, evaluate and map your plot, design your landscape, and identify and procure the materials you will need to execute your plans.

Figure 3.1. A large, stone gathering area was part of my garden plan from the start.

CHAPTER 3

PLANNING YOUR FOREST GARDEN

After I attended the permaculture workshop, I was itching to embark on planting my own edible forest garden. I had been market gardening for seven years, and I was ready for a new venture. But as I started reading more about the subject of forest gardening, I realized that a great deal of planning and preparation would have to take place before planting could begin.

When envisioning a forest garden, it's important to begin by studying the land and making note of its diverse features. All aspects of your site will impact the design of your landscape and the choice of plants you install. The slope of the land, the spots where rain pools on the surface and where water flows, the direction and intensity of the prevailing wind, the places where frost hits first and where it is held off, the movement of sun and shadow, the position of existing structures, the nature of your soil—all of these details are important. A thorough examination of your plot requires observing it during all four seasons. It could take an entire year before you complete a full-scale map of your plot noting all these elements. Meanwhile, you will also be formulating your goals for the use of your garden, which will help you decide what amenities, such as play or sitting areas, privacy screens, or water features, to include in your landscape design. On top of that are decisions about what plants to include in your initial planting.

A forest garden planting is quite different from a traditional vegetable garden or an annual flower bed where the plants live for only one season. You will be installing perennial plants that might live for several years or even a few decades into the future. The relative permanence of these choices, their potential cost, and the effort and disruption involved in replacing or moving them after they are planted make it well worth your while to approach your forest garden project with a good deal of forethought. Understanding as best you can the growth pattern and ideal habitat for each plant from the start will increase the odds that you'll choose a planting spot where it will grow successfully. This research, too, takes time.

Not every decision you make will work out as planned. This is to be expected, as you can see from the numerous examples I share with you throughout this book. Rome wasn't built in a day; nor is an edible garden established in one year. It is normal to revise and even transform your design as time passes. Your diligence will be rewarded as the fruits of your labor (literally) increase each growing season.

Figuring Out Your Garden Goals

Obviously, producing food is one goal of an edible forest garden, but what kinds and amounts of edibles do you want to grow? Do you want to have small quantities to harvest and eat for that day's meal, or are you looking to can and freeze or otherwise preserve larger amounts of produce for future use or to give as gifts? And what else are you hoping for? Would you like your forest garden to be a shady retreat where you can sit outside and enjoy nature, a place for your children to play, a space to entertain guests or throw a party? All of the above? Clearly the size of the plot you have available may limit some of these choices. But if you are clear about your goals from the start, with a little ingenuity, you may well be able to work all of them into your plan.

Since Dave and I were already operating a farm business, I decided that one goal for the forest garden would be as a profit center. This would take the form of U-pick, as well as using the space as an event venue such as for weddings and family reunions. It could also be a site for gardening workshops, tours, and other events. Thus, right from the start, I realized that a large gathering area would be part of the design. Since the public would be visiting the garden, it was also important to make the design as accessible as possible. As I mentioned in chapter 1, I also wanted to create a section of the garden using only hand labor to serve as a teaching venue and an inspiration to gardeners without the resources or space to bring in machinery to clear land or create garden features.

I also decided that I would like to test the limits of what I could grow in my climate, with its very cold winters, and my heavy clay soil. As I pored over plant catalogs, I read about many appealing trees and shrubs that were better suited to warmer winters and better-drained land, including almonds, peaches, and cherry bushes. But they sounded so appealing that I wanted to try them anyway, realizing from the start that they might not endure. My intention was to match these plants with microclimates in my garden that would be most conducive to their survival. These microclimates might already exist in my plot or, if not, I would create them to increase the likelihood that these plants would be successful. I was also betting that climate change would temper our winters enough to allow these plants to persist. If my bet proved correct, I could be the only gardener in my area harvesting peaches!

Woody trees and bushes can be expensive. If your budget is limited, you may wish to stick with plants that have proven to do well under your climate and soil conditions. In addition, fruit and nut trees can take years to become productive. If you are interested in quicker results, you may wish to stick with perennials like many berry bushes, herbaceous plants, and ground covers that you can begin harvesting in a year or two. In any case, the goals you set and decisions you make at the start will continue to evolve over time as you observe the outcomes of your initial choices.

As I developed my forest garden, I conducted experiments to see whether some of my goals were achievable. For example, to test my hypothesis that I could grow edible plants even in shady woods, I decided to plant a bed in a wet spot in the shade of several existing trees. To create the bed, in early

Figure 3.2. Guests enjoy sitting on the large patio in my forest garden.

May I piled up branches trimmed from locust trees and covered them with topsoil. I wanted a water feature to attract frogs nearby, so I dug a shallow pond by hand. As I dug the pond, I piled the excavated mud and sod (upside down) on the adjacent developing planting mound. Then I added a thick layer of leaves, more topsoil, and wood chips. I planted the mound with some leftover kale seedlings to hold the soil in place while I pondered what perennials I wanted to place there.

By the end of June, the contents of the mound were composted enough to install permanent plants. I interplanted shade-adapted berry plants and a variety of herbs, edible ground covers, and nitrogen fixers. (See "Mounds Built on Saturated Ground" on page 291 for a list of the specific plants.)

The plants all established themselves that first summer, and then the top growth died back as winter arrived. The following spring the kale was gone, which I had expected, but all of the perennial plants leafed out. Most thrived over the course of the growing season. I concluded that the experiment was a success, and I was ready to build my garden into the woods.

Observing the Land

Learning how to study the land is a valuable skill for any gardener. That includes noticing and evaluating microclimates, drainage, slope, areas of light and shade, direction and strength of prevailing wind, and the nature of your soil.

As soon as the land was cleared of brush and trees, I began studying and mapping the garden. This process involved observing the land at different times of day over a period of months, noting the type and location of all the features I observed, and creating a rough sketch. I quickly learned that if I didn't make a note about an observation, I soon forgot what I had noticed. Even if your plot is small and you only intend to plant edible shrubs, it is still important to have a detailed knowledge of the nature of your plot from the beginning.

By the way, this habit of noting your observations is a great one to maintain as you implement your plan for your garden and watch it grow and flourish over the coming years. Record your observations about the weather, the growth of your plants, any pest or disease problems, the times when flowers

Figure 3.3. This experimental mound built in a shady spot is one of my best experiments. Just three years after planting, success! The mound is populated with thriving plants.

Figure 3.4. In early spring, the tall trees in an existing hedgerow cast long shadows across the landscape. Observing how light moves across your plot over the course of the day and the seasons can be an enchanting experience.

bloom and fruits ripen, and the times of year when you perform certain maintenance tasks. You can look back at your notes to anticipate times when your garden will need attention. Reviewing your observations can also help you puzzle out the cause of problems and give you clues about steps you can take to solve them.

While I was still studying the land, I also planned the landscaping and the first plantings for the Enchanted Edible Forest. The process was a challenge but exciting at the same time. It was like solving a math problem with many variables. I also spent time observing and planning when I decided to expand the garden in 2015.

One of the first pieces of information to collect about your land, if you don't already know, is what hardiness zone you live in. Our farm is in USDA Hardiness Zone 4. Hardiness zones are defined by the average minimum winter temperature. The average low winter temperature in Zone 4 ranges between –20 and –30°F (–29 and –34°C). The lower the zone number, the lower the average minimum temperatures, so Zone 3 ranges from –30 to –40°F (–34 to –40°C), for example. Plants that can't tolerate the average low temperature in a given zone, or that require a colder winter than a zone provides, will likely not survive. The converse is also true: Plants that can't tolerate an extended and extremely hot summer season or that require a longer, warmer growing season than your zone provides will probably not do well, either.

Most nursery catalogs and reference books note the hardiness zone range for each plant they describe. You can find your zone by searching the term "USDA Hardiness Zone Map" on any web browser. The interactive map allows you to search your location by town name or zip code. Since I garden in Zone 4, I shy away from plants that are hardy only to Zone 6, for example, because they probably wouldn't live through our winters. Even though my focus is on cold-hardy plants, however, much of the information in this book about planning, planting, and maintaining a forest garden will be helpful to you regardless of what hardiness zone you live in.

It takes time to fully understand some of the environmental aspects of your property. For example, the position of the sun and its angle change throughout the year. It can take a full year of observations to discover all the variations in sun and shade patterns as the seasons progress. My garden is at the 44th latitude north. In the summer, the sun comes up in the northeast and sets in the northwest, and our land receives up to 16 hours of daylight. In the winter, the sun rises in the southeast and sets in the southwest, and daylight can be as short as eight hours. At the spring and fall equinoxes, the sun rises directly east and sets directly west, and daylight is 12 hours long. The angle at which the sun's rays strike the ground changes significantly from the winter to the summer, closer to perpendicular in summer and at a much lower angle in winter. Buildings, walls, trees, or other objects on the property may cast significant shade at certain times of the day or year. Noting these patterns is important to help match plant to site, and also because they affect microclimates.

NOTE MICROCLIMATES

A *microclimate* is a small area that differs from its surroundings with respect to temperature, moisture, shade, wind, or other factors. For example, the area adjacent to the south side of your house, if unshaded, will be an extra-bright and warm microclimate that will receive direct sun for most of the day as well as heat and light reflected off the building. As a result of all the sun exposure, the soil there is likely to be dry. In contrast, the area on the north side of your home will be shaded for most of the day, and so the soil there is likely to be cooler and moister than on the rest of your property. In addition to microclimates that may already exist, there are several ways that you can enhance or even create a particular microclimate within your garden. In the pages ahead, I describe several ways I created or enhanced microclimates in my garden, and suggest how you can do the same in yours.

IDENTIFY SLOPES

Studying and recording the lay of the land—the degree and directionality of slopes on your site—is important. In *Creating a Forest Garden*, Martin Crawford offers the rule that for every degree of south-facing slope, you gain two growing days, and for every degree of north-facing slope, your garden loses two growing days. This rule of thumb applies at all temperate latitudes and is critical for positioning plants.

Slope is important to enhance ripening and for pushing the envelope on hardiness zones. For example, for fruits or nuts that ripen late in the season and are not considered generally cold-hardy, the goal is to maximize the probability that the crop will reach full ripeness before the growing season ends. I wanted to grow varieties of American persimmon, pawpaw, and Asian pear that are late to ripen and rated as borderline hardy in Zone 4. South- or southwest-facing slopes are best for these kinds of crops because they receive the greatest intensity of the sun's warming rays, thus enhancing the growing season. And even in winter, greater exposure to sunlight on these slopes may somewhat mitigate the intensity of the cold.

A north-facing slope, though, is ideal for early-flowering plants such as apricot trees. Apricots are native to areas where spring temperatures increase slowly and gradually over several weeks' time. In my region, spring warms erratically; there are often a few days to a week of 70°F (21°C) temperatures in March and then a return to subfreezing temperatures. Such a hot spell can stimulate apricot flower buds to become active, and then they are killed by the sudden cold. On a north-facing slope, plants have less direct exposure to the sun in late winter and early spring. The ground warms more slowly than on a south-facing slope, delaying the awakening of the tree. Even though air temperature partially influences bud break, the ground temperature is also significant. Thus, you can prolong dormancy of plants and trees like apricots by paying careful attention to the directionality of slope when choosing a planting site.

ASSESS DRAINAGE

Some plants thrive only in well-drained soil, while others can tolerate water-saturated soil. To make informed decisions, study how water behaves on your site: Where does water puddle on the surface? Where do streams or rivulets run following heavy rainfall or during snowmelt? Where is the soil surface always dry? It is often a good idea to dig some test holes to investigate the condition of the soil a few inches below the surface. This can reveal clues about the amount of moisture or lack thereof in different parts of the garden. To determine the quality of drainage, you can dig a hole 1 to 2 feet (30–60 cm) deep, fill it with water, and observe how long it takes for the water to be absorbed. If it drains down within three hours, your soil is well drained but may dry out quickly in the absence of precipitation. If water takes more than 12 hours to percolate into the earth, you will have a drainage problem. Even if your plot is relatively flat, there may still be differences in the condition of the soil from one area to another that are worthy of note. This information is key to planning the proper fit between the moisture needs and preferences of plants. The better the plant is suited to the habitat, the more successful it will be.

In my garden I discovered some wet areas where the water table sits above ground level for most of the year. I could tell this was the case because water remained pooled on the surface for several months. In three spots I observed transient aboveground streams. At very wet times of the year, such as in spring during snowmelt or rainy periods in fall, one of them ran the whole length of the garden, almost 200 feet (60 m), flowing to a low spot at one end. When it rained, water drained into the garden from the road that runs parallel to its northeast side and also ran off from the pasture bordering the garden's southwest edge. Siting plants, other than a water plant like watercress, in a stream of running water would not make any sense. I ultimately decided to construct culverts (as I describe in "Building a Culvert" on page 58) to channel the water underground, so I could increase the diversity of plants in the soil above. I also dug ponds in spots where standing surface water persisted almost all year

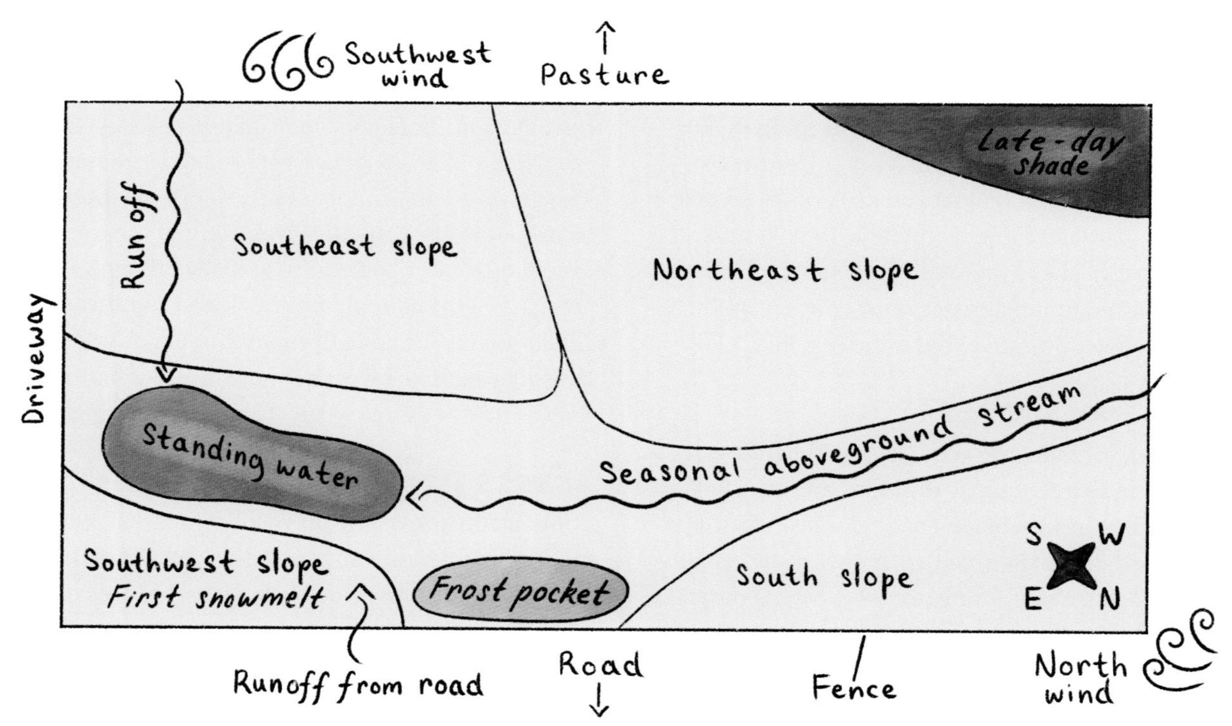

Figure 3.5. This sketch of the initial lay of the land in the Enchanted Edible Forest isn't to scale, but it captures important features that I noticed as I observed my future garden site and thought through my garden plan. Creating a sketch like this is helpful for planning a garden of almost any scale. Illustration by Zoe Chan.

long. Solutions to drainage problems may differ depending on the size of your plot and your vision for your garden. Rather than install a culvert to channel water underground, you may wish to do a little dredging to create a shallow trench lined with pebbles to better channel an aboveground flow. In addition to digging ponds and building culverts, techniques such as building raised beds or hügelkultur mounds described in chapter 4 may best suit your setting.

It is also instructive to observe what happens to water as the seasons progress. In my garden, as the snow melted in spring, the snow cover on the southeast corner of the garden melted first. It was a significant difference from the rate of melting in the rest of the garden. That area had a slight southwest slope, and in my soil exploration I found a lot of underground stone in that area. These observations revealed a microclimate that was somewhat warmer than the rest of the garden, especially early in the season and probably late in the season as well. I chose borderline Zone 4 plants for this spot hoping the extra heat would help the plants survive the winter and lengthen the growing season enough to produce ripe fruit. In light of the rock close to the surface, I also built up the layer of soil above the rocks to provide more growing area for plant roots, and I chose plants that had shallow root structures.

STUDY LIGHT AND SHADE

Become aware of any natural features or built structures that block sunlight from reaching parts of your garden. Note the time of year the shade occurs and at what times of day. This information is critical to inform decisions on where to place plants. For example, in a tropical environment where the sun is intense, shade is necessary for many plants to thrive. In a temperate environment like USDA Hardiness Zone 4, most fruiting plants require several hours of direct sun each day. *Full sun* is often defined as six to eight hours of direct sunlight per day. Many annual vegetables and some fruit trees do best in full sun. Other fruiting plants need some shade; direct sun from dawn to dusk is too much for them (gooseberries and red and pink currants, for example). Noticing that one section of my garden is shaded in the afternoon by scrubby pasture plants as well as tall trees helped me select appropriate plants and trees for that area.

DECIPHER THE WIND

Wind can affect the growth of plants and can also be dehydrating. Noting the seasonal changes in wind direction, speed, frequency, and intensity is crucial. Planting a barrier such as a windbreak of trees and bushes can help reduce the effects of wind. The barrier can contain deciduous plants and trees that lose their leaves in fall, because the growing season is when wind can have the most impact on plant growth. I was not aware of the importance of windbreaks at first, and planted my plum patch in a location where the trees received the brunt of the prevailing southwest wind. During the first two years after planting, their branches became skewed toward the northeast due to the frequent intense wind affecting their growth. On the other hand, some plants (such as grapes) need excellent air circulation to ward off fungal disease and others (such as raspberries) benefit from good air movement to help reduce insect pest problems. Placing them in a spot where the wind blows is beneficial.

EVALUATE YOUR SOIL

To help you understand the characteristics of your soil, I recommend taking soil samples and sending them to a soil testing lab for analysis. The lab you choose will provide instructions for preparing the samples. A comprehensive test that addresses both nutrient levels and also soil characteristics such as organic matter content is most helpful. When I first had my soil tested, I learned that the organic matter level in certain parts of the garden was high. Over the years, the brushy growth and trees had deposited a significant amount of organic matter on and into the soil in the form of leaves and roots. The lab tests results provided information on the kind and quantity of nutrients. The results also showed that the pH, a measure of the acidity or alkalinity of the soil, was between 6.0 and 7.0 (7.0 is neutral), which is within the range where most plants grow best. I concluded that acid-loving plants like blueberries and cranberries would not do well. As a result, I decided not to plant acid lovers in the garden to avoid the labor and cost of regularly adding acidifying amendments. If your soil has a pH lower than 6.0, there are many edible and supportive plants that can do well under your conditions. On the other hand, many gardeners choose to "sweeten" their soil by adding lime or compost. Your local cooperative extension can help you determine what kind and quantity of amendments are best suited to your situation.

Determining your soil's organic matter content is important because organic matter retains moisture while at the same time improving drainage. When a soil lacks organic matter, watering can become a major undertaking. For example, orchardist Stefan Sobkowiak says that at Miracle Farms, his permaculture orchard in southern Québec, the native soil was "sandy sand," completely lacking in organic matter. (Sand drains effectively but does not hold water.) Stefan installed a drip irrigation system to adapt to these conditions. He ran the drip lines along the fruit tree rows and watered every day to ensure that the trees had enough water.

Heavy clay soils that are low in organic matter can have the opposite problem: Clay traps and holds moisture effectively, but drains poorly. Thus, it is very possible to overwater plants in clay soil.

If your soil is lacking in organic matter, there are steps you can take to improve it. One choice is to plant cover crops such as clover. You can also spread copious quantities of organic mulch or compost on the soil surface, which will become incorporated into the soil over time. You can build raised mounds using plenty of organic materials. And as always, choosing plants adapted to your conditions, including your soil type, is a good practice.

Planning Your Landscape

As you study your plot and note its various features, you'll probably start dreaming up ideas for the design of your landscape. To capture your ideas, it's very helpful to create a to-scale map of your planned garden space, including all its existing features. Once you have an accurate map, you can begin to fill in details.

CREATING A TO-SCALE MAP

Here's the basic method to follow to create a map that serves as a two-dimensional model of your undeveloped garden space.

Step 1. Collect the materials you'll need, including a tape measure at least as long as the longest dimension of your plot, graph paper, tape, a ruler, pencils, and a drawing compass.

Step 2. If you haven't yet measured the overall length and width of your garden area, do so now. Then figure out what the size of your map will be if you create it on the scale of 1 (graph paper) square = 1 square foot.

Step 3. Set up your base plan to scale by drawing the boundary lines of the garden space to scale. My original garden is 100 feet × 200 feet (30 × 61 m). In order to make a map 200 squares wide and 100 squares high, I first had to tape together several sheets of graph paper.

Step 4. Next, collect data about the location of each landscape feature within your garden space (such as trees or tree stumps, shrubs, the flow of surface rivulets, slopes or rock outcroppings, and any human-built structures). One way to precisely map a feature is to triangulate its distance from two known locations, such as a fence post or tree along the boundary line of the garden (I

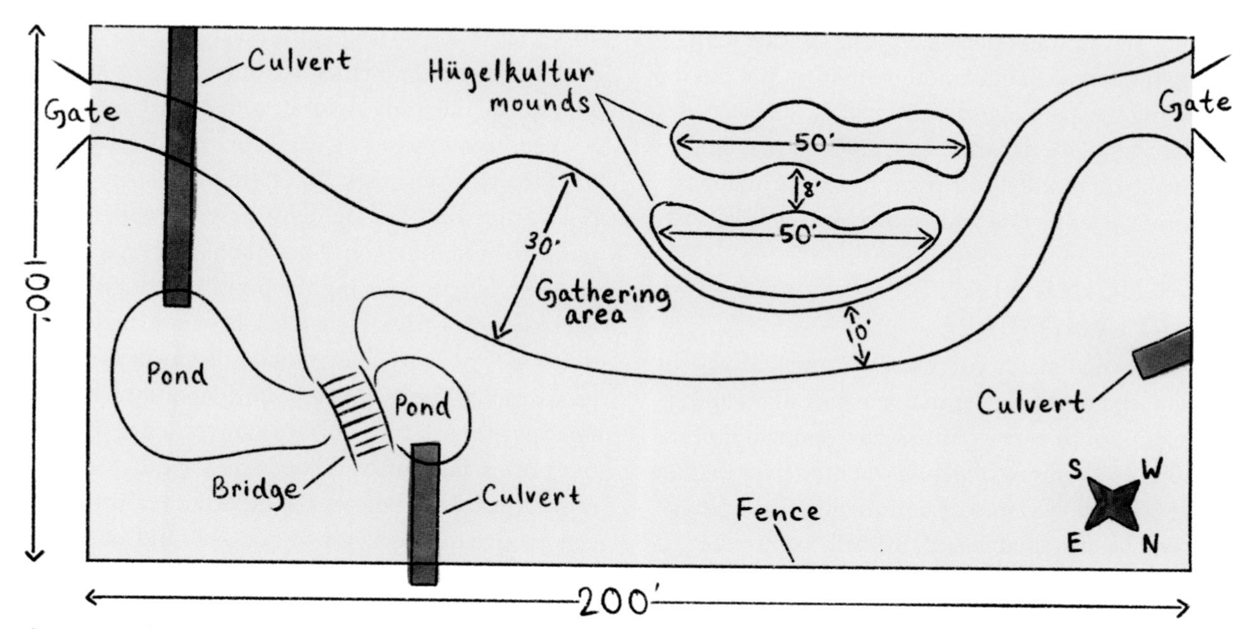

Figure 3.6. This sketch shows (not to scale) the initial landscape design for the original ½ acre (0.2 ha) of the Enchanted Edible Forest. Illustration by Zoe Chan.

explain how to do this in step 5). For each feature you want to include, measure its distance from two permanent features that you've already marked on your map. Be sure to record all of these measurements on paper as you work.

Step 5. Return to your base map and use the drawing compass to draw arcs on your map at the measured distances for a particular feature from the two permanent objects. The point where the two arcs overlap marks the precise spot of the feature. (This is triangulation.)

Step 6. Add notes on the map about aspects of the landscape such as slopes, and sketch in the contour line where two slopes meet. This is similar to the information you noted on your first informal sketch (see figure 3.5 on page 32), but now you will be drawing in those features at an accurate scale.

Step 7. On a calm day, take your map outside and double-check the accuracy by remeasuring distances on the ground and comparing them with what you've drawn on your map. Frequent checks of this kind while developing a map are essential so that you don't compound errors as you continue to add information to the map.

DESIGNING LANDSCAPE FEATURES

Once you are satisfied that your drawing accurately represents the existing features of your plot, you can begin to plan the landscaping and figure out where you want to place key plants such as fruit trees. First, decide on the major elements of your landscape design, including wide access routes, gathering spaces, drainage improvements, ponds, and planting mounds. These landscape elements will reflect the goals you set for your garden as well as what you learned through observation.

Draw a sketch indicating the landscape features you wish to include, taking into consideration the characteristics of the land that you have already entered in your to-scale map (see figure 3.6). Next, determine the exact placement and dimensions of each planned element in the garden itself, by using grade stakes to mark the boundaries of each feature.

Finally, record these details on your to-scale map, using the same triangulation method as you did for mapping the existing features.

Your map is a dynamic tool, and you may find that you will modify your plans (and your map) as circumstances dictate and your vision evolves.

ARRANGING PLANTS IN YOUR LANDSCAPE

Deciding what plants you want to grow, and where to plant them, is another major aspect of planning your garden. In part 2 of this book, you will find detailed descriptions of plants you may want to grow, while part 3 delves into designing plant groupings that will be sustainable and mutually compatible. The rest of this section outlines basic factors to consider when choosing and arranging your plants.

While I was observing and mapping my land, I also began to compile a list of plants I wanted to include. This process involved reading about the growth habits and habitat preferences of all kinds of perennial edible plants: tree fruits and nuts, berry bushes, perennial vegetables, ground covers, and vines. My sources included books and nursery catalogs (and I've listed many of my favorites in the resources section at the end this book). I also attended workshops at the sites of other edible gardens, including the edible forest garden of Jonathan Bates and Eric Toensmeier in Holyoke, Massachusetts, and horticulturist and author Lee Reich's garden in the Hudson Valley of New York State.

When it comes to deciding on plant placement, you'll be balancing multiple factors, such as preferred habitat, future growth, and mature canopy size. Keep in mind that a garden should be planned for four basic dimensions: length, width, height, and *time to maturity*. This set of dimensions is one reason working with a to-scale map of your landscape is essential. For some overstory trees, you may need to think ahead as much as 50 years because that's how long it will take them to reach mature size. Then back up to the present and calculate planting distances so that when the trees or bushes are mature, they will flourish with proper spacing.

A common mistake gardeners make is planting trees too close together. When trees are spaced too closely, the roots of neighboring trees compete for moisture and nutrition. The canopies compete for space and light, reducing the overall productivity. Even though a fruit tree *whip*, a baby tree that has not yet formed any side branches, looks too small to be spaced 20 feet (6 m) away from another whip, this spacing will turn out to be correct once the trees reach their mature height and width. Most nursery catalogs indicate the expected mature canopy size of select varieties of trees and/or the proper planting distances.

I saw an example of the importance of canopy size and tree spacing during a tour of a nursery in the Southern Tier of New York State. The nursery manager pointed out a walnut tree stand and explained that the nut trees, planted about 10 feet (3 m) apart, were appropriately spaced to produce timber but would not produce a good crop of nuts. He said that walnut trees needed to be spaced 40 feet (12 m) apart for good nut production. Tree spacing should reflect your goals for your forest garden.

As the lay of my land was quite varied, it contained a number of distinct habitats that allowed me to incorporate plants with different habitat preferences. I planned the placement of most of the trees and shrubs for the entire ½ acre (0.2 ha) at the outset. The first step was to figure out the appropriate spacing for those with the largest canopies—the trees for the overstory and understory. For a small forest garden that is designed around a single tree, it is possible to plan and plant all seven layers of the garden at one time. You can also take this approach in a larger garden if you plan to plant in small segments at a time.

In my garden, I took the approach of planting all the woody plants across the entire ½ acre first. I concentrated on choosing locations for the trees

A Looming Threat

As I was planning the Enchanted Edible Forest, I learned from my local cooperative extension of an emerging invasive pest that was wreaking havoc in conventional berry plantings, a small fruit fly called spotted wing drosophila (SWD). Possessing a saw-like protrusion that can cut through the surface of unripe berries, this fruit fly lays its eggs inside immature fruit of crops such as blueberries and raspberries. As the fruit ripens, the larvae hatch and begin to mature. When the fruit is ready to harvest, the larvae sometimes emerge as tiny, squirmy white worms. Worse, they can be hidden inside the fruit unseen until you take a bite (the larvae won't harm you, but it's not an appetizing experience).

I was concerned about this invasive fruit fly infesting my newly planted berries. I also believed that our state's cooperative extension service would soon determine best management practices for this pest. Therefore, although I had already planned their locations, I decided to delay planting most of my berries for a year to give the researchers time to develop these management strategies so I could incorporate their recommendations. (See "Red Raspberry" on page 175 for a summary of organic management practices for this pest.)

with adequate spacing to allow for the width of their mature canopies, and placing the shrubs in appropriate habitats. Later I filled in around those plants with herbaceous plants and ground covers.

I also kept in mind the permaculture principles I described in chapter 2. I tried to maximize diversity; in most cases, I avoided placing two trees of the same species side by side. To maximize solar absorption for all plants, I placed the tallest trees, such as black locust and pear, to the north, northeast, and northwest, so they wouldn't shade other plants. I did not place any tall trees on the sunny side of the ponds. I wanted to be sure the water would absorb the maximum amount of solar radiation, creating the warmest possible microclimate for surrounding plants. I planned and planted a tall, dense windbreak along the north corner of the fence to shield the garden from north winds.

To the best of my ability, I attempted to match plants with their preferred habitats. For example, I planted trees and shrubs that preferred or could tolerate wet soils in the lowest spots. Drought-tolerant plants, or those requiring the best drainage, were placed in the highest. I sited shade-adapted bushes to the north side of trees, whose eventual canopies would shade them. I planted raspberries just inside the fence, where the edge of the adjacent scrubby woods mirrored the forest edge habitat where raspberries are found in nature. The preferred habitats of the plants you choose to incorporate need to be kept in mind as you plan their locations in your garden.

ADDING PLANTS TO THE MAP

In terms of your master map and plan, there are a few ways to work in stages as you map out specific plantings in your garden-to-be. One way is to draw the planting plan on sheets of tissue paper overlaid on the master map of the garden. You can also use a drawing compass to pencil in circumferences

Figure 3.7. Here is a close-up of one small part of my master garden map, with paper circles depicting a grouping of plants.

that represent the diameters of the mature canopies of trees.

The way I find the easiest to plan the arrangement of trees and shrubs is to cut out paper circles with diameters representing the mature canopies of the trees I want to include. Be sure to label each circle for the tree it represents. Then move the paper circles around on the base map like a puzzle to see how they best fit together. With this strategy, there is no erasing and redrawing; you can temporarily affix the circles in place once you have a suitable arrangement.

When I am satisfied with the layout, I can proceed to planting. I also use paper circles to make a permanent record of my planting choices, as shown in figure 3.7. In addition to the name of the tree, I make note of the nursery it came from and the date it was planted. Where circles overlap, I place the circle representing the taller plant on top.

There are a variety of software programs that can be used to plan a garden, but I have not used one because my frustration tolerance for learning to use new computer software is almost nonexistent. If you are comfortable with a computer-based approach, that may be the planning method of choice for you.

SOLVING LANDSCAPE PROBLEMS

As you evaluate your site and begin to plan your landscape features and plant arrangements, you may well identify problems that your site presents. Figuring out solutions for these dilemmas can be both challenging and gratifying.

New plantings can benefit from protection from strong winds. This was certainly an environmental factor I had to contend with when I began planning my garden, and strong winds are becoming more common in many regions as climate patterns continue to become more erratic. If your location lacks natural windbreaks, it is helpful to include one or more in your garden design.

In *Creating a Forest Garden,* Martin Crawford provides a detailed discussion of considerations in planning a windbreak. Basic principles include:

- Choosing fast-growing species that will grow tall enough to serve the intended purpose.
- Planting trees and bushes close enough so they merge and become dense as they grow.
- Making sure there are no gaps.
- If your windbreak design includes more than one layer of plants, placing the tallest, most vertical plants in a line perpendicular to the prevailing winds, and the shorter, bushier ones downwind from there.
- Protecting plants from weed competition to increase the speed of their establishment.
- Understanding that root competition from a densely planted windbreak will make it difficult for other plants to thrive close by.

These principles were my guide as I planned four windbreaks in my original ½-acre garden: one in the north corner to protect a persimmon tree, another to block the northwest wind, another along the entire northeast side of the garden, and a fourth to block the southwest wind. I chose a variety of

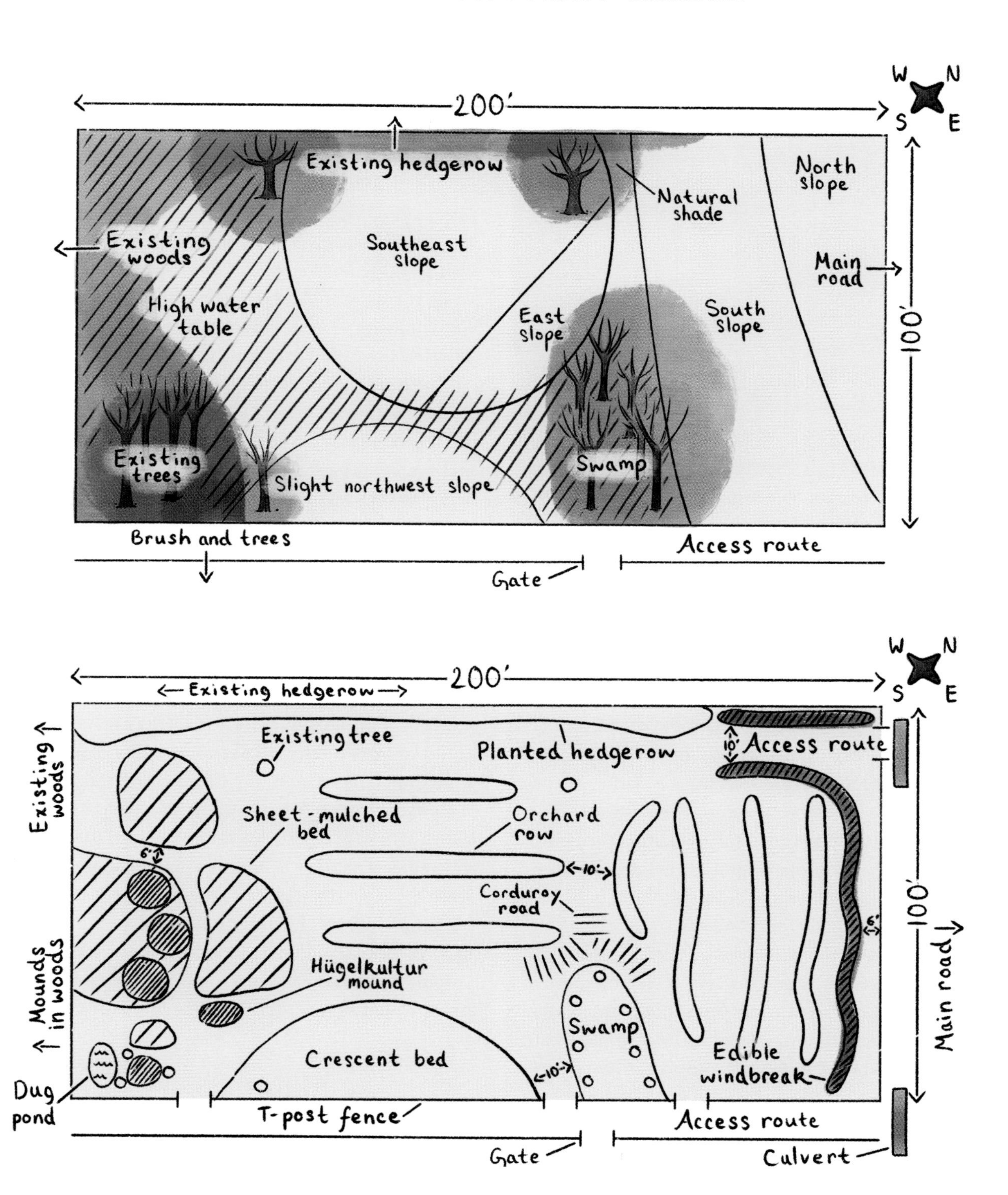

Figure 3.8. The first rough sketch for the expansion of my edible forest garden shows the location of existing trees, slopes, and saturated ground. The second sketch shows my initial design for this area. Illustrations by Zoe Chan.

Figure 3.9. For a windbreak to be most effective, the plant stems should be dense right to the ground, as seen in this hedge comprising hazelbert and sea buckthorn.

species of trees and shrubs for these windbreaks, as I describe in chapter 14.

When I expanded the garden beginning in 2016, I didn't anticipate major problems with wind, because the plot was almost completely surrounded on three sides with natural windbreaks including a savanna-like area to the southeast, a stand of dense woods to the southwest, and a dense natural hedgerow to the northwest demarcating the boundary between my property and my neighbor's, all effectively blocking the wind along those borders.

However, I had other issues to contend with in this new area, including root competition from the dense hedgerow, large expanses of sloping ground, and saturated soils. I came up with design solutions for all of these issues. I decided that I would create a raised bed (to reduce root competition with the existing trees) parallel and adjacent to the natural hedgerow along the northwest side of the plot. There I would plant a second hedgerow populated with useful plants of my choosing. I would also plant edible windbreaks in strategic spots where natural protection from wind was lacking.

When I visited Stefan Sobkowiak's permaculture orchard in southern Québec, Canada, in the spring of 2016, his planting inspired a design solution to deal with the expanses of sloping ground. At Stefan's orchard, the fruit trees are arranged in long rows, interplanted with berry bushes, perennial vegetables, and herbs. I realized this model would work well on these large slopes. Widely spaced, fairly straight rows would make maintenance efficient. A riding mower or even a tractor could navigate easily between the rows for routine mowing.

The final challenge was how to deal with the areas of saturated ground. While there were no aboveground flows of water, several areas had high and even aboveground water tables. The ground was also very wet on the contour where some slopes met. One clump of trees surrounded a "swamp" where the water from spring snowmelt and fall rains reached a depth of up to a foot. In the sections where the water table was often aboveground, I determined I would either create raised beds, build up the soil by sheet mulching with thick layers of organic matter, or create hügelkultur mounds. I decided that the areas where slopes met could be turned into access routes rather than trying to plant there. Along the circumference of the swamp, I would dig in mushroom totems, which would benefit from submersion in the wet ground (see chapter 12 for information about growing mushrooms).

Along the way of planning the landscape and solving these problems, I identified the materials I would need to execute my plans and figured out how I would source them.

Sourcing Materials

As you are planning your landscape, you may want to begin sourcing the materials you will need to execute your plans. I needed to source a variety of materials, including stakes, compost, leaves, wood chips, mulch material, gravel, and large limestone slabs.

GRADE STAKES

When it's time to start implementing your plans on paper in the actual landscape, it is helpful to have lots of stakes on hand to demarcate access routes, planting beds, and other features. I used grade stakes, which are inexpensive 1-by-2 inch (2.5–5 cm) wooden stakes 2 feet (60 cm) long, pointed at one end and squared off at the other, available in bundles at home centers and lumberyards. They are easy to hammer into the soil and to pull out and reposition. In addition to grade stakes, I used sticks gathered from the woods and bamboo stakes from potted nursery plants. For better visibility at a distance, you can paint the tops or mark them with colorful flagging tape.

I used the stakes to indicate the outlines of hügelkultur mounds, sheet-mulched beds, access routes and pathways, patios, and ponds, as well as to mark spots where I planned to plant trees and shrubs.

COMPOST

It's hard to have too much compost on hand when you're starting a forest garden. In my situation, I was particularly aware that I would need copious amounts of organic matter to amend my heavy clay soil and to build up the soil in the wet spots. We do not routinely produce compost in bulk on our farm, so I needed to find another source of supply for creating my forest garden. Fortunately for me, there is a resort community of four hundred cottages on Wellesley Island where most of the cottage owners view yard waste as disposable "garbage." This "refuse" accumulates in large mounds on the community's commonly held property, and our town's buildings and grounds crew routinely trucks it off the island. The town's maintenance supervisor was more than happy to deposit this semi-composted organic material on my property rather than cart it to a dump on the mainland. Thus, I acquired a huge quantity of vegetable matter with which to amend my developing land. As noted in "Let Gravity Help" on page 22, I deposited a large pile of this material on the highest point of the garden where it continued to compost in place while I removed quantities of it to apply thick mulch over the surface of my planting beds. When it was all used up, in its place on the crest of the hill I built a semicircular stone patio. I deposited another large quantity of this free organic matter directly on top of the existing ground just inside the northeast garden fence to build a 200-foot-long (60 m) mounded raised bed there.

You might not live next to a colony of cottages, but think creatively about where you might find an inexpensive source of compostable organic material. For example, often restaurants have quantities of table scraps, which make good compost and which they may be happy to give you in bulk.

LEAVES

As with yard waste, most landowners on our island don't know what a treasure fallen leaves can be. Every fall they spend hours raking and bagging these leaves for disposal. To organic growers, dry leaves are like gold; they contain abundant nutrients accumulated by trees' deep and expansive roots throughout the growing season. Once they became aware of our need, our neighbors began bringing these bagged leaves to us, rather than hauling them to the dump. In this fashion, over the years, we have obtained invaluable mulch material at no cost. Sometimes these leaves arrive whole and sometimes they have been shredded. Either way, I spread them directly over the ground to provide mulch, to build raised beds, or as a component in hügelkultur mounds. If you don't have neighbors willing to bring their leaves to you, in the fall homeowners often place their leaves in bags by the curbside. If you get there before the garbage truck, these are free for the taking and make excellent mulch and compost. Just be sure they don't contain any weed seeds or live invasive plants that could contaminate your planting.

WOOD CHIPS

Wood chips are also ideal material for mulch. Wood is a fantastic insulator; when placed on the ground around a newly planted seedling tree, it conserves moisture and keeps the ground cooler in summer and warmer in winter. Its presence prevents soil erosion and germination of weed seeds in the underlying ground, and delays incursion by surrounding plants. As wood chip mulch gradually decays, it provides nutrient-rich organic matter to the soil below.

The best kind of wood chips are those produced by chipping freshly cut branch tips and leaves, which are the most nutrient-dense parts of a deciduous woody plant. This type of material is called *ramial* wood chips, and arborists often produce them when they chip up the debris after removing an unwanted stand of brush. They are also available in our area every seven years when the power company clears brush and tree branches intruding below transmission lines.

In March of the first year of my forest garden project, I contacted every arborist in our local area by phone or email to ask whether they expected to work on Wellesley Island during that season, and if so, would they consider dumping their chips on our property. Most readily agreed, to avoid the time and expense of carting them off the island and dumping them in a landfill.

Figure 3.10. Truckloads of wood chips provide mulch for garden beds and hügelkultur mounds. Over time the piled chips gradually compost in place, forming a rich humus.

Caution: Be sure that the chips you use are primarily from deciduous trees, not conifers (pine or cedar) or black walnut. Chemicals in chips from these species produce complex natural substances, such as juglone in black walnut, that may hinder the growth of other kinds of plants.

OTHER MATERIALS

Every project has certain unique features, and you may well need materials beyond the basics discussed above. For example, you might need fencing materials (fencing is discussed in more detail in chapter 4), cardboard, or stone for hardscaping. The more you can stay alert to potential ways to scrounge materials, the better, as described in "My Search for Stone."

One thing I knew I would need on an ongoing basis was additional mulch material to suppress weeds. For this purpose, organic gardeners can use newspapers or brown (but not white) cardboard. At the time I first began my forest garden, though, my organic certifier prohibited use of newspaper with colored inks and of any type of cardboard, so I was forced to search for another material. My solution was a biodegradable mulch paper that comes in 4-foot-wide (1.25 m) rolls. I ordered ten 500-foot-long (150 m) rolls of this material. These rolls of paper worked well to cover the 50-foot-long (15 m) hügelkultur mounds just after construction, when I wasn't yet ready to plant them. I was concerned that weed seeds in the soil would germinate and grow, presenting competition for the plants I intended to install sometime later. The paper mulch, held down by leaves or wood chips, shielded the beds from sunlight, preventing this unwanted outcome.

A few years later, I learned that my certifier had changed the rule, and cardboard was now acceptable in organic operations. I found a local appliance business that was more than willing to let me

My Search for Stone

My landscape plan called for a 10-foot-wide (3 m) access route traversing the entire 200-foot (60 m) width of the garden. At the heart of the garden, the access route would widen out into a patio area. As I envisioned events like weddings taking place on the patio, I realized that trucks occasionally might need to cart in a tent, dance floor, or catering equipment. I wanted both the access route and the patio to be able to sustain the weight of large vehicles. Coincidentally, the backhoe contractor who was creating the access route had just dug a foundation for a house that involved excavating large slabs of 6-inch-thick (15 cm) limestone. With no other takers for the stone, the contractor offered it to me. I gladly accepted it, and all I had to pay was the cost of having it trucked to my farm. It turned out to be enough stone to complete half the access route and gathering area.

Where could I find more of this material? I called local quarries and stone contractors and came up empty. One stone contractor told me he had been searching for similar material for two years without success. A year went by, and I was getting desperate. If I couldn't source similar material, my project would be in limbo indefinitely.

That third summer, I attended a lawn party at a friend's camp. I decided to announce to the crowd my desperate need. In response, the hostess said, "Hop on my four-wheeler; we are going to visit my neighbor." We rode next door where, to my amazement, the owner had bulldozed a private quarry in his backyard containing the exact type of stone I needed, and the owner didn't want it because it wasn't suitable for the rock wall he wanted to build. He and I struck a deal, and I had the stone trucked to my garden. That stone was sufficient to complete the access route and gathering area, with some left over for future projects. If you let random people know what your needs for materials are, you are likely to learn of a place to find them.

Figure 3.11. We ran out of limestone with only part of the access route and half of the gathering area completed.

Figure 3.12. The biodegradable mulch paper I used to cover new hügelkultur mounds is good for deterring weeds between rows of annual vegetables as well.

remove their discarded large-appliance boxes to save them the trouble of disposing of them. There is so much that gardeners can get for free while being eco-friendly in the process!

Plans Are Made for Changing

Before I move on to the topic of putting your plan into action, let me caution you again: Be sure to allow plenty of time to study your land and plan the landscaping and initial planting. Foreseeing problems like poor drainage and access and remediating them in your initial plan will avoid unnecessary frustration, labor, and expense in the future. Periodically step away from the planning process, let some time pass, and then return. You may find that you've developed new insights and creative solutions while your mind was focused elsewhere.

Be realistic. No one should expect themselves to foresee every potential problem or mistake without the benefit of experience. This was certainly my case. As I began to implement my initial plan, there were several points at which I decided to modify it because I realized that what I wanted to do ran contrary to the natural forces at play.

As described above, my original landscape plan called for a gathering area. As it was being constructed, however, I decided on the spot to make it bigger, so that it could accommodate larger events, expanding it from 15 feet (4.5 m) wide to 30 feet (9 m) wide. This change required cutting into the hillside, which created a vertical dirt face 1½ feet (45 cm) tall at its highest point along that side of the oval patio area. Over the next two years, I observed this small "cliff" beginning to erode. The remedy was to build a stone retaining wall to hold the slope in place. Although this involved a good deal of additional time and expense, the result was far superior to what I had originally planned.

Another change in plans occurred when I dug a hole at a spot where I planned to plant a cherry tree. Unexpectedly, the hole filled with water. I had not anticipated that the water table was so close to the surface in this part of the garden. The wet area extended about 300 square feet (28 sq m), so I couldn't simply dig another hole close by—it would fill with water, too. From my research, I knew that cherry trees require particularly good drainage, and that the tree would likely die if planted there. Other plants would probably struggle as well. My remedy was to fill in the hole and then cover the entire area with soil being dredged out to create the adjacent pond. I was able to raise the ground level by about a foot. Then I planted the cherry tree in this newly raised ground, where it has done fine.

Some changes in plans are a long time unfolding. In the fall of 2014, I had the opportunity to visit the well-established forest garden in England designed and planted by author Martin Crawford. After touring that garden and returning to my own garden, I realized I should have planted a windbreak to block the prevailing southwest wind. Unfortunately, I had already planted some bushes in the spot where the windbreak was needed, but even so, I modified my original plan, transplanted the bushes elsewhere, and installed the windbreak.

I share another story about changes made after creating the original plan in "My Tamarack Travesty" on page 115, where I describe how I changed the composition of a windbreak hedge when it became apparent that it wasn't fulfilling its intended purpose.

My final example entails the interaction of my designs with the vagaries of nature. My original plan called for two ponds with a narrow strip of land between. I hired a backhoe operator to dig the ponds, and the excavated areas filled up with water over time through natural runoff. Because the ponds were situated on a slope, however, I observed that the lower of the two ponds was full, while the upper pond remained very shallow: Being continually drained by the slope, its water surface rested about 18 inches (45 cm) below the intended shoreline. The solution appeared to be to install a weir, a dam designed to raise the water level. We installed the weir, and it worked as designed. It worked as designed, that is, until a family of muskrats moved in and tunneled *under* the weir, thus draining the pond. After unsuccessfully attempting to shore up the damage, I was resigned to installing a lining in the upper pond as the only viable solution. Could these problems have been anticipated? Possibly, if I had consulted a landscape architect, or if I had dug only one pond. My plan to dig two on a slope was my imprudence.

No matter how simple or complex your plans are, the process of executing them will be full of learning opportunities, which is part of the pleasure of undertaking a forest garden. Chapter 4 is full of tips to help you implement your plan successfully.

Figure 3.13. A change in plans to enlarge the gathering area in the Enchanted Edible Forest necessitated an added amenity: a curving retaining wall that adds aesthetic interest as well as built-in seating.

Figure 4.1. Decked with rugosa roses, the limestone gathering area in my Enchanged Edible Forest leads to a decorative trellis at the head of the bridge, a perfect setting for a wedding photo.

CHAPTER 4

IMPLEMENTING YOUR PLAN

With a detailed map of your site in hand and your plans for developing your landscape complete, it is time to put your plans into action. In this chapter I provide instructions on methods for preparing land for planting and for constructing a variety of features that you may choose to include in your garden. I describe in detail land prep techniques (such as frost seeding, sheet mulching) and how to build raised beds and hügelkultur mounds. I also discuss ways to capture, hold, and channel water. Finally, I delineate structures you can install to encourage or exclude wild creatures.

If your plan calls for creating hardscape features such as a patio that might require use of heavy machinery and large quantities of materials, it is probably best to complete this work before any soil

Figure 4.2. Excess stone was used to create a small patio, spiral staircase, and stepping-stone path; later a wooden railing and toolshed were added amenities.

improvements you intend to make, because the presence of trucks, tractors, and other equipment on the garden site is likely to compact the ground.

My initial landscape plan included significant hardscape changes. I knew that a large backhoe would be needed to dig out the ponds, build the large mounds, and create the wide access routes. And right from the start, I also envisioned a wooden bridge with a trellis between the two ponds. These major construction activities continued off and on over three years.

As your garden landscape begins to take shape, there may be opportunities to add features you hadn't envisioned during your original planning stage. It's fun to follow your creative muse when this happens. For example, the main access route in my garden was constructed with a bed of gravel made from crushed limestone (called crusher run) topped with thick slabs of limestone. There was gravel and stone left over, which inspired us to build some additional patios and spiral staircases. The ideas to add a barrier railing at the deep end of the large pond, benches for seating, and a toolshed came later as well.

At this stage you may not have finished making choices and placements of all the plants you want to include. No worries. This process will be ongoing as you prepare more plots for planting as described below, and as your vision expands over time.

Sheet Mulching

Sheet mulching is a simple technique of creating a clear planting area by covering the ground surface with layers of organic material. Before laying down sheet mulch, though, it's important to cut down and clear out the top growth of any unwanted shrubs and trees as best you can and mow any perennial ground cover plants down as close to the soil surface as possible. Once you've prepared the area, simply gather large slabs of plain brown cardboard, such as large-appliance boxes, and lay them flat on the ground, overlapping the adjoining pieces by at least 8 inches (20 cm). Arrange them as much as possible in the desired shape of the bed. Temporarily secure the cardboard with stones, bricks, heavy tree limbs, or whatever you have on hand. Spread a layer of seed-free organic matter at least 6 inches (15 cm) deep on top of the cardboard. Tree leaves, grass clippings, vegetable garden waste, and compost all serve this purpose well. Top the whole bed with wood chips to hold down the organic matter against the wind. The cardboard serves as a layer that blocks light to the plants below, and they will slowly die out before the cardboard breaks down. The organic matter piled on top gradually decays along with the cardboard, creating an enriched bed for planting. Tip: I have found it best to remove the securing stones or branches as I pile organic matter on top of the cardboard. The cardboard under them tends to decay quickly, allowing perennial weeds to emerge before enough time has passed for the intact cardboard to snuff them out.

Within a few weeks or months, depending on the season and the nature of the original ground cover, this method creates a clean and fertile bed ready for planting edible perennial plants. (Without the topping of organic materials, the cardboard might take as long as two years to decay.) This technique can also be used to eliminate competing plants under a newly planted windbreak.

EXTRA-THICK SHEET MULCHING

If the sod is heavy and full of perennial weeds or there are woody plant roots in the underlying soil, a double or multiple layer of cardboard can be better than a single layer in order to snuff out woody roots and persistent perennial weeds. Building up extra-thick layers can also help in wet soil areas to raise the ground level. In one part of my forest garden, the soil is saturated for at least eight months of the year, and even longer during an unusually wet summer season.

To make this area more palatable for plants, I decided to sheet-mulch with two layers of cardboard

Figure 4.3. Cardboard sheets overlap and cover a 4-foot-wide (1.25 m) area adjacent to an orchard row. This technique prepares the ground for planting additional shrubs and ground covers.

Figure 4.4. A double of layer of cardboard is arranged around shrubs on an existing bed overlaid with 12 inches (30 cm) of organic material to raise the soil farther above the water table and, hopefully, eliminate perennial quack grass.

covered with copious quantities of leaves and wood chips. In some locations, successive applications of this sheet mulching have gradually raised the land enough so that plants comfortable with wet soil can do well. You may also need to re-sheet-mulch an area to eliminate persistent perennial weeds. A downside of this extra-thick sheet-mulching technique is that it forms an extremely desirable habitat for voles and mice that enjoy ingesting the bark of woody plants.

TARP COVERING

A variation on sheet mulching that works well but doesn't require adding organic matter involves covering the area with a thick piece of black plastic such as a silage tarp. As with the sheet-mulching method, it is essential to first cut down and clear out the aboveground portions of unwanted shrubs and trees, and mow the herbaceous plants close to the soil surface. Then lay the tarp and secure it by burying the edge in a trench or weighing it down with large stones or sandbags. Within a few weeks or months, depending on the season, when you remove the tarp you will have a sod-free bed in which to plant.

Letting Animals Do the Work

If you have livestock animals, they can be of great help in clearing a site. Using animal power made sense to us because of the size of our plot and the fact that it was totally overgrown with scrubby brush and scattered trees, and because we had the animals!

As I described in chapter 1, the kickoff of our land clearing was the goat wedding. Because there was so much vegetation on the site, though, the newlyweds weren't able to make much of a dent. Fortunately, Dave had an entire herd of goats and cows to help out.

On our farm, we use a technique called interspecies grazing. The goats and cows graze together in the same pasture because they complement each other. Goats are primarily browsers, eating the tips of tree branches, leaves of woody plants, and seed heads of tall herbaceous plants. The cows are primarily grazers, preferring to eat grasses and other plants that grow closer to ground level. Together, they clean up and fertilize a scrubby area pretty well and don't compete in terms of the forages they prefer. As an extra bonus, the cows act as guard animals, protecting the goats from predators like coyotes. After the animals were in the garden area long enough to eat a lot of the greenery both on the ground and on the branches, Dave moved in a mama pig with her brood of 13 piglets.

Pigs are wonderful at rooting out and eating nutritious taproots of forbs, broad-leaved weeds such as dandelions. They also happen to just love quack grass, an invasive perennial grass. They stayed in the garden during late fall and winter then through to early spring, doing an excellent job of cleaning up the ground cover plants while leaving behind some saucer-like hollows, lasting evidence of their work. After that Dave used a bush hog and chain saw to level the scrubby bushes and cut down the box elder, ash, and elm trees.

Pigs can be useful in an annual vegetable garden as well. Pigs will enjoy eating up the remains of crops

Figure 4.5. Our herds of goats and of Red Devon and Belted Galloway beef cows did the lion's share of the work of consuming unwanted vegetation on the site of the edible-forest-to-be.

Figure 4.6. Piglets Oscar and Mayer—here keeping an eye on Dave and Splash Gordon—did a fine job of preparing a plot that we later used for production of annual vegetables.

such as squash and corn. They can also be used to prepare a vegetable plot for planting. One fall when Dave and I first purchased two piglets to raise for meat, we fenced them inside a sod-covered ¼-acre (0.1 ha) plot, where we left them until mid-spring.

Over the course of six months or so, they did a fine job of rototilling the ground with their snouts, consuming the weeds and grass and fertilizing as they went. That next spring, followed by a light tillage with a rototiller, the ground was prepared to plant with annual vegetables.

Frost Seeding

Once the land is cleared, the next step is to attend to the soil. In my plot the soil was heavy clay, and I was concerned about the roots of fruit trees being able to thrive in this dense, water-retaining medium. Although there are many ways to prepare soil for planting, the easiest and most expeditious method is to use plants to do the job for you. Seeding with a cover crop can address many soil inadequacies: Plants can be used to add organic matter, improve drainage, suppress weeds, scavenge nutrients, and prevent erosion.

To incorporate a cover crop with minimal labor, I recommend the frost-seeding method, a technique in which seeds are sown over bare ground during a time of year when frosts occur. For my cover crop I chose to use clover, which both fixes nitrogen and has deep roots. During the spring in my region, daytime temperatures rise above freezing, but often dip below freezing at night. If you scatter seeds of cold-hardy crops such as clover on bare ground, the freeze–thaw cycle causes the soil particles to expand and contract. This acts to suck the seeds into the ground, where they have good contact with the soil. When the soil temperature reaches the right level for germination, the seeds readily germinate, and the roots of the growing clover penetrate and break up the clay. When a cover crop is mowed, the plants automatically prune their roots. The severed roots decay, which contributes organic matter to the soil. The open spaces left after the roots decay help improve drainage and aeration.

The pigs did such a good job clearing the site that it was primarily bare ground, ready for frost seeding. You can also create bare ground for frost seeding by sheet mulching or tarping. In late March 2013, when the winter snow had melted, I used a hand-cranked seeding device to scatter clover seeds over the ½-acre (0.2 ha) plot. (Alternatively, seeds can be easily scattered by hand.)

That same spring, I started planting edible plants. Unfortunately, the clover had not yet had enough time to do its job. The clay was sodden and heavy, and when I dug into it, it stuck to the shovel in a giant clump that had the appearance and consistency of pottery clay. Nonetheless, I persevered and I did manage to plant some fruit trees. The following year, I could see a remarkable change, thanks to the clover! Now when I dug, the clods were only ½ to 1 inch (13–25 mm) in diameter so the trees had spaces for their roots to move through the ground.

Forming Raised Planting Beds

My solution for some of the saturated soil areas in my gardens was to make raised planting beds in the shape of small mounds. If you have access to a sufficient quantity of topsoil, compost, or manure, you can build a raised mound by simply laying a cardboard footprint (to discourage plants from intruding into your mound), then piling the soil or organic material on top in the desired shape, size, and height. If you don't have an adequate supply of material, which was my case, then you can use the following method.

The first step is to rototill or otherwise work up the ground to pulverize sod and perennial weeds and to create loosened soil. (Dave did this step for me.) Next, rake the loosened material into piles in the desired shapes. If you have some additional topsoil available, use it to increase the overall bed height. This process of mounding up the soil creates air spaces and also enhances drainage due to the pull of gravity. Cover each mound with a layer of wood chips at least 2 inches (5 cm) thick to prevent erosion and discourage weed growth. Finally, sheet-mulch the areas between the mounded beds to prevent weeds from growing there and to prepare those areas for ground cover plants of your choice.

To learn more about how I incorporated planting mounds in the Enchanted Edible Forest, refer to "The Crescent Bed" on page 290.

Including Hügelkultur Mounds

Incorporating hügelkultur mounds in a garden can help aerate the soil, increase growing area, create additional microclimates, and add aesthetic interest. Building a mound is an excellent way to compensate for heavy clay soil and to create growing space if your soil is very thin or rocky. Hügelkultur mounds are similar to the simple raised mounds I describe above, but they involve more materials and

Figure 4.7. Shadows highlight a series of small mounds built in the crescent bed to create planting beds raised up above the high water table.

layering. The mounds have a base of logs and branches, which are covered with other organic materials. The wood slowly decomposes, adding organic matter that provides nutrients to the plants growing in the mound. The wood also acts as a water reservoir, actually pulling water up from the soil below as it rots. When there is a dry spell, the roots of plants on the mound find the water that's been held by the decaying wood, thus making the structure self-sustaining.

These versatile landscaping elements will expand your garden's diversity and productivity in multiple ways. And you can size them large or small to fit the scale of your landscape. Mounds can increase growing space by providing up to three times more surface area than flat ground, and by creating more space under the ground surface for root growth. Increasing the growing area is advantageous when available land is limited.

Mounds create and exaggerate microclimates, including slopes, and establish varied habitats such as full sun, shady, high and dry, and low and moist. By lifting plants above frost pockets, exposing more ground surface to the air and the sun's rays, and creating warmth with heat produced within as the organic matter decomposes, mounds can extend the growing season for species planted on them. Mounds also provide viable growing space in areas where the water table is at or above ground level. Their height doubles as a noise and visual barrier, creating greater privacy. Mounds make the garden accessible, lessening the need to stoop over while tending or harvesting, and facilitating harvesting from a sitting position in a wheelchair or golf cart. Mounds can add aesthetic value, as well, when designed with gently curving boundaries or in other interesting shapes.

BUILDING A HÜGELKULTUR MOUND

Some sources prescribe a very precise methodology for building hügelkultur mounds, but I find that whatever organic matter I have available eventually

Figure 4.8. Step 2.

decays and forms a rich humus. I have built several of these mounds, each composed slightly differently depending on the materials at hand. Here is the basic method I use.

Step 1. Clear the ground of weeds and brush and mow the sod. Lay some large sheets of cardboard on the ground to delineate the mound's footprint. This footprint should be in the size and shape of the base of the completed mound you envision, extending slightly beyond the intended base.

Step 2. Build a substructure of wood to define the eventual shape of the mound in the center of your cardboard footprint, leaving ample space on all sides for the addition of organic matter (see step 4). For this step I have used large sections of dead tree trunk, semi-rotted logs scavenged from the forest floor, pieces of large dry tree branches, piles of thin branch trimmings, and combinations of all of the above. Arrange the wood so it rises to a peak at the center, with sides sloping gradually down in all directions. One technique is to pile the thickest pieces close together in the middle, and then add smaller pieces graduating smoothly down on all sides. The height can be anywhere from 1 to 3 feet (30–100 cm), and half or less of the planned finished height of the

Figure 4.9. Step 4.

Figure 4.10. Step 5.

mound. Be sure that no pieces of wood stick out, or they may eventually protrude out of the completed mound.

Step 3. Inspect your work to be sure that the pieces of wood are stacked together tightly, leaving minimal spaces where rodents could reside inside the mound and nibble on your plant roots from below. If you have some loose soil handy, sprinkle it into any gaps in your wood structure. Or stuff the gaps with dead twigs, as an added deterrent to rodent infestation.

Step 4. On top of this basic wood structure, pile alternating layers of whatever organic materials you have available: grass cuttings, weeds, vegetable garden waste such as pumpkin vines, wet or dry leaves, reeds, cattails, compost, manure, and any kind of soil you have on hand. It is helpful to include some fresh, green organic matter in addition to the brown for quickest decomposition. Overall, the goal is to create a total thickness deep enough to install your plant cover in. The organic matter should be at least 1 foot (30 cm) thick over the entire wood base, and as much as 3 or 4 feet (1–1.25 m) thick, depending on how much soil is in the mix. The less soil, the thicker this layer of organic material needs to be. Over time, the force of gravity, as well as the impact of rain and snowfall, and the process of decomposition, will reduce the thickness of these layers by at least half.

Step 5. Top the mound with a layer of material at least 2 inches (5 cm) thick that does not contain any viable weed seeds. This is important to prevent future weed problems. I usually top my mounds with a thick layer of wood chips, which serve to hold down leaves against the wind and prevent the rain from eroding the underlying soil layer. Compost or grass clippings can also work well for the top layer, but be sure they are free of weed seeds.

To expedite the decay of the organic material, you can thoroughly moisten each layer of organic matter as you build the mound, but that can be very time consuming. I usually build mounds late in the growing season and let nature provide the moisture in the form of fall rains and winter snowmelt. By spring, the organic matter is sufficiently decayed, and planting can commence. I often plant annual vegetables in a mound during the first year. This serves the purpose of getting living roots into the mound, which is important to support the microbial community in the decayed organic matter. It also gives me time to observe the microclimates of the mound and decide what perennial plants would do well there.

Figure 4.11. Three sisters (corn, squash, and beans) on a new mound: The strategy of planting annual vegetables on a new hügelkultur mound gets roots in the ground, produces an edible crop, and buys time to plan a more permanent planting.

As you can see from the description, building hügelkultur mounds is a labor-intensive undertaking, making it a great group project. I have had lots of help building mine, including WWOOFers, college groups, interns, friends, and a hired hand. Perhaps you could invite your friends over for a party to build yours!

USES OF HÜGELKULTUR MOUNDS

I have built hügelkultur mounds to increase growing space, enhance microclimates, overcome adverse growing conditions, and add aesthetic appeal. I even built one partially enclosing a sunken patio to create a unique sitting area. They are an inexpensive, versatile addition to any growing space. Here are examples of a few ways I have incorporated them.

In my original garden, my backhoe operator built two 50-foot-long (15 m) mounds parallel to each other on a slight northeast-facing slope with soil he dredged out to create the two ponds, as shown in figure 3.6 on page 35. Loosening and piling up the soil improved drainage for the fruit trees I planned to plant there. The mounds also emphasized the north-facing angle of the hillside, making it steeper than the existing slope. This provided a suitable microclimate for plantings of apricots in that space, extending their dormancy and increasing the probability that they would bear fruit. An unanticipated benefit is the way these large mounds mute noise and screen out views of the surrounding area, creating unplanned private spaces. Mounds of any size can serve these functions for you, as well as create more growing area if your space is limited.

When I expanded my garden, there was an area where my efforts to grow plants proved unsuccessful despite repeated thick sheet mulching, even when I chose plants allegedly comfortable in wet soil. My solution was to construct several large hügelkultur mounds to provide some well-drained land for planting. Those mounds are now teeming with diverse perennial plant life. This technique of building mounds also works if your soil is too thin to support a variety of plants because bedrock is close to the surface. Building one or more hügelkultur mounds can provide an ample soil base and a water sink that will even support good-sized trees.

Once I realized that my shady mound experiment was a success (as I described in "Figuring Out Your Garden Goals" on page 27), I was ready to build my hügelkultur mound garden in the woods. That fall, I was lucky to have four energetic young WWOOF volunteers in residence on my farm. It was during the COVID-19 pandemic, when many young people chose to take a "gap year" from their college studies rather than attend courses remotely. I gave my volunteers the task of building five

A Word to the Wise

I was absent from the scene while my four WWOOF volunteers were building the bases of some hügelkultur mounds. When I returned, I noticed they had stuffed some of the spaces between the logs with green leafy twigs. It turned out that the stems were willow brush. Willow is one of the most prodigious rooters among hardwoods. In fact, a traditional way to promote rooting in cuttings is to soak them in willow water—water that has been infused with pieces of willow branches to create an effective rooting medium.

If the willow twigs stuffed into the base of the mound took root, it was likely that willows would take over the mound, outcompeting anything I planted. At my direction, my WWOOFers reluctantly deconstructed their mounds to remove the invasive willow.

Whatever materials you choose to build your mounds with, *never, never* include live twigs or branches or clumps of invasive weeds such as quack grass. These "prohibited" materials may well invade your mound with woody growth or weeds that will be impossible to remove once established.

Figure 4.12. Before and after: These mounds in the woods add aesthetic interest and a feeling of serenity to the landscape in addition to providing viable growing space.

hügelkultur mounds in the woods to the southwest of the existing garden. The woods are fairly dense here, creating a good deal of shade, and the water table is above the ground for much of the year.

Dave used his chain saw to remove some dead and tilting trees to eliminate potential hazards, improve access, and allow more light to enter. Fall rains came, raising the water table above the ground between the mounds. To allow dry access, we created "stepping-stone" paths with thick sections of tree trunk laid in gentle curves between the mounds.

I was so pleased with the first collection of mounds that we built an additional four farther into the woods. On a roll, I decided we would construct

Figure 4.13. Four mounds in the swampy woods create growing space above standing water.

four more inside the "swamp" in a different section of the garden.

If you have an area of woods where you would like to grow perennial food plants, building mounds could be your best option. Not only will they get your plants above the water table and beyond root competition, but if your site is populated with trees such as walnut or pine that can deter the growth of other species, the plants atop your mounds might escape that negative influence. Observe your woods and note the spots where some light penetrates, perhaps just inside the forest edge or in small natural clearings in the interior. These could be ideal spots to build your mounds. If no such spots exist, you may need to clear some brush or cut and trim up some trees to allow some light to enter, still leaving most of the woods intact. In chapter 14, I provide plenty of ideas for what to plant on your mounds in the woods.

Creating Ponds, Culverts, and Swales

Features that hold and channel water are enhancements that you may well want (or need) to include in your garden design. Some can be installed by hand, while others will require professional expertise and equipment.

For example, a shallow, hand-dug pond can be a home, breeding habitat, and water source for frogs, turtles, water snakes, birds, and beneficial insects. A large pond is an amenity that can serve the same purposes, but it's best to consult a professional on the design of a large pond, and hiring heavy equipment will likely be required to construct it. In my garden I dug a small, shallow pond by hand, and my contractor built two larger ponds with his backhoe.

BUILDING A CULVERT

A culvert is a structural pipe placed below the ground surface to allow water to flow through. (The diameter and length of the pipe you need will be a function of the amount of water flow and the channeling distance.) Once built, you can use the surface of the culvert as a pathway or a kind of raised planting bed.

I highly recommend you consult a professional for advice before embarking on a culvert project to determine what materials and equipment you will need to effectively manage your water flow. You can obtain professional advice from your local Soil and Water Conservation District agent, a Natural

Resources Conservation Service (NRCS) conservationist, a civil engineer, a landscape architect, or an experienced excavating contractor.

In my original garden I incorporated both large and small culverts. To channel the considerable drainage from the pasture under the wide, stone access route, I relied on my backhoe operator's expertise to design and build a sizable culvert. With advice from Dave, I constructed two 10-foot-long (3 m) culverts by hand to divert small intermittent streams underground. My goal was to create drier soil conditions above, allowing me to choose from among a wider variety of species to plant in these locations.

If you decide to install a culvert yourself by hand, the first task is to figure out the quantities of materials you will need. This includes calculating what length of pipe is required and of what diameter. (Traditionally made of pottery, modern drainage pipes are typically plastic but sometimes metal.) The pipe should be long enough to span the area under the ground where you intend to channel the water, and wide enough to channel the water at peak volume, such as during a heavy downpour or rapid spring snowmelt. To determine the dimensions of the pipe, I recommend consulting someone who can offer expert advice. You'll also need geotextile fabric of sufficient length and width to underlie the culvert. Geotextile fabric is a synthetic woven fabric designed to allow water to flow through but to prevent movement of solid materials like stones and gravel and even to prevent infiltration of mud. You'll need a sufficient quantity of

A Cautionary Tale

Working with contractors was one of my big learning experiences in the first years of developing the Enchanted Edible Forest. In the midst of the work to install the stone patio and the main access route through the garden, I had to hire a new backhoe contractor because my original contractor left the business.

The new contractor, who used a machine called a skid steer, did well with the first tasks he took on, so I felt comfortable being away from home on the day when he was completing the limestone access route. When I returned that evening, I was appalled at what I saw. The contractor had left a foot-wide strip of bare crusher run between the limestone slabs of the route and the adjacent sod. I knew immediately that it would be impossible to add more limestone to cover this gap without disassembling the entire day's work. I cursed myself for not being present while this last stretch of access route was completed. If I had been there, I would have caught the problem when it could still be remedied. I learned an important lesson from this mishap: When working with a contractor, it's essential to communicate one's vision explicitly rather than assume the contractor understands it implicitly.

Now there was nothing to be done but cover over that long stretch of bare crusher run with wood chips and hope that they would decompose and turn to soil. But what could I plant into wood chips that would thrive above a layer of rock below? Luckily, in the course of my first few years of growing perennial plants in the garden, I made some observations that could resolve this dilemma, as I explain in "Two Counterintuitive Discoveries" on page 166.

gravel to underlay, surround, and cover the culvert pipe by at least 6 inches (15 cm). If you plan to install plants over top of the culvert, instead of the gravel layer on top of the pipe you'll need enough soil to cover the pipe by at least 6 inches, and possibly more than that to support the root growth of the plants you want to install there.

To execute the construction, you will need a shovel to dig with, a wheelbarrow or cart to transport materials, and a level to determine the slope of the pipe as you install it. Once you have assembled your tools and materials, you can begin the construction:

Step 1. Dig a concave depression in the soil deep and wide enough to lay the fabric, a 6-inch (15 cm) layer of compacted stone, and then the culvert pipe at the proper depth to channel the water underground. Keep in mind that the bottom of the pipe opening at the end where the water enters has to be placed below the level of the existing water flow.

Step 2. Lay down the geotextile fabric. The fabric should extend at least 1 foot (30 cm) past the ends of the pipe at either end, and be at least 2 feet (60 cm) wider than the pipe on both sides.

Step 3. Cover the fabric with a layer of stone at least 6 inches thick.

Step 4. Place the culvert pipe in position. The pipe must be laid on a slight slope sufficient to channel the water downhill, usually a slope of about ¼ inch (6 mm) per foot (30 cm). Build up or remove stone as needed to create the proper slope, sliding the pipe back and forth until it is nestled in the bed of gravel at the proper angle. Use a level to determine that the slope is adequate.

Step 5. Cover and surround the pipe with more stone, leaving the ends open. If you expect heavy foot traffic over the culvert, add at least 6 inches of stone above the pipe; if vehicles will be driven over the culvert, at least 12 inches of stone should be added. Grade the stone gradually on both sides of the pipe so the surface of the stone does not have a sharp bump. If you intend to plant above the culvert, cover the pipe with at least 6 inches of soil, suitably graded, in lieu of the gravel overlay.

Step 6. Place large rocks above and at the sides of the open ends of the culvert to prevent other materials from falling down and blocking the openings.

CREATING SWALES

A swale is a shallow ditch with gently sloping sides dug along a contour on a hillside to capture and hold runoff. In permaculture practice, it is generally paired with a berm or mound on the downhill side into which the water collected by the swale is gradually absorbed. Plants and trees installed on the mound benefit from the captured moisture as well as organic matter that tends to collect in the swale over time. The trick to creating an effective swale is to have its base run exactly level so the runoff is captured and remains in the swale until it is absorbed rather than flowing out or pooling on either side of the swale. If you are considering including a swale in your design, consult a specialist (see recommendations in "Building a Culvert" on page 58) to help you develop your plan. A modest project may be doable with hand labor; or you may need a backhoe or other piece of equipment to execute a large one. As I described in chapter 2, I ended up having to fill in a swale that also served as a path after an accessibility expert pointed out that it might be a hazard for a person in a wheelchair.

Installing Paths and Corduroy Roads

As I've described, I decided to include a variety of access routes in my garden to accommodate everything from people strolling on foot to a large tractor. One of the widest tracks, which was built by a contractor, has a stone surface. Many of the access routes have a covering of native sod or perennial cover crops, which I maintain by mowing; others are covered with wood chips. By and large, walking paths created themselves as the result of repeated trampling underfoot, though I did create some defined paths using stepping-stones and cross-cut sections of tree trunks. You may wish to use any or all of these surfaces for your paths and access routes.

In the area where I expanded the garden, some sections of access routes were covered with surface water for most of the year. For those areas, none of the coverings noted above were suitable. I knew that traversing these flooded sections with a golf cart or tractor, even a wheelbarrow, would leave deep ruts at best. Worse, the reduced traction could cause a vehicle (or boots) to become "stuck in the mud," requiring emergency extraction.

The solution to this difficulty is an old farmer's trick called a *corduroy road*. This is a pathway constructed with closely spaced logs, placed perpendicular to the direction of traffic. Over time, the logs become embedded in the ground, providing a hard surface over which to walk, pull a garden cart, or drive a vehicle, even when covered with water. The surrounding soil prevents oxygen from reaching the logs, delaying their decomposition. On our farm, these constructs last a good 10 years before needing repair or replacement.

Figure 4.14. In one part of this corduroy road, sod has filled in between the logs, but the surface is still solid enough to drive over.

If you need access over an area covered with water, a corduroy road could be your solution, too. To build such a structure, use logs of equal diameter cut to length to accommodate your intended use: roughly 3 or 4 feet (1–1.25 m) wide for foot traffic, a wheelbarrow or garden cart; 6 feet (2 m) wide for a golf cart; and 10 feet (3 m) wide for a tractor. Lay the logs side by side, tamping them down in the mud or digging out enough sod so they lie firm, level, and packed side by side as tightly as you can place them.

Structures for Protection and Sustainability

Built structures, both simple and complex, can attract and sustain a multitude of helpful native creatures or be designed to exclude others. Amphibians, including newts, salamanders, toads, and frogs, consume snails, slugs, and other destructive critters. Snakes eat mice and slugs. Shrews and moles are carnivores relying primarily on insects for sustenance. Simple assemblages of natural materials such as piles of mulch, leaf litter, brush, rocks, and rotting logs provide shelter for many of these beneficial creatures. Birds are another wonderful garden ally, and adding birdhouses to your garden can help to keep birds foraging there.

Other types of structures provide protection against four-footed creatures, including deer and woodchucks, that like to feast on edible crops.

BIRDHOUSES

Although birds do eat fruit and berries, their primary food sources are insects and rodents, and in the balance they are a benefit in an edible garden. As they forage, birds fertilize the landscape with poop rich in nitrogen and phosphorus.

In *Edible Forest Gardens*, Dave Jacke and Eric Toensmeier provide extensive lists of wild birds and the locations where they scavenge for prey. Here are some of the wild birds that live in my region and where they forage for food:

Baltimore orioles. Insects on foliage at all levels.
Chickadees. Insects gleaned from the lower canopy.
Eastern bluebirds. Flying insects and invertebrates on the ground.
Nuthatches. Insects in tree bark.
Owls and hawks. Ground-dwelling mammals, including rabbits, mice, rats, voles, chipmunks, and squirrels.
Robins. Insects on the surface of the ground.
Tree swallows. Insects in flight.
Warblers. Insects in the overstory and understory canopies.
Woodpeckers. Insects in bark and the wood of trees.
Wrens. Insects in the lower canopy and on the ground.

I learned my lesson the hard way about the importance of encouraging birds to patrol pests in my garden. One year we had a scourge of caterpillars. I first noticed that some leaves on a branch of a precious apricot seedling were missing. On closer inspection, I saw several caterpillars chomping on the surrounding leaves. I knew that chewing caterpillars could wreak havoc, completely defoliating even large forest trees. I could not permit this in my garden. I grabbed each caterpillar between my fingers and squeezed. Then I inspected other nearby trees and bushes and used the same technique to exterminate every pest I found. I continued these executions daily as new caterpillars marched in from the surrounding woods.

Within a few days I noticed itchy, red raised bumps all over my body. At the same time, I was becoming irritable, overreacting to even the mildest frustrations. I described these symptoms to a friend, and she immediately identified the cause. It seems that caterpillars carry histamine in their skin, an inflammation-causing chemical that occurs naturally in humans but can be toxic in large doses. My symptoms were from a poisonous dose of histamine, absorbed through my skin from the multitude of caterpillars I crushed with my bare fingers. Once enlightened, I resumed squashing the pests, this time between two small rocks or twigs to avoid direct skin contact. Meanwhile, I installed a sticky insect barrier called Tanglefoot to trap these intruders as they climbed up my fruit tree trunks.

Within a couple of weeks, I won the battle. The onslaught of caterpillars abated, leaving little damage. My efforts, however, took up a lot of time and were disagreeable to say the least. There had to be a better way.

I thought back to the time when I visited Miracle Farms in Québec. There, I was struck by the hundreds of birdhouses throughout the 5-acre (2 ha) orchard. I was surprised and confused when orchardist Stefan Sobkowiak told us what he did when he saw a caterpillar nest on the branch of a fruit tree: nothing. He explained that the birds residing in all those birdhouses dispensed with these pests before they moved on to a second branch.* This was news to me—I had thought that if I didn't eliminate caterpillars right away, they would move on to defoliate the entire tree. As counterintuitive as Stefan's method seemed, I decided to try his approach.

I installed six birdhouses with openings sized to attract the eastern bluebird. (In general, the size of the opening in a birdhouse will determine which birds will and will not take up residence within it.) Each birdhouse is affixed to a metal pole and protected from below by a stovepipe baffle. Snakes and rodents will attempt to steal eggs and young chicks. The freestanding pole and baffle are designed to prevent this pilfering.

Though I haven't seen bluebirds nesting in the houses, both wrens and tree swallows moved in to

* Stefan Sobkowiak (Miracle Farms proprietor), in conversation with the author, May 2016.

Figure 4.15. An empty caterpillar nest in a mulberry seedling: Birds devoured the insects before they could defoliate most of the leaves.

Figure 4.16. A tree sparrow delivers food to her chick inside a birdhouse designed to attract bluebirds.

raise their young. Young birds require a lot of protein. During the time in late spring when these young birds are maturing in their nests, their parents gather insect pests in the surrounding garden to ensure their brood is well fed. Many native birds, including wrens, sparrows, chickadees, warblers, and robins, capture caterpillars to feed their young. Now when I see a caterpillar nest on a branch of a tree, I do what I learned at Miracle Farms—nothing. Within days, beneficial birds remove the caterpillars and feed them to fledglings in nests inside my birdhouses. In fact, if I had thought to make the connection, I had already seen the potential of birds to solve pest problems in my own vegetable garden, as I describe in "Birds and Brussels Sprouts" on page 64.

I also put up two larger houses, one affixed to a telephone pole overlooking an open area to attract a kestrel (a small hawk), and one nailed to a tree at the edge of the woods facing the garden to attract a small screech owl. Both birds specialize in consuming small rodents like mice and voles. Neither has yet chosen to take up residence in one of my houses. I have the same vacancy rate in a bat house I mounted on a telephone pole inside the garden. I am still hopeful that one day these houses will be occupied by the intended critters.

I had better luck a couple of years later when I installed two birdhouses designed to attract chickadees. These year-round residents occupied both immediately and built nests of moss. They raised their broods and vacated in time for wrens returning from the south to move in, build their nests, and raise their families in the same houses.

It is useful to clean out the nests from the last season's brooding well before winter's end because some birds, such as chickadees, take up residence early in spring. For the same reason, if you intend to install new birdhouses for chickadees, it's best to do so by year's end. These birds begin scouting for nesting sites on warm days in late January and February. For returning migraters like wrens and tree sparrows, new houses should be up by mid-March at the latest.

You can also attract beneficials by setting up a perch for hawks and owls. These carnivores like to

Birds and Brussels Sprouts

In my vegetable gardens, I always used to keep kale and brussels sprouts covered with row cover to protect them from pests, especially imported cabbage worms. These green caterpillars hatch from eggs laid by white-winged butterflies, and they can devastate members of the cabbage family by consuming their leaves. The row cover works well to exclude the butterflies from kale plants so they cannot lay their eggs on the foliage. And I harvest kale frequently, so I'm likely to spot any caterpillars that do appear. In the case of brussels sprouts, though, I find it hard to keep the row cover secured over the plants, as they grow tall. Wind gusts frequently upend the cover, giving the butterflies access. And I don't harvest brussels sprouts until late in the growing season, so I don't check on those plants as frequently.

One year, when I checked to see how far along the sprouts were, I lifted the row cover and a white cloud of butterflies, hundreds of them, rose above the plants. Every stem was stripped clean of leaves. Butterflies must have gotten under the row cover early in the season and deposited eggs. The hatchlings then feasted on the plants, pupated, and emerged as adult butterflies, all the while incubated and hidden by the row cover. That crop of brussels sprouts was completely ruined.

The next year, I did an experiment. I covered one row of brussels sprouts with a row cover for the entire season. I covered another nearby row for a few weeks while the plants became established and then removed it. Later that season, I lifted the row cover from the first row and compared. In the covered row, a third of the leaves were gone due to an infestation of cabbage worm. The row without the row cover had no damage. My conclusion—birds dispensed with any eggs or worms in the open row. But birds couldn't get under the fabric on the row I had left covered, allowing the cabbage worms to nibble at will. From then on, I left brussels sprouts plants uncovered once they were established and let the birds deal with the cabbage worms for me.

Figure 4.17. Three robin chicks await insects gleaned by their parents from the brussels sprouts planted nearby.

scan the territory for prey from a tall (10–12 foot / 3–3.5 m or higher) perch with an unobstructed view in all directions. A pole with a small platform affixed to the top is not difficult to build, and it can be installed on the crest of a hill or in a field with no taller structures nearby. Hawks will use it during the day and owls at night. I have not erected one because the telephone poles at each end of my garden serve this purpose for our native red-tailed hawks.

Some wild birds will nest only in the natural landscape, not in a birdhouse. Providing a diversity of plants varying in height increases the odds of their nesting in your garden. It is also helpful to provide perches for smaller birds in both edible forests and vegetable gardens. For example, a few sunflowers scattered about can serve as outposts from which songbirds can survey the garden for insects. Later in the season these helpful aviators will feed on the sunflower seeds. The more structures you install to attract beneficials, the healthier your garden will become, and the less work and frustration you will experience in the long run.

EXCLUDING DEER

Because deer are nocturnal, they do most of their damage under the cover of darkness, browsing on the leaves and tender shoots of almost every plant. Our island is overpopulated by native deer to the point that the understory layer of the woods does not regenerate because every new seedling is browsed without mercy. I knew these ruminants would present a threat to the plants in an edible forest, which is why I insisted on fencing the plots before I planted a single plant there. If your area is populated by deer, you will need to plan protection from them at the outset.

Heavy-Duty Electrified Wooden Fence

We knew that we would need to exclude not only the deer, but also our cattle and goats that would be grazing in a nearby pasture. Thus, we decided on a heavy-duty electrified wooden fence to enclose the original garden. This barrier required skill and heavy machinery to install. Dave chose to work with rot-resistant native locust and red cedar logs. He used a post pounder attached to a tractor to drive the posts into position. He laid a strand of barbed wire along the ground just outside the posts to deter burrowing mammals. Next, he enclosed the wooden framework with welded wire, secured to the outward-facing side of the posts. The final step was to attach two strands of high-tensile wire, one high and one low, to the posts, and electrify them.

The next spring, when I began planting trees inside the fence, I baited the top strand of high-tensile wire with "sandwiches" of peanut butter wrapped in strips of aluminum foil clipped to the wire by wooden clothespins. Deer are curious. When they smell the peanut butter, they explore it with their noses. The aluminum carries the charge from the electrified wire to give them a strong shock at this highly tender spot, and they quickly learn to avoid the fence. I replenished the bait that fall as well as in the spring and fall of the following year. By then, two generations of deer were deterred. Except for a single day when I neglected to

Figure 4.18. This heavy-duty electrified wood fence is one way to exclude deer from your garden. Though it does the job, it is an expensive installation requiring expert skills.

close the gate for the night, I have had no problems with deer incursions into this fenced garden area.

T-Post Fence

When I expanded the garden to a second ½ acre (0.2 ha), Dave had neither the time nor the ambition to build another heavy-duty fence. That was okay with me because I wanted the new site to exemplify how an edible forest garden could be constructed without the use of heavy equipment. We decided to build a T-post fence using only simple hand tools. Truth be told, this fence has gone through three distinct iterations: electric wire, fishing line, and electric fence tape.

Aluminum Electric Wire: The earliest design employed two 14-gauge aluminum electric wires, one about 1 foot (30 cm) above the ground and the second 3 feet (1 m) up. In early spring, we also baited the top wire with peanut butter sandwiches. That worked to deter deer as long as the wire remained electrified and the deer could see it. This fence was bordered on two sides by brushy woods. It intercepted a deer run, a path regularly traversed by deer probably for generations. As soon as the surrounding foliage appeared in spring, it obscured the fence. Then the deer, following their familiar path, simply stepped over the fence and happily entered the garden.

The first evidence of this incursion was the absence of leaves on some recently installed Asian pear trees. This sign was troubling. I rebaited the fence right away, trimmed what I could of the obscuring foliage, and hoped for the best. This appeared to work for a while, but as the trees and brush leafed out further, the deer returned again to strip my Asian pears. After this happened for a third time that season, I knew I had to provide some structure besides the fence to protect the Asian pears and other young trees. I came up with the idea of wrapping tomato cages with chicken wire and installing them over the seedlings. (See “Tomato Cages” on page 80 for a description of these cages.)

Fishing Line: The next spring, I visited the Cornell University Agricultural Experiment Station’s

Deer Netting

We also use fencing to keep deer out of our market garden. We constructed our first fence for excluding deer from our vegetable beds using cut saplings about 9 feet (2.75 m) long for the poles to support 7-foot-tall (2.25 m) deer netting. The black plastic netting looks like a delicate, nearly invisible, small-squared web. We inserted the poles in the ground at a slight outward slant. Then we wove a thin piece of rope through the top and bottom edges of the netting. We attached the netting to each pole, stretching it taut between them. Deer never breached the barrier, probably because they couldn’t see it, and it must have made them recoil when their noses touched it. Unfortunately, it did not deter the dogs, who punched through the bottom in several places as soon as it was installed. This reduced the life of the netting but seemed not to hinder its purpose of deterring deer. Once we acquired farm animals, we replaced this fence with a heavy-duty electrified one to keep goats, cows, and pigs from intruding. If you don’t have farm animals, just the netting could keep deer away from your edible garden.

research farm in Willsboro, New York. I wanted to learn about an experiment being conducted to determine what combination of warm-weather cover crops worked the best to suppress weeds in annual vegetables (it turned out to be millet and hairy vetch). While there, I noticed a thick web of fishing line strung around the open sides of a high tunnel (an unheated greenhouse). I asked the lead scientist, Mike Davis, PhD, what the fishing line was for. He informed me that it was a terrific deer deterrent.

As soon as I arrived home, I purchased enough colorless fishing line to wrap around the posts encircling my garden six times, with each strand spaced 3 to 6 inches (8–15 cm) apart. I secured the line as tautly as I could and then waited to see what would happen. Sure enough, this technique kept the deer out. I imagine the deer recoil from this invisible obstacle in the same way that we humans do when we unexpectedly come in contact with a cobweb while walking in the dark.

Some of the fine fishing line I installed began to droop, leaving gaps. I purchased heavier-weight line to reinforce the barrier. With occasional repairs and reinforcements, this barrier worked for about a year. Then the deer apparently figured out how to make their way over or through the lines. I was back to square one. Thank goodness for the tomato-cage barriers that still protected my trees despite the nightly raids.

Electric Fence Tape: Around this time, I read an article in a grazing magazine that described a type of electric tape as a useful deer deterrent. It seems that in contrast with a thin strand of electrified wire, the inch-wide white or yellow electrified tape is visible in the dark. Curious deer approach it to explore with their noses, and zap! They quickly learn to avoid it. I purchased a roll of this tape in a bright yellow color (though I now believe white would have worked as well), strung it around the fence, and turned on the juice. Lo and behold! As long as I kept the undergrowth trimmed, this technique appeared to be a permanent fix. Since it was installed, the deer entered the garden only once in the winter when, overnight, a branch fell over the fence, shorting out the power. I noticed their tracks in the snow the next day and repaired the damage. The next night, they stayed away.

Figure 4.19. The yellow electric tape strung between metal T-posts is my best solution so far to deter deer.

If you have a small area to protect, multiple strands of heavy-duty colorless fishing line surrounding your plot may well do the job. If you can design your plot to be long and narrow, this technique will work even better, as deer are reluctant to enter an area that appears confining.

For a larger garden, I strongly recommend trying the electrified tape. A small solar panel on a post can supply sufficient power to energize your tape.

BARRING BUNNIES

When we first began farming, there were no wild bunnies on Wellesley Island. We were told they had been wiped out by a combination of predators and disease. Because there were no rabbits, there was no need to install structures to exclude them.

Then, years later, the rabbits appeared. At first, it was a delight to see a lone cottontail hopping about our front yard before one of our dogs chased it into the brush. However, there is a good reason for the expression "breeding like rabbits." Soon there were telltale signs of their growing numbers and boundless appetites: pea seedlings and lettuce sets reduced to nubs in the annual vegetable gardens and woody stems nipped off at an angle close to the ground in the Enchanted Edible Forest. Clearly, we had a problem.

The T-post fence around the expanded garden still allowed natural predators like foxes to enter, so they were able to keep the rabbits there under control. The vegetable and original edible forest garden enclosed by heavy-duty fences were another matter. The fence excluded foxes and coyotes, but not rabbits, which could fit through the spaces in the welded wire.

I have found that floating row cover, the lightweight fabric I use in my vegetable garden to protect plants from chilly nighttime temperatures, will exclude rabbits in places where it is practical to drape over young plants. Where floating row cover is not feasible, adding barriers onto the existing fencing can exclude rabbits from your garden in the three following ways. Use metal material for these barriers because rabbits can chew through vinyl, and be sure that the lower vertical openings are no wider than 1 inch (2.5 cm) to prevent baby rabbits from squeezing through. With some modifications, these barriers can also exclude groundhogs.

Barrier in a Trench

Dig a 6-inch-deep (15 cm), 8-inch-wide (20 cm) trench around the perimeter of the outside face of the existing fence. Lay a length of 3-foot (1 m) galvanized metal chicken wire against the outside of the fence and down into the trench, with the bottom edge extending outward 8 inches to form an L shape at the base of the trench. This underground barrier prevents the bunnies from burrowing under the fence. If you need to exclude groundhogs, which are preeminent burrowers, make the trench at least 1 foot (30 cm) deep and 1 to 2 feet (30–60 cm) wide, and use chicken wire that is 4 feet (1.25 m) tall (or taller). Refill the trench with soil and secure the chicken wire to the fence, stretching it as tight as possible using plastic zip-ties. This is probably the most foolproof of the three methods but can be quite labor-intensive for a large plot.

L-Shaped Barrier

There's no trench involved with this method. Using 3-foot (1 m) chicken wire or a suitable substitute, fold the wire so the bottom of the L lies flat on the ground surface, extending at least 1 foot (30 cm) outward from the base of the fence (for groundhogs, 2 feet / 60 cm). Use garden staples to secure the chicken wire to the ground. Fasten the wire upright, stretched taut, to the fence posts using plastic zip-ties. This is less laborious than the first method, but still fairly rabbit-proof. A downside is that the aboveground footer could become entangled in a lawn mower or string trimmer used close to the ground outside the fence.

Hardware Cloth Reinforcement

For this barrier, use a stiff metal fencing material such as hardware cloth or welded wire with 1-inch (2.5 cm) openings (at least at the bottom) and at least 2 feet (60 cm) tall. Place the lower edge flush with the ground (with the narrowest openings at the bottom) and secure it to the fence with plastic zip-ties. This method involves the least labor but runs the risk of the bunnies or groundhogs burrowing under the fence. I recently installed a bunny barrier using this method. So far, so good.

As you develop your edible landscape, plan to incorporate some built structures as simple as a rock pile or birdhouse or as complex as a durable fence to welcome helpful critters and keep out the pesky ones.

Now that your plot is protected, landscaped, and furnished with ample access, and you have prepared and built your chosen planting beds, the next steps are to install, nurture, and protect the plants that will occupy your edible forest garden, and that's the focus of the next chapter.

CHAPTER 5

PLANTING AND TENDING YOUR GARDEN

In this chapter I offer detailed instructions on how to get your new plants off to the best possible start and provide continuing care. Because of their large size, extensive root structure, and long life spans, planting and caring for woody shrubs and trees is a bit different from tending an annual vegetable garden. And even though a forest garden will require less labor than an annual vegetable garden of equivalent size, there are still seasonal maintenance tasks that need to be done each year to preserve and enhance the plants you install. In appendix 3, I provide a chart that lists management and maintenance tasks throughout the year.

Below I offer planting guidance and discuss routine tasks that will help your plants prosper, including weeding, watering, and pruning. I also describe some more structural installations to protect plants from animal pests.

Ordering and Storing Plants

Once I decided on locations for particular trees and shrubs in my forest garden, I began ordering plants. I source my plant materials from nurseries all over the United States and Canada, and I even ordered some saffron bulbs from a company in the Netherlands that specializes in them. Nurseries generally supply woody stock as bare-root plants or as potted plants. Bare-root plants are grown in the ground at the nurseries and then dug up while dormant. The soil is washed off the roots, and the plants are stored under controlled conditions until it is time to package them for shipping. Potted trees, shrubs, and other perennials may be growing in natural soil or in a potting mix, and they are usually already leafed out when they are shipped.

I generally purchase young trees and shrubs because they tend to be the least expensive choice, plus they are most resilient and easiest to plant. The downside is that you have to wait longer for young plants to gain substantial size after planting.

CREATING A PLANT NURSERY

It often happens that an order of plants arrives at your doorstep when the timing is not right to plant them. You might not have prepared the planting site yet, or—as often happens in my case—you may order so many plants that there is not enough time to plant them all on the day they arrive. In these instances, it is helpful to create a plant nursery, an

Figure 5.1. These bare-root tamarack seedlings (*left*) are being heeled in on the shady north side of my house, covered with soil and watered thoroughly (*right*), then covered with mulch.

area set aside where you can temporarily keep bare-root plants alive and protected until you are ready to plant them. A semi-shaded area with decent soil is ideal, such as a spot shaded most of the day by your home or garage or a large tree.

You can use a technique called *heeling in* to temporarily cover plant roots and keep them moist and sheltered from temperature extremes or damage. If your site is covered with sod, first remove the sod from the area where you want to heel in your plants. With this technique, rather than preparing a full-fledged planting hole, you will dig a shallow trench in which you can lay the roots of bare-root plants on an angle, rather than upright. Try to fashion a trench in which one side is at about a 30-degree angle from the soil surface; make the trench deep and wide enough to accommodate all of the roots of the plants you want to protect. Lay the plants with the roots along the gently sloping surface; it's fine to set the plants quite close together, as shown in figure 5.1. Cover the roots with soil and then water thoroughly. After watering I like to mulch the area with a thick layer of leaves or wood chips to keep the ground cool and preserve soil moisture. In this manner, you can preserve plants for future planting whether that occurs in a few days or months, or even the following year. That said, it is best to plant woody seedlings in their permanent locations before they leaf out, because the seedlings will have a better chance of adapting to their new digs if they are transplanted while still dormant. Moreover, when they leaf out, new shoots will begin growing toward the light, which means they will be growing sideways when you replant the tree upright. If I notice leaves starting to grow on heeled-in plants, I drop everything else and transplant them right away.

For potted woody plants, I dig a hole deep enough to contain the pot. I then place the pot in the hole so that its rim is level with the surface of the surrounding soil. I use some of the removed soil to fill in around the sides of the pot. Sinking the pot in this manner protects the plant's roots from winter cold and moderates the temperature of the soil in the pot in the heat of summer, making frequent watering less necessary.

The nursery is also a place where you can allow naturally spreading or self-seeding herbaceous plants to self-propagate. Simply plant a few "mother" plants in the nursery and let them grow there until they multiply. Then you can transplant the "babies" into your garden beds and leave the mother plants to continue producing more transplants.

Figure 5.2. One apple mint plant installed in my nursery has created an abundance of new plants to transplant into my garden.

Planting Bare-Root Trees and Shrubs

When a bare-root tree arrives, open the packaging and inspect the tree to make sure it is alive and in good condition. You can tell if a bare-root tree is alive by gently scraping off a tiny section of bark low on the trunk. If the tissue underneath the bark is green, the tree is alive: if it's brown, it likely is not.

You can plant a bare-root tree as soon as the frost leaves the ground—the sooner the better. As the ground gradually warms, your tree will emerge from dormancy in the most natural way. It is important to keep the roots moist and maintain the plant in a dormant or nongrowing state until planting. If you are unable to plant it immediately, but will within a week, moisten the roots if they appear dry and rewrap them in the packaging materials from the nursery to preserve the moisture. Then place the tree in a cool, unlighted space like a dark cellar or even in the refrigerator until you are ready to plant. If planting will be substantially delayed, it is best to heel in the tree as described previously.

On planting day, first gather your tools and supplies:

- Shovel with a pointed digging head
- Three pieces of plastic sheeting or cardboard 3 to 4 feet (1–1.25 m) square (for placing soil on)
- Garden hose (or other source of water)
- 5-gallon bucket
- Brick or rock
- Trunk guard
- Wood chips or other mulch, sufficient to create a 4-inch-deep (10 cm) layer about 2 feet (60 cm) in diameter around each tree after planting

PROTECTING THE ROOTS

Take precautions to protect tree roots from injury during the planting process. The roots are vulnerable to injury from dehydration, the sun's rays, and being crushed or broken. To protect the roots from possible injury, the best conditions for planting are when the sky is overcast and the air is cool, moist, and calm. Another good time to plant is in the evening when the sun is low, the air has cooled, and the wind has died down. Try to avoid planting when the weather is hot, dry, sunny, or windy. To ensure that the plant is adequately hydrated, it is best to soak the roots in a pail of water for at least 2 hours, but not more than 24 hours, before planting. Make sure all of the roots are submerged in the water. When necessary, weigh down the roots with a brick or rock to keep them submerged, making sure that the weight doesn't break or crush any roots. To prevent roots from drying out during the planting process, diligently minimize the time the roots are exposed to air or the sun's rays. I can't overemphasize this point: Depending on the ambient conditions, even a few seconds of exposure can be enough to permanently damage fine root hairs.

DIGGING A HOLE

Most nurseries provide instructions on how to plant their stock, and such instructions often suggest incorporating amendments or improved soil in the planting hole. I disagree with these recommendations. I generally follow the recommendations of Bill and Diana MacKentley, former owners of St. Lawrence Nurseries in Potsdam, New York, in the nurseries' planting guide; they add no amendments in the hole. In keeping with permaculture principles, I believe it is best for plants to learn to meet their nutritional needs from their new natural environment rather than from external sources.

After you've read the instructions, it's time to start digging. Your goal is to dig a hole wide and deep enough to encompass all the roots of your bare-root plant. The same procedure holds for planting a bare-root shrub as for a tree. In both cases, you will size the hole to accommodate the root mass.

Step 1. If the planting site is covered by sod, first mow or use a string trimmer on the sod, then dig it out and place it in one pile atop a piece of plastic sheeting or cardboard. (Later, you will place the sod in the bottom of the planting hole.)

Step 2. Dig the hole, placing the soil on a separate piece of sheeting or cardboard from the sod. You'll want to place the rich, crumbly topsoil in its own pile, and place any chunks of rock, clay, pure sand, or dense subsoil on a third piece of sheeting or cardboard.

Step 3. When you think the hole is the right size, remove your tree from the water bucket and lower the roots into the hole. Check the trunk: You should be able to see a line and change in color, which indicates the level of the soil surface at which the plant was growing at the nursery. Unless otherwise instructed, your goal is to set the plant roots at that same depth, so that the color change on the trunk lines up with the soil surface in your planting bed. One way to check this is to lay a board (or a long tool handle) across the hole; the line on the trunk should be at the same level as the board. Then quickly put the roots back in the bucket of water. If needed, adjust the size of the hole by digging out more soil or adding soil back into the hole.

If your soil is clayey, score the bottom and sides of the finished hole with your spade or a pitchfork. The indentations you make in this way break up any slick (a compacted surface that may form on the inside surfaces of the hole during the digging process), which is hard for tree roots to penetrate.

If you are planting in clay-type soils that hold moisture and are poorly drained, the sides of the hole should be vertical. This is because, after planting, the soil returned to the planting hole will soak up precipitation more effectively than the surrounding undisturbed soil. With clay soil, the goal is to minimize the surface area of disturbed soil to limit the amount of water that is absorbed close to the plant roots, to avoid potentially suffocating them. You may also wish to build a small mound in advance or as you plant to raise the roots higher above the water table to improve drainage.

Sandy soil usually drains very well but doesn't retain water well. Thus, for sandy soil, dig a wider, saucer-shaped planting hole. It will provide a larger surface area to absorb water during future precipitation events. In this situation, or in a more arid climate, you may also want to dig a deeper hole and plant the tree a few inches below the surrounding soil surface, creating a concave saucer around the trunk. This technique will place the roots closer to the groundwater, and the concave surface of the hole will collect even more water during precipitation events.

PLANTING THE TREE

Now that you're ready to plant the tree, here's what to do next:

Step 1. If you removed sod and set it aside, place the sod green-side down in the hole. Chop it up with your shovel and tamp it down firmly with your feet or the shovel. As the turf decays, it will provide a boost of nutrients to the adapting plant.

Step 2. Use your topsoil to build a small mound in the middle of the hole. Many trees have roots that can be spread out like an upside-down cone over the sloping sides of the mound.

Step 3. For this step, it's handy to have a helper who can hold the tree in place while you fill the hole with soil. But if you're working alone, first remove the tree from the bucket and spread out the roots over the mound. Then immediately start covering the roots with additional topsoil. You will have to use one hand to hold the tree upright at the correct height and use your other hand to move the soil until there is enough soil in the hole to support the tree. If you have a helper, they can hold the tree, leaving you free to apply the soil with both hands.

Figure 5.3. This row of newly planted cherry trees is ready to have trunk guards installed to protect them from bark-chewing rodents. Then more mulch will be added to retain moisture and keep weeds away.

Figure 5.4. The cherry tree seedlings you plant today will look like this after just a few years.

Step 4. Once the roots are covered with topsoil, place some of the poorer-quality soil and chunks of clay on the outskirts of the hole to help support the plant. As you work, keep rechecking to make sure the tree is positioned at the intended depth, not too deep or too high. Also check that the graft union, if there is one, will sit above the final soil level when the hole is completely filled in, unless otherwise instructed by your nursery.

Step 5. When the hole is about two-thirds filled, tamp down the soil with your hand or foot to ensure all the roots come into good contact with the soil. Tamp moderately—too much pressure can crush roots or squeeze all of the air out of the soil. Keep in mind that wet soil compacts more easily than dry soil, so if you're working with wet soil, be gentle when you tamp it down.

Step 6. Begin to add water to the planting hole. The goal is to be thorough but to apply the water gently enough to avoid washing soil away from the roots.

Step 7. Once the soil is saturated and water begins to pool in the hole, replace the remaining soil, placing the better-quality soil closest to the trunk.

Step 8. At the outskirts of the hole, use some of the poorer-quality soil to build a small circular mound; this mound will keep water from escaping from the planting area.

Step 9. Slowly apply more water, until the water pools on the soil surface before it seeps in.

Step 10. Place a trunk guard (also called a tree guard or trunk protector) around the base of the trunk and surround it with a layer of mulch at least 4 inches (10 cm) thick and 2 feet (60 cm) in diameter.

For mulch, I usually use wood chips, but you can use straw, grass clippings, compost, or any other weed-seed-free organic matter. Before I spread the mulch, I sometimes place a piece of cardboard on the ground around the base of the tree to further deter weed pressure. Shape the mulch into a "doughnut" around the trunk. The mulch layer should be very thin adjacent to the trunk and thick at the outer rim of the doughnut. Mulch left touching the trunk can trap moisture, which can lead to fungal disease, or provide habitat for destructive insects or rodents.

POST-PLANTING CARE

Keep your tree well hydrated, especially throughout its first growing season. Many sources suggest watering on a weekly basis if your garden doesn't receive at least an inch (2.5 cm) of natural rainfall a week. This advice is probably sound if you have sandy or loamy soil. With heavy clay, however, it is

Keeping Your Perspective

Over the years I have lost countless plants. Some met their demise through choices I made, such as pairing a plant with an unsuitable habitat, over- or underwatering, replacing it when it was still alive, allowing too much competition from other plants before it became established, or choosing a plant incompatible with my hardiness zone. These misjudgments account for some losses, but not all.

There are many reasons a plant may die through no fault of your own. A nursery may inaccurately classify a plant's hardiness zone range, claiming that the plant can tolerate harsher or warmer winters than is actually true. The nursery stock may be defective due to poor root development, disease, or damage before shipping. Poor packaging may expose the roots to air or cause other damage in transit. Exposure to extreme heat or cold during shipment can also injure plant tissue. The wrapping around bare roots can become dislodged, or the soil enclosing the roots of potted plants can dry out with lethal results. Once a plant is installed in your garden, rapid changes in temperature, extreme wind, or excess rain may suddenly kill it. Soil that is too acidic, basic, rocky, sandy, dense, wet, or dry can cause problems as well. Too much or not enough sun, shade, or wind can lead to a plant's demise, as can competition from neighboring or native plants. Predation by insects or rodents can be deadly.

With experience, you will learn to avoid or prevent some of the above circumstances that can cause plants to die prematurely. But there will always be factors outside your control that just need to be accepted.

Figure 5.5. These aluminum labels represent just some of the perennial plants that have perished under my watch.

all too easy to overwater, saturating the soil and excluding all air, which can unintentionally suffocate the roots. My practice is this: Once or twice a week, I stick a finger down through the wood chip mulch surrounding a newly planted tree to feel the soil below. If the soil is cool and moist to the touch, there is no need to water. If it the soil feels at all dry, it's time to water. With my heavy clay soil topped by wood chip mulch, a good initial watering at planting time is sufficient to sustain most newly planted trees for the entire growing season, even in a drought year, but my soil is unusual in its ability to retain moisture. It is best to check the moisture level in the soil around your newly planted trees once or twice a week all season long and water whenever the soil feels dry.

It is important to keep competing plants from encroaching on a newly planted tree until its roots are well established. Early in my learning curve, I was careful to keep the area around newly planted trees free of weeds, but I made an exception for clover, the remnants of the frost seeding I had done to prepare the overall garden site. I allowed the clover to spread over and into the mulch surrounding the seedling trees, believing that clover would be beneficial because of its nitrogen-fixing capacity. Over time, however, I observed that clover seemed to compete with the new plants for moisture and root space. The new trees failed to grow as robustly as those without the infringing clover. Based on this observation, I began restricting the clover around new plantings, and results improved. Once a tree is fully established, it is fine to allow clover or other beneficial plants to cover the ground close to the trunk.

Planting Potted Trees and Shrubs

If you order potted trees, check with your supplier to find out where the plants were stored before being shipped. If they were kept in a heated greenhouse or another location with ambient temperatures significantly warmer than your outdoor conditions, you probably need to harden off the plants. Hardening off involves gradually exposing plants to the outdoors by putting them outside for a short time the first day and gradually increasing exposure time over the next few days until the plants are left outdoors for a full 24 hours.

When a potted tree is adequately hardened off and you are ready to plant, water the root ball thoroughly. The best way to do this is to submerge the potted plant in a bucket of water in an upright position until air bubbles cease to rise to the surface. Then remove it from the pot and inspect the roots. If the roots are *root-bound*—encircling the outer surface of the root ball—trim them with a scissor or clippers until you can spread out the still-attached roots to examine their expanse. If you plant trees or shrubs in the root-bound state, the roots will likely continue to grow in this manner, circling round and round within the planting hole, and your plant will struggle to survive.

Once you've trimmed the roots as needed, place the root ball back in the pot or otherwise protect the root ball from the air and sun while you are digging a planting hole to size. Place the root ball in the hole to make sure the depth and width of the hole are adequate. Then plant as described previously, spreading out any loose roots to encourage their outward growth. If the root ball is a solid mass of fine roots, you needn't build the mound of topsoil to spread the roots around. Just place the bottom of the root mass on the bottom of the hole and fill in around it. Water it well, then apply a trunk guard and mulch.

Transplanting Trees and Shrubs

As your garden develops, there will be occasions when you need or wish to move an established plant. You may decide that a tree would do better in a new habitat that you think will be more conducive to its growth. Or you may want to relocate a shrub that is becoming too crowded or shaded by neighboring plants.

A Transplanting Technique

During a walk in the farthest reaches of our farm one early summer day, I discovered a cluster of basswood seedlings growing in the brush near a mature basswood tree. I decided to try transplanting some of them into my edible forest. The narrow stems of these young trees were no more than 8 inches (20 cm) tall, with just a few leaves, so I assumed they were about two years old. When I dug down to extricate the roots, I learned otherwise: These roots were thick and deep, characteristic of a much older sapling. I surmised that these plants were probably stunted aboveground by repeated browsing by deer and/or rabbits, while their roots continued to expand. I was able to dig up six of these seedlings, but only by cutting the taproots about 6 inches (15 cm) below the ground. Because of the root damage, I was skeptical about their survival.

Considering how to promote their viability, I recalled a mulberry tree I had ordered that arrived leafed out in a pot. The supplying nursery had trimmed away the upper half of each of its leaves. I later confirmed with the nursery that this was done to reduce transpiration, the loss of water through the leaves. I decided to try this technique with the basswoods. I removed about half of the leaves from each plant entirely, then snipped away part of each remaining leaf. Then I planted, watered, and mulched each tree, hoping for the best. I checked on them about every other week until they became dormant in fall. To my amazement, their leaves remained stiff and green—they were alive! I resolved to use this technique during future summers whenever I needed to install leafed-out bare-root plants.

When transplanting, the goal is to capture as much of a plant's root structure as you can, retaining as much of the soil surrounding the roots as possible. This will give the plant the best chance of establishing itself successfully in its new location. Here's the technique:

Step 1. Make vertical cuts with a shovel in the soil around the plant at a distance about half the height of the bush or tree. These cuts should overlap so you create a continuous vertical slice encircling the roots. If the plant is 1 foot (30 cm) tall, a circle with a 6-inch (15 cm) radius is probably sufficient. If it is 2 to 4 feet (60–125 cm) tall, a radius ranging from 9 to 12 inches (23–30 cm) should be adequate, and so on. The depth of the cuts also depends on the size and root structure of the plant. For shallow roots, cuts 6 inches deep may capture most of the roots. If the plant has deeper roots or a taproot, dig deeper to capture as much of the root mass as possible.

Step 2. Begin digging down deeper and toward the tree. Your goal is to extricate the roots in a bowl-shaped mass with the soil intact. You may encounter roots that you need to cut through using the point of the shovel or a pruning implement.

Step 3. Once you have cut through all the roots you encounter, spread a damp cloth on the ground nearby.

Step 4. Extricate the root ball from the hole, wrap it in the damp cloth to shield the exposed roots from the air, and transport it to the new location.

Depending on the size of the tree, a root ball can be quite heavy; you may need help to move it.

Step 5. At the new site, dig a hole to match the size and shape of the root ball you have extricated.

Step 6. Remove the cloth and lower the root ball into the hole. The surface of the root ball should be level with the soil surface surrounding the hole.

Step 7. Use the excavated soil to fill in any gaps between the root ball and the sides of the hole.

Step 8. Apply some water to settle this soil, and add more soil again until it is level with the surrounding surface.

Step 9. Use more of the removed soil to build a raised circle around the tree to form a saucer to hold water.

Step 10. Water the tree just until the water begins to pool.

Finish by applying a trunk guard and mulch as described previously in the instructions for planting a bare-root tree (step 10, page 74).

Figure 5.6. This volunteer cedar tree chose to grow nestled between red currants and raspberries in the shade of an apricot tree.

Sometimes soil does not adhere well to the roots, but instead falls away, leaving the roots bare. In this case wrap the roots in a wet cloth and follow the instructions for planting a bare-root tree (page 72). If the roots are bare and the plant is leafed out, it may benefit from the technique described in "A Transplanting Technique" on page 77.

Weeding and Watering

Weeding and watering are two tasks that diminish considerably as your forest garden develops. At first, sheet mulching or other techniques to remove weeds as described in chapter 4 are necessary to create clean planting beds. Then mulching and spot weeding reduce competition with young plants until they become established. Rather than pulling out weeds, I recommend deadheading them: Cut off the fading flowers and immature seed heads of weed plants. This limits their spread without disrupting the soil. Deadheading also works to prevent herbs or other herbaceous garden plants from unwanted self-seeding. Once your ground is covered with plants you intend, there's much less need to weed and mulch. Rather than weeds, volunteer tree and shrub seedlings may appear amid your plantings. These you can remove or leave, depending on your judgment. In my garden, a few cedar seedlings have spontaneously germinated from seeds either deposited by birds or present in the wood chip mulch. I have chosen to let them grow because they make excellent habitat for nesting birds and because, being evergreen, they add a touch of color to the winter landscape.

Woody and herbaceous plants need to be watered sufficiently during planting and for a period thereafter while their roots are becoming established. Once the roots are able to source water independently, there is little need for supplemental water. When I planted my first trees and shrubs, I read that a specialized drip watering system for these plants would be beneficial. Dave and I purchased and installed a system, but this investment of capital and labor turned out to be unnecessary—I never needed to use it. With mulch and ground cover plants keeping the soil cool and preventing evaporation, my garden has turned out to be self-sustaining with regard to moisture. One exception is some brambles that have needed additional watering during droughts to ensure adequate berry production. Regardless of the type of soil you start with, over time, as your ground remains undisturbed and organic matter accumulates on and within it, your need to water will correspondingly diminish.

Pruning

Pruning is the practice of selectively removing branches from a tree or bush. In nature, "pruning" occurs randomly when branches are broken or damaged by a windstorm, a fire, or the nibbling of wild ruminants. In a garden or farm setting, though, productivity and health of plants improves when limbs are systematically clipped. Thinning branches improves air circulation to ward off disease, lets in light to hasten ripening, encourages blossoms and can result in fewer but larger fruits. Removing dead, damaged or diseased limbs helps prevent infestations from spreading to healthy growth. Pruning also renews a tree or bush by encouraging growth of new branches that fruit more reliably than older ones. If your space is limited or you prefer not to climb high to harvest fruit, you can prune to manage the size of your trees and shrubs. Pruning can also result in plant forms that are more resilient and aesthetically pleasing or in dense growth of a hedge. Removing lower limbs of a tree can allow more light to bathe plants below.

LEARNING TO PRUNE

Each type of tree or shrub you plant may require slightly different pruning techniques. Permaculturist Sepp Holzer does not prune any of the plants on his extensive permaculture farm in Austria. That is an option for you as well, especially if you have a good deal of acreage and prefer to let your forest garden remain more natural. But if you would like to actively enhance the health and productivity of your plants, it is wise to apprise yourself of pruning basics. Most domesticated fruit trees benefit from pruning to enhance the quality of their fruit. I am admittedly a novice in this endeavor, though I tried my hand at pruning a few apple and peach trees the last few seasons. I learned that this enterprise is as much an art as a science, and takes much practice to master. I also observed that trees are resilient and will in most cases survive even the most egregious mistakes of the beginner.

Shaping and maintaining trees and bushes when they are young promote the best form and growth from the get-go. In addition, pruning cuts heal more quickly when only young, thin branches are removed. In part 2 you will find detailed descriptions of pruning techniques for some types of berry bushes and brambles along with some suggestions for caring for grapes in northern climates. I recommend you seek expert advice to learn how best to maintain the plants you choose, and do so early on. Excellent references for pruning techniques can be found in the resource sections of some nursery websites, on YouTube, and in some of the references sited in the selected bibliography of this book. Your local cooperative extension office can also be a resource.

I recommend that you familiarize yourself with basic pruning principles and, if possible, apprentice yourself to a more experienced friend or neighbor to learn specific techniques.

THINNING FRUIT

In addition to pruning, many fruit trees benefit from annual thinning of immature fruit. Thinning can remove diseased and distorted fruits while enhancing the size and quality of those remaining. It is also a way of helping a young tree by allowing it to direct most of its energy into expanded growth rather than fruit production. Each type of fruit tree will benefit from a slightly different thinning practice; consult your nursery or another source to determine the timing and technique best suited to your chosen trees.

Dealing with Pests

Whether your garden plot is in a city, suburb, or country setting, you are likely to have to contend with some types of pests. Deer, mice, woodchucks, and rabbits are examples of wildlife that can molest your plants. To exclude deer, Dave built a heavy-duty electrified fence, as described in chapter 4. However, there are less elaborate ways to protect your plants. Below I describe a few simple structures that do the job.

TOMATO CAGES

Despite a perimeter fence, my garden expansion experienced regular deer incursions when the fence was not adequate to deter them. In the early days of our market garden, we used cone-shaped wire tomato cages to hold sprawling tomato plants upright. Since then, we've switched to permanent trellises to support the tomato crop. Pondering what to do to protect my nascent trees, I thought: *What if I turn a tomato cage upside down so the wide opening is at the bottom, wrap it with chicken wire, and place it around a young fruit tree? Would that work to keep the marauding creatures at bay?* I constructed a few of these tepee-like barriers and carefully lowered each one into place over an immature Asian pear or one of my other valued trees. Then I secured the base of each cage with three or four garden staples pushed firmly into the ground.

Over time, I observed the caged plants. I saw signs that deer were indeed entering the garden at night to feed, including their telltale hoofprints and occasional leaves and stem tips missing from some bushes and ground cover plants. But there was no further damage to the enclosed trees. Satisfied that my invention was effective, I constructed more of the tree cages. From then on, I enclosed every newly installed sapling with a modified tomato cage, leaving it in place until the tree was taller than the deer could reach. I continue to do this with all new plantings as extra insurance against deer in the event that the barrier fence fails.

TRUNK GUARDS

Even before I planted my edible forest garden, I had read about the importance of protecting the trunks of trees and shrubs from girdling by rodents during the winter. The planting guide from St. Lawrence Nurseries written by Bill and Diana MacKentley recommended using hardware cloth (which is a material made of stiff, welded wire with ½-inch / 13 mm openings in a square grid) rolled into cylinders to encircle the trunks. I purchased sufficient quantities of this material cut to size to protect all the trees I planted the first year.

Figure 5.7. An upside-down tomato cage wrapped with chicken wire protects a young tree from deer in the Enchanted Edible Forest.

Instead of installing the trunk guards around most of the trees during the planting process, I waited until fall to begin the procedure. That year, cold weather arrived early, and when the temperature dipped below 40°F (4°C), it was painful on my bare hands to bend the tips of the metal wires to secure the cylinders in place. I gave up after installing just a few protectors, hoping for the best. The next spring, the melting snow revealed my imprudence. Most of the unprotected trunks were fatally girdled by the furry pests, while the protected trunks were unscathed. I resolved to never put off protecting my valued trees again and began installing the guards as part of my planting procedure described earlier in this chapter.

Figure 5.8. Trunk guards protect the trunks of these cherry trees from rodents scavenging under the snow.

Shrubs were another matter. The hardware cloth guards could not encircle their spreading branches, so I left them unprotected. For several seasons, no problems arose. Then one year, the rodent population must have exploded because, after the snow melted in spring, the evidence of rodent malice was clear. Multiple stems of nearly 100 of my shrubs were irreparably chomped. Luckily, most of the shrubs (with the exception of the Meader cherries and one beach plum) recovered by sending up new shoots from their roots. Nevertheless, I resolved to protect them in the future.

The staff at my local hardware store recommended I try surrounding the entire base of a multi-stemmed shrub with a mantle of metal screening material. That fall, I purchased a 4-foot-wide (1.25 m) roll of aluminum screening and cut it in 1-foot-wide (30 cm) strips. I wrapped a strip around the base of each bush, and where the ends of the strip overlapped, I used clothespins to secure

Figure 5.9. Classic examples of girdling: Under the cover of snow, rodents gnawed through the bark on this fruit tree from a point below the graft union to several inches above it (*left*). Spring melt reveals the devastated stems of a berry bush chewed by rodents (*right*).

Figure 5.10. A metal screen wrap protects a multi-stemmed berry bush from rodent nibbling under the snow.

them at top and bottom edges. This protection worked remarkably well. One year when I wrapped most of my shrubs in this manner but left the stems of my elderberries unprotected, the wrapped shrubs were unharmed while almost every elderberry trunk was girdled mercilessly.

Mice and voles do their damage beneath the snow cover, so it's important to ensure that the hardware cloth guards and screen wraps you make are tall enough to exceed the average snow height in your region. We rarely get more than a foot of snow on Wellesley Island. After a deep or drifting snowfall, I scout the garden to scoop away any snow that covers the tops of these wraps, both inside and outside the barriers, so they will continue to serve their purpose.

Propagating Berry Bushes

Propagating berry bushes yourself is a fun and satisfying way to expand your garden at no cost or to make a garden gift to pass on to friends and neighbors. Here are four methods that are easy to implement.

LAYERING

The basic technique of layering is to lower the middle section or tip of a stem to the soil surface, clip it firmly to the ground, and cover a short length of the stem with a handful of soil or compost. Once it sets down roots at the tethered location, the resulting young plant can be severed from the mother branch, dug up, and potted or transplanted to a new location.

In my garden, I have observed blackberries, currants, gooseberries, jostaberries, and goji berries layering on their own. I usually discover these fortuitous plants in autumn after the leaves have fallen, when they are more visible. I mark their locations with bamboo stakes, so I can locate them in early spring to pot up or transplant.

I like to experiment with layering intentionally. In the spring, while pruning my berry bushes, rather than remove a low-lying branch, I wrap a piece of brightly colored flagging tape around a horseshoe-shaped landscape staple and use the staple to secure the branch to the ground. The flagging tape marks the spot. Then I drop a handful of semi-composted wood chips on top. Come fall, or even early spring of the following year, I sever the new plant from the original branch, dig it up, and plant it elsewhere or give it to a friend.

HARDWOOD CUTTINGS

Taking dormant hardwood cuttings is a great way to propagate many kinds of fruiting shrubs. Using this method, you can generate many new plants with little effort.

In late winter or early spring, cut off dormant sections of pencil-thick, year-old stems 6 to 10 inches (15–25 cm) long. Be sure each cutting has at least three buds. Cut the top end off each cutting at an angle just above the uppermost bud, and cut the base of the cutting straight across just below the bottommost bud. That way you can tell which direction they were growing and be sure to put the squared-off ends into the rooting medium or soil. The cuttings can be preserved in the refrigerator until you are ready to plant them.

When you are ready to plant, dip the lower half of the stems in a commercial rooting medium or willow water. Insert the cuttings 1 or 2 inches (2.5–5 cm) apart, and deep enough to submerge two buds, in pots filled with potting medium. Water them well and place them in a shaded area outdoors that receives natural precipitation. Then wait. By summer, the cuttings that root will be flourishing and ready to be transplanted. I have had success ranging from 0 to 100 percent in propagating shrubs in this manner: none with blackberries or clove currants, nearly 100 percent with jostaberries and elderberries, and varying degrees of success with gooseberry, highbush cranberry, honeyberry, autumn olive, sea buckthorn, grapes, and most varieties of currants.

DIVISION

Any suckering plant can be propagated through division. First, locate a young sucker some distance from the base of the original plant. Next, take a sharp shovel and dig down vertically between the mother plant and the sucker to sever the root that connects them. This root may be as much as several inches below the soil surface. Continue making cuts with the shovel in a circle around the sucker, far enough from its base to capture most of its roots. Then use the shovel to gently scoop the root ball out of the ground, retaining as much soil around the young roots as possible. Immediately replant the division in a pot or in a new location, water well, and mulch. Raspberries, aronia, elderberries, juneberries, pawpaws, and persimmons lend themselves to this method.

SEED

The specific techniques for propagating by seed vary depending on the plant you wish to propagate. That said, there are several general considerations. First, determine whether your seed needs a period of stratification (exposure to cold) and for how long. Second, find out whether the seed needs to be scarified, soaked in boiling water, or otherwise abraded to allow it to absorb water. Then do some research to determine the soil temperature most conducive to germination.

Once you have that information and have prepared the seeds properly, go ahead and plant them in potting soil (preferably organic). Be sure to plant them at the proper depth. A good rule of thumb is a depth equal to between two and three times the diameter of the seed. If you're planting in a pot or flat, be sure there is sufficient depth of soil below the seed to allow room for roots, particularly taproots, without stunting. In the case of a taprooted plant, if you plan to repot it into a larger container once it has a few leaves, a pot 4 to 6 inches (10–15 cm) deep is probably sufficient. If a plant has a spreading root structure, 2 or 3 inches (5–7.5 cm) deep is plenty. For a discussion of how I successfully propagated a tree from seed, see "A Propagation Story" on page 138. Beware that propagating woody plants from seed can be tricky, and a low (or even nonexistent) germination rate is

Figure 5.11. These potted dormant currant cuttings were planted in March and placed outdoors. By late spring they had rooted and were leafing out and beginning to grow new shoots.

not unusual. It's a good idea to plant more seeds than you think you need. In the best-case scenario you will have extra shrubs or trees to give away or expand your forest garden even further.

Propagating Perennial Herbs from Seed

Perennial herbs are relatively easy to propagate from seed as long as you plant them at the right time under the right conditions. There are advantages to planting herb seeds indoors as opposed to outdoors. Under conditions of light, soil temperature, and moisture that you can control indoors, you are more likely to be successful. Plus, you can determine in advance how many plants you will have to transplant outdoors once they are ready. Sowing seeds directly outdoors in the garden is a bit of a crap shoot. You may get so many you need to thin them out, or few or none, or they may be overcome by neighboring plants before they are mature enough to fend for themselves. When I am particular about the outcome, I start them indoors; otherwise I just toss them on top of mulched ground in a habitat that I believe they will like: shady and moist for shade-loving plants such as sweet cicely, sunny and dry for drought-tolerant ones like chamomile and echinacea. Over the years I have learned which herbs are likely to do well just thrown on the ground by observing which ones self-seed and where, *and so will you*!

It takes about six months for seedlings of woody-stemmed herbs to reach sufficient size to be planted outdoors. This includes lavender, oregano, thyme, and sage. The best time for sowing these herbs is in early to midwinter, so they are ready to move outside once the soil warms. More succulent herbs like chives and sweet cicely can be sown indoors in early to mid-spring. Alternatively, many herbs can be direct-seeded outdoors once the soil warms.

To start seeds indoors, begin with a 4 × 4 or a 4 × 6 inch (10 × 10 or 10 × 15 cm) flat with or without individual cells, and fill it with sterile potting soil. Many herbs are tiny and frail when they first emerge as seedlings. At this stage, they are vulnerable to potentially fatal fungal infections. Non-sterile soil often harbors fungal spores that can infect and kill your tender plants.

To sow tiny herb seeds such as thyme and oregano, drop a small bunch onto the soil and use your finger to gently press them against the surface. Or you can sprinkle the seeds over the entire surface of the soil in the container, and use the flat of your hand to press the seeds into close contact with the soil. It's important to be gentle to avoid compressing the soil too much, which could prevent oxygen from penetrating the mix, or make it so dense that the tiny seedling roots will not be able to penetrate. Larger seeds, such as sage and lavender, can be planted to a depth of two to three times their diameter. Make about 12 to 20 equally spaced holes of that depth with your finger and drop one seed in each hole, covering them with potting mix. Once the seeds are placed, water them gently so as not to wash them out, until water drips from the holes at the bottom of the container. I use the spray nozzle attachment at my sink, using a very gentle flow. I hold the nozzle flat rather than at an angle to the soil surface so the spray doesn't unintentionally move the seeds or soil around. As I water I move the nozzle continuously above the soil so that no water pools in one spot. After watering, place the container under a fluorescent or alternative bright light for 14 hours each day, ideally in an environment where the ambient temperature is about 70°F (21°C). Make sure to keep the soil surface evenly moist through gentle daily watering. I sometimes lay a piece of plastic wrap over the soil to help hold moisture in during dry winter days.

Herb seeds can take up to three weeks to germinate. Once seedlings emerge, be sure to keep the soil thoroughly moist (but not saturated) until roots become established, a week or two after the first leaves emerge. After that, it's safe to allow the soil surface to dry out between waterings. Once the seedlings have developed a few leaves, it is best to bottom-water. To do this, fill a shallow pan or flat with water and submerge the bottom of the vessel containing the seedlings in the water. The potting soil will absorb water through the holes in the bottom of the vessel. This watering method does not wet the stems or leaves, and thus helps prevent damping off, a fungal disease that thrives in wet conditions. Damping off fungi can infect tender

Figure 5.12. Your reward for your tender care of your herb seedlings is seeing them flourish in the garden.

tissues of seedling stems at the soil surface, often killing the seedlings.

After about eight weeks of growth, your herb seedlings may be getting crowded in their containers. It's time to separate and replant them individually in small pots. (I use empty yogurt cups with holes punched in the bottom). Gather all your materials before you begin—you will need your pots, a flat tool (such as a tongue depressor), potting soil (can be the same as your germination mix), and a watering can. Water the flat containing your seedlings thoroughly. When the soil is wet, it is more likely to cling to the roots of the plants as you remove them, keeping the roots from drying out as you move them from the flat to a pot. To remove the seedlings, insert your flat tool near the edge of the flat and gently lift up on the soil, teasing a small cluster of seedlings away from the rest. Once you've removed a cluster, you can either separate them into individual plants or smaller groups, doing your best to keep the soil around the roots intact. Use your judgment: The larger, stronger plants can be separated by softly clasping them with two fingers at the base of the stem and gently pulling them apart, whereas the smaller, weaker ones may fare better if left in clumps. For example, in the case of herbs, I would likely separate sage into individual plants but leave thyme in clumps.

Gently hold a seedling or cluster of seedlings by the leaves or stems and suspend them inside one of your empty pots, lowering it in until the location where the stem met the soil in the original flat is about ¼ to ½ inch (6–13 mm) below the lip of the pot. Hold the plant in this position with one hand and use the other to sprinkle potting mix down around the roots until they are covered and surrounded with soil. Tap the pot gently on a surface to settle the soil, then add a little more potting mix to fill in where the soil settled, leaving the ¼- to ½-inch space to the top of the cup. This space will let you water without washing the soil out. Make sure all the roots are covered with soil but the stem is not. Water the transplant until water seeps from the bottom of the pot, then let the surface soil dry out before rewatering. After a few days, the herbs can be moved to a location that doesn't stay uniformly warm, such as an unheated room in your house that is warmed by the sun during the day and cools down at night. You will just need to make sure that the location is well lit.

When the weather begins to warm in the spring, it is time to start acclimating your plants to the great outdoors. Begin by setting them out on a warm, calm day in a spot with indirect light, for no more than two hours, and then bring them back indoors. Over the next two to three weeks, gradually expose them to direct sun, a greater range of temperatures, and longer durations outdoors, working up to a full 24 hours. If you have a cold frame—a transparent, enclosed structure built close to the ground that retains the sun's heat—place the plants inside, vent it during the day, and close it at night. These young perennials can withstand light frost, but cover them with row-cover fabric or bring them inside if heavy frost or a freeze is predicted.

By late spring or possibly even earlier, depending on your location, the soil and ambient temperatures will be warm enough to plant your herbs into your garden beds. Be sure to water them well, apply mulch, and check them regularly while they are settling in to make sure they have enough moisture. Once they are established, natural precipitation should meet their needs. While you may not be successful with every perennial herb you attempt to grow from seed (I certainly have not been), some will do well. Once planted in your garden, they will provide you with flavorful harvests for years to come.

We've come to the end of the overview of forest garden design, planting, and care. Now it's time to dig in deep on deciding what you want to grow. In part 2, "Plants for Every Level," I share my experiences growing a multitude of interesting and nourishing plants for an edible forest, from trees to ground covers and all the layers between.

— PART 2 —

PLANTS FOR EVERY LEVEL

Figure 6.1. Although pawpaw is borderline hardy in Zone 4, I took a chance on it, and six years after planting, it flowered!

CHAPTER 6

CHOOSING PLANTS

Now that you understand the nature of your plot, have considered what landscaping and infrastructure you may wish to incorporate, and are acquainted with the basics of plant care, it's time to choose your plants. This process takes due consideration. Perennial trees and shrubs can be pricey, and planting them and getting them off to a good start takes significant time and effort. Understanding the habitat preferences of plants will require research. When you choose plants that match your hardiness zone and the habitats you have available, you are most likely to be successful. At the same time, taking a calculated risk with a specimen that you really want to try, even if it may not be happy under your conditions, keeps things interesting.

Some gardeners like to populate their gardens exclusively with native plants. If that is your case, you will find many native plants to choose from in the chapters that follow. I incorporated many natives in my garden but chose to have greater diversity, incorporating a large variety of species, including non-natives and even some plants that are not hardy to my zone.

Perennial plants are different from annual vegetables and flowers in ways that are important to take into account as you choose plants you wish to grow. For example, since perennials live for years, rather than just one growing season, you need to know whether they can tolerate the average lowest winter temperatures in your region. It's also important to understand their habitat preference, root structure, pollination requirements, and eventual height and width so you can choose a site that meets their needs. It's also wise to learn about and anticipate maintenance requirements such as annual pruning and protection from pests. And although some perennials may provide a harvest within the first year or two after planting, others may grow slowly and take years to become productive, so you need to have the patience to wait and watch.

About Plant Hardiness

Do you know what hardiness zone you live in? If not, refer back to "Observing the Land" on page 29 to find out how to check on your zone. Knowing your zone is important in choosing plants for your plot. Most nurseries list a range of USDA hardiness zones for each plant they offer for sale. For example, a nursery may say that a particular cultivar of raspberries is hardy in zones 4 through 8. (A *cultivar* is a cultivated variety of a plant.) As you compare listings from nursery to nursery, though, you'll discover that the ranges listed for a particular cultivar often differ from source to source. In the plant descriptions in the following chapters, I have indicated my best estimate of the hardiness zone range of the plants I describe.

If you live in a region with cold winters and a relatively short growing season, it is often best to source plants acclimated to those growing conditions. For example, if you garden in the Upper Midwest, look for nurseries that are located in Zone 4 or 5. On the other hand, if you live in the warm and humid Southeast, with relatively warm winters and an extended growing season, choosing plants from nurseries located in Zone 7 or 8 may maximize your chances of success. With experience, you will determine which suppliers offer stock that does well under your conditions.

In the future, climate change could lead to milder winters and warmer summers overall, while also extending the growing season. Hardy plants that require an extended cold, dormant period may not do well. More-tender plants could possibly thrive and be productive in hardiness zones beyond what their present-day ratings indicate. I am already seeing evidence of such changes.

Given this reality, I decided to take some chances and push the envelope on hardiness zones by growing some plants adapted to warmer winters. In the worst-case scenario, species that are listed as hardy only to Zone 5 or 6 will not survive for the long term in my garden. In the best case, though, they will survive and even thrive, and I might be the only gardener in the region who is growing and harvesting those fruits and nuts. For example, I decided to try my hand at growing Asian pears, even though they are not native and also are challenging to grow in Zone 4 (see "My Asian Pear Adventures" on page 92). If you want to gamble on some plants that are not rated as hardy in your zone, select the hardiest varieties of the species you can find, and plant them in beneficial microclimates that will enhance the possibility of success.

Understanding Fruit and Nut Trees

When I began researching the Enchanted Edible Forest, I was excited to learn about some familiar and some completely unfamiliar fruit and nut trees to include in the design. Because our winters are so cold and the soil in my garden is heavy clay, I knew I would be taking a big chance with those that preferred a less harsh climate and required excellent drainage. But I decided to attempt growing most of them, nonetheless.

If you are new to growing fruit and nut trees, too, it's important to be patient with them. Some will not flower or fruit until many years after planting; nut trees may take more than 10 years to produce a crop of nuts. Also, some may die back above ground level during a harsh winter, but their roots will still be viable. I've lost count of the number of times during the first few years of my forest garden project when I hastily replaced a tree I thought was dead, only to find live roots after I dug it up. If you cut away the dead branches, the tree may well re-sprout from its roots and live on for many more years.

My experiences over the past decade have taught me some important things about how fruit and nut trees grow and reproduce. I share those learnings here to help you maximize your success with the fruit and nut trees you decide to include in your edible garden.

ROOT DEPTH

Unlike some trees that grow deep taproots, many types of fruit trees have shallow root systems. It's crucial that they receive adequate moisture and be protected from competition from other plants while their roots establish. Most also require excellent drainage.

Nut trees often have deep taproots that take time to establish before the young trees begin to add significant height, so again, be patient. Because of their taproots, deep soil is advantageous for many of these trees to prosper. And when your trees do produce a crop of nuts, harvesting them before they are snatched by wild creatures such as squirrels and chipmunks can also be a challenge.

HOW GRAFTED TREES GROW

Many fruit trees as well as some kinds of nut trees sold by nurseries are grafted plants. This means that part of a branch or *scion* of the desired cultivar is attached to the roots from a different cultivar or species, referred to as the *rootstock*, in such a way that the rootstock and scion grow to form one tree. The scion conveys the qualities of the desired fruit or nut, while the rootstock can convey hardiness and possibly disease resistance and determines the eventual size of the tree. Only trunks and branches that emerge above the graft union (the spot where the scion is grafted onto the rootstock) will be true to the characteristics of the named cultivar. Stems that sprout from below the graft union will have the characteristics of the rootstock and will not produce the expected quality of fruit. When you purchase a grafted tree, check what type of rootstock it is growing on, and be sure that rootstock is suitable for your conditions.

A *seedling tree* is an ungrafted fruit or nut tree that has grown from a seed. In my experience, some seedling trees may be hardier in certain instances than grafted cultivars of the same variety, but the quality of their fruit or nuts can be less predictable.

Nurseries offer many fruit and some nut tree varieties in *standard* (full-sized), *dwarf*, and *semi-dwarf* forms—the mature size depends on the rootstock on which the trees are grown. In Zone 5 or colder areas, it is best to plant standard trees, which have expansive, hardy rootstocks that are more resilient in face of subzero temperatures. If you desire a smaller tree, keep a standard tree rigorously pruned to fit your space. Or you can choose a natural dwarf fruit tree, which is a variety that naturally remains smaller than other trees of the same type. For example, Evans sour cherry is a natural dwarf cherry variety that can be grown as an ungrafted tree. It reaches only 10 feet (3 m) tall and wide at maturity while most other tart cherry varieties grow twice as large.

POLLINATION

In order for fruit and nut trees to produce a crop, the flowers need to be pollinated. How does the

Figure 6.2. Butterflies pollinate a number of tree species as well as shrubs and herbaceous plants. This one is attracted by the oregano flowers.

My Asian Pear Adventures

Although I had never tasted an Asian pear or seen an Asian pear tree, after reading about them in books and catalogs, I knew I wanted to grow them. Asian pears are borderline Zone 4 at best. I chose a spot with a south-facing exposure where I thought they would be happy. I chose two cultivars, rated as hardy for Zone 4, and planted them. The following spring, both trees were dead. I noticed that the bark on their trunks was ripped and shredded. A little research revealed that Asian pear trees are particularly vulnerable to bark splitting in spring due to expansion and contraction of the trunk when the afternoons are warm and suddenly turn cold as the sun sets. This phenomenon, called *sunscald*, kills the bark cells and can eventually lead to the death of a tree. The remedy is to paint the trunks white, so the heat of the sun is reflected during the day, and the resulting temperature differential is not as steep when the trunk cools as the sun sets. (Incidentally, painting the trunks white is a good practice for all fruit trees, not just Asian Pears.)

Figure 6.3. Asian pear in spring bloom and fall color.

I was also concerned about the quality of the soil in which I had planted the trees. It was heavy clay, unremediated, and even if the trees had lived, I wasn't sure they could ever thrive in those soil conditions.

I decided to replant in the same spot but with a modified plan. I built a crescent-shaped south-facing raised bed to loosen the soil and improve drainage and to concentrate the heat of the summer sun to enhance the ripening of the fruit. I painted the trunks of the young trees white right after I planted them. Two years later I harvested a few pears from those trees. Then we had a very severe winter. Both trees were winter-killed above the snow level, but the roots and the trunk just above the graft were still alive. Both trees regrew from above their grafts. They flowered the next year, but a late frost killed the blossoms and no fruit formed. The next winter they were again winter-killed above the snow cover. They regrew, and once again bore some fruit the second year. I am hopeful that these trees will survive and ultimately thrive as the climate warms and the winters perhaps become less severe. Meanwhile, I am thrilled when a two-year span of favorable weather yields a harvest of deliciously sweet, crisp, and juicy Asian pears.

pollen move from flower to flower? Depending on the species, it can be transported by wind, water, insects, birds, bats, or butterflies. For example, pawpaw trees are pollinated by flies!

The conditions necessary for pollination also vary from species to species. A species or variety of fruit tree can be *self-fertile*, meaning that the pollen produced by its own flowers is sufficient to make fruit. But some species and varieties require cross-pollination, which means that the trees need to be pollinated by another variety of the same species that blooms at the same time.

Still another consideration is what type of flowers a plant bears. Many fruit trees have what are called *perfect* flowers, which means that each individual flower includes male and female reproductive structures (pistil and stamens). However, there are plant species in which some flowers bear only female reproductive structures and others bear only male reproductive structures (squash plants are an example of this in the vegetable garden). And with trees, some species bear male flowers and female flowers on separate plants—there are male trees and female trees.

Be sure you check a tree's pollination requirements as you decide what to grow. Unless they are billed as self-fertile, most fruit and nut trees need another compatible tree placed within at least 50 feet (15 m) for pollination. If the species has male and female trees, at least one of each sex is needed. And even if a tree is self-fertile, pollination can be enhanced by having another compatible tree nearby.

CHILLING PERIOD

Most fruit and nut trees require a "chilling" period, which is a minimum period of time (number of hours) during dormancy when they must be exposed to cool temperatures; the standard chilling temperature ranges between 32 and 45°F (0–7°C). This chilling exposure is what triggers the plant to produce flowers and fruit the following growing season. The length of the chilling period varies by species and cultivar.

Beware that as the climate warms, a cultivar that historically has been productive of fruit in your region may no longer bear well because of insufficient chilling hours.

POME FRUITS VERSUS STONE FRUITS

Pome fruits are characterized by a core containing a few seeds enclosed in a tough membrane. This center is surrounded by an outer layer of edible fruit. Apples and pears are familiar examples. Plants in this family require good drainage and are prone to infestations by insect pests and diseases. Most authorities prescribe a spraying regimen to keep these trees healthy and productive. As I have a personal aversion to the chore of spraying, even with "organic" materials, I knew I was taking a big chance by planting pome fruits and that they might never produce palatable fruit. I was also taking a chance because several of the pome fruits I planned to grow were rated for warmer climates. I dove in despite these concerns. How did things turn out? To find out, read the entries about pear, apple, medlar, and quince trees in chapter 8.

Stone fruits are so named because in the center of each fruit is a pit as hard as a rock. These seeds, also called *drupes*, can be attached to the surrounding flesh or not (clingstone or freestone). Mangoes and olives are stone fruits that grow in warm climates. In chapter 8, I describe my experience with stone fruits like cherries, peaches, and plums that grow in more temperate habitats. Stone fruits require excellent drainage and are prone to fungal diseases, as well as some of the same insect pests as pome fruits. Some, such as apricots and peaches, are also particularly susceptible to late frosts killing flowers or flower buds and thus preventing fruit formation. If your property is in a frost pocket or is otherwise prone to late frosts, choose the latest-flowering cultivars of apricots and peaches you can find to help avert this problem.

Plants with Other Useful Functions

In addition to selecting plants that produce food products that interest you, it is important to populate the layers of your garden with supportive plant species that will make your garden resilient and self-sufficient. Even though some may not produce food, these plants take care of their food-producing neighbors by providing nutrients, attracting beneficials, and deterring pests.

NITROGEN-FIXING PLANTS

After water, nitrogen is the element needed in greatest quantities for plant growth. Nitrogen makes up 78 percent of the Earth's atmosphere but cannot be used by plants in its gaseous form. In order for plants to take up nitrogen, it must be chemically "fixed" in a usable form. Plants that can facilitate nitrogen fixation by soil microbes are many and varied, and they may occupy any of the layers in a forest garden. These plants include clovers, which can be part of the ground cover and herbaceous layers, shrubs such as goumi and Siberian peashrub, understory trees including redbud and green alder, overstory trees such as black locust, vines including wisteria and groundnut, and root crops such as licorice.

Besides their ability to provide usable nitrogen, all nitrogen-fixing plants supply other benefits such as food, medicine, and aesthetic appeal. They are also pioneer plants, moving in after a disturbance to prepare the ground for other kinds of plants that follow. Their leaves and branches are particularly rich in nitrogen and thus form a rich mulch and offer nutritious fodder for animals, especially in times of drought. Some nitrogen fixers provide edible leaves, berries, seeds, pods, flowers, or roots. Beneficial insects and birds are attracted by their flowers, branches, fruits, and seeds. They can also serve as trap crops for pests and as nurse plants for other species that need additional sustenance or protection.

Plants in the legume family (which includes the clovers and the locusts) form nodules on their roots where bacteria live in a beneficial relationship with the plant. The bacteria take sustenance from the plants in the form of sugars and starches. In return, they convert nitrogen from the air into nitrogen compounds that plants can absorb through their roots. Thus, these bacteria provide nutrients to the plants in which they reside as well as to surrounding plants, assisted by the web of fungi in the soil.

Here in part 2 I've included some nitrogen-fixing plants in each chapter. And in part 3, you will find guidelines for including nitrogen fixers in plant groupings.

NUTRIENT ACCUMULATORS

A garden becomes self-sustaining when all the nutrients required for health and growth are contained within it. Intentionally incorporating plants that either convert or scavenge for these nutrients can eliminate the need to provide amendments sourced externally. The deep roots of these nutrient accumulators selectively pull up minerals from layers of the soil that plants with a shallow root structure cannot access directly.

When the nutrient accumulator drops its leaves in the fall or when it dies to the ground, the plant matter decays, and the nutrients accumulated in its tissues become available to nourish other plants. The elements that nutrient accumulators concentrate in their leaves and fruits also add to their nutritional value as food plants for people. These natural amendments may include three of the nutrients most needed by plants—phosphorus, calcium, and potassium—as well as nutrients needed in smaller quantities, including iron, selenium, magnesium, copper, cobalt, and silicon (see appendix 1).

Nitrogen-fixing plants are nutrient accumulators, too, and in addition to providing nitrogen for the benefit of other plants, several nitrogen fixers accumulate other essential nutrients. For example, licorice, lupines, clovers, and vetches scavenge for phosphorus, the nutrient required by plants in the most quantity after nitrogen. Black locusts accumulate not just nitrogen, but also calcium and potassium.

BENEFICIAL ATTRACTORS

Plants can attract an array of beneficial insects and animals, including pollinators, pest-devouring insects, and birds. Flowering plants that provide food for pollinators can ensure that these beneficial insects are abundant when needed to pollinate the food-producing plants in your garden. Many types of insects provide pollinating functions, including bees, beetles, ants, flies, and butterflies. Other beneficials such as ladybugs, parasitic wasps,

Figure 6.4. Japanese beetles congregate on the leaves of thimbleberries, which grow wild in my region and produce a raspberry-like fruit. This keeps the beetles away from my raspberries!

Figure 6.5. Nutrient accumulators can serve multiple functions. This comfrey plant, which is a nutrient accumulator par excellence, also attracts pollinators like this bumblebee.

lacewings, and spiders dine on garden pests. Plants that attract all of these helpers include various types of herbaceous plants, fruiting trees and shrubs, ground covers, overstory trees, and vines.

PEST DETERRENCE

The first line of defense against pests and disease is healthy plants. Plants that are receiving the right amount of light, water, airflow, and nutrients will be in the best position to resist debilitating influences. Surrounding plants can also provide some protection from pests and disease.

Any plant with a strong odor can serve as an *aromatic pest confuser*. To my knowledge, there is a dearth of scientific research proving the benefits of these plants, but anecdotal evidence suggests that they are helpful. These include members of the onion family (chives, garlic) and culinary herbs (mints, oregano, sage, lavender). The idea is that when the pest such as a plum curculio enters the garden looking for a plum tree and encounters a mass of, let's say, oregano, it is overwhelmed with the smell of the herb, becomes confused, and loses the scent of the plum tree. Aromatic pest confusers can theoretically disorient the pests and thereby make it harder for them to infest their preferred host bushes or trees. In addition, herbs like lavender exude oils that have anti-fungal properties, so may help guard against fungal disease, particularly in fruit trees. The flowers of herbs and members of the onion family are also attractive to pollinators and other beneficial insects.

In addition to their utility as attractors and confusers of insects, certain plants can serve as *trap crops* to attract pests away from more valuable plants. In my garden, I have discovered several of my plants already serving this function. For example, I noticed Japanese beetles huddling on the leaves of thimbleberry (see figure 6.4). I have seen aphids congregate on the stems of two types of nitrogen-fixing plants—hairy vetch and wild blue lupine. Also, if you plant fruit-bearing shrubs such as elderberry and juneberry along the borders of a garden, they may divert birds from feeding on ripening fruit crops within the garden.

Layer by Layer

In the chapters that follow I describe in detail many of the plants I grow in my garden, organized by permaculture layer: overstory, understory, shrub, herbaceous, ground cover, root, vine, and fungi. Not surprisingly, the plants in each layer do not conform to strict boundaries—for instance, many plants in the herbaceous layer serve very well as ground covers and a vine may also offer edible underground tubers. I have placed plants in the layers I consider best representative of their growth habits. Root crops are integrated into various chapters, and chapter 12 covers both vines and fungi.

Each of the plant descriptions includes a quick-reference summary of basic information: the plant's native range, mature height and width or canopy, soil conditions, sunlight needs, and hardiness range. In addition, you'll see one or more colorful *icons* next to the plant name; these icons indicate some of the key functions or benefits the plant provides.

For each plant, I also convey what I have learned and observed to help you become acquainted with the plant's characteristics and, I hope, benefit from my experience. As you peruse the descriptions, consider which may suit your personal aesthetic and the habitats in your garden. Plan to include some that provide the sustaining functions described above. Choosing plants to suit your tastes and your garden's habitats is one of the joys and challenges of creating an edible landscape.

Functions and Benefits

These icons represent the major categories of functions and benefits that plants provide in an edible forest garden. Look for these icons in the plant descriptions throughout chapters 7 to 12.

- NF Nitrogen Fixer
- FR Fruit
- N Nut
- MT Multipurpose Tree
- PV Perennial Vegetable
- H Herb
- EF Edible Flower
- B Berry
- GC Ground Cover
- CF Cut Flower

CHAPTER 7

THE OVERSTORY

In this chapter I describe a variety of useful and aesthetically appealing overstory trees—trees that grow 30 feet (9 m) tall or larger. You may not have space for trees of this size in your garden, but if you do, they can enhance your landscape while providing other important functions. Nitrogen fixers, fruit and nut trees, and what I call multipurpose trees can occupy this layer. If you have a relatively large plot, it is advantageous to incorporate overstory nitrogen fixers because their expansive size allows them to provide this essential nutrient to a large volume of plants, while some of them have thin canopies that allow a good amount of light to filter through to the plants below.

It is important to carefully plan the placement of overstory trees, which may take decades to reach their full height and girth. In order to space trees properly, you need to consider what their function will be. If you want a tree to produce food, it should be given enough room to accommodate the full width of its mature canopy. On the other hand, if the tree will be used for timber, wind protection, wildlife value, or aesthetic appeal, spacing can be considerably closer. And if your plot size is limited,

Figure 7.1. Tamarack is a fast-growing multipurpose tree whose branches occupy the overstory layer.

some of the trees described below can be cut back so you can maintain them within dimensions that suit your space.

MT

American Basswood

Tilia americana

Native Range: Eastern North America
Height: 20 to 80 feet (6–25 m)
Canopy: 20 to 50 feet (6–15 m)
Soil Conditions: Adaptable
Sunlight: Full sun to deep shade
USDA Hardiness Zones: 3 to 9

American basswood, also known as American linden, is a large overstory tree that grows rapidly in full sun or much slower in deep shade. In full sun, mature trunks can reach 5 feet (1.5 m) in diameter and the canopies can be tall and broad. In shade, the trunks remain slender and the trees form an elegant, vase-like shape, as shown in figure 7.2. Basswood leaves are deeply veined and asymmetrically heart-shaped, growing many inches in diameter in deep shade. The young leaves are edible and make a tasty substitute for lettuce in salads. The fragrant, yellowish white flowers, which dangle from the tree in bunches, attract bees and can be steeped for a soothing tea.

American basswood trees grow wild in the understory of the woods on our farm, but I have also planted a few seedlings and transplants in my garden. I first planted two purchased seedlings in a moist, shaded spot in a hedge along the garden's border. Once they grow taller than I can reach, I plan to pollard them alternately, following a procedure outlined by Martin Crawford in *Creating a Forest Garden*. The first year I will cut one back so the new tender growth is easy to reach, similar to the pollarded mulberry pictured in figure 7.14. Once that tree regrows out of reach, I will cut back the second. That way the tender leaves on the fresh shoots will be continually accessible for harvest.

Figure 7.2. This wild basswood (*center*) grows in the shade of a mature oak (*left*). Even during the leafless season, the basswood's elegant, vase-like silhouette makes it attractive.

FR American Persimmon

Diospyros virginiana

Native Range: Southeastern US
Height: 20 to 80 feet (6–25 m)
Canopy: 20 to 50 feet (6–15 m)
Soil Conditions: Tolerates different soil types but prefers well-drained soil
Sunlight: Full sun or partial shade
USDA Hardiness Zones: 4 to 9

You may be familiar with Asian persimmon (*Diospyros kaki*), whose fruit is sometimes found in grocery stores or specialty produce shops. American persimmon is a different species that grows naturally in most of the southeast quadrant of the United States, from southern Connecticut to Florida and eastern Texas to southern Indiana. This native, fruit-bearing tree has been cultivated in North America for as long as humans have occupied the land. Native seedlings, which can grow to 80 feet tall and live for well over 100 years, require one male and one female tree for pollination. Some grafted cultivars, selected for the quality of their fruit, grow only 15 to 20 feet (4.5–6 m) tall (thus occupying the understory) and are self-fertile. Persimmons have a suckering habit, but the resultant new growth will not be true to the grafted cultivar. Despite their growth potential, persimmon seedlings can be pruned to maintain the desired height.

These trees are among the last to leaf out in the spring. In the summer, the persimmon's dark, oblong leaves and spreading canopy make it an attractive specimen tree—singular trees that create focal points in the garden. During the fall, the leaves turn yellow, pink, red, or purple, and the fruit ripens to a deep red-orange. The fruits hang on the tree like bright Christmas ornaments through early winter, and the tree's charcoal-to-black checkered bark solidifies its three-season landscape appeal.

American persimmon fruits are slightly oval orbs between 1 and 2½ inches (2.5–6 cm) in diameter,

Figure 7.3. An American persimmon seedling in fall.

with a calyx (a whorl of sepals) where they attach to the tree. Persimmon fruits require exposure to frost to become edible; they are astringent until they soften. Though I have yet to taste one, these fruits allegedly have a texture like custard and a complex flavor, and are delectable eaten fresh, frozen, dried, or processed into pies, puddings, and other concoctions.

Early on, I incorporated two cultivars, Yates and Meader, in my garden, both of which have survived to this writing. Later, I also planted several seedlings, which so far seem to be more tolerant of bitter cold than the grafted cultivars.

I planted the Yates and Meader persimmons on well-drained slopes facing south and southwest, respectively. Each has been winter-killed down to the snow, but above the graft, at least once. These resilient trees both grew back the next season. Both cultivars matured enough to produce inconspicuous white flowers below their branches, but they have not borne fruit to date. I am patient and hopeful. Even when fruits do form, it is possible that they will meet the same fate as my medlar, turning black and inedible before ripening due to the not-quite-long-enough growing season (see the "Medlar" entry on page 131). However, I am betting that the warming climate will continue to extend temperate fall weather, eventually enough to allow fruits a chance to mature.

Black Locust

Robinia pseudoacacia

Native Range: Appalachian Mountains
Height: Up to 80 feet (25 m)
Canopy: Up to 30 feet (9 m)
Soil Conditions: Tolerates wide range of soil types, drought-tolerant
Sunlight: Prefers full sun
USDA Hardiness Zones: 3 to 8

Black locust is believed to be native to the Appalachian Mountains but has naturalized to much of the United States and Canada. This tree grows extremely fast; I planted 12-inch (30 cm) seedlings that soared to 15 feet (4.5 m) tall in three years. With this rapid growth rate, the locust can provide a quick windbreak or trellis for vines. Its hard, rot-resistant wood is excellent for fence posts and firewood, and young leafy stems provide nutritious forage for ruminants. These vigorous trees can be chopped down to the ground (coppiced) or pruned hard, leaving only the trunk and branch stumps, and will regrow.

I love black locust's four seasons of landscape appeal. In mid- to late spring, its branches are adorned with blooms that look like large bunches of white grapes. The blossoms are attractive to pollinators and edible for people. When I stand under one of these blooming trees, the buzzing of bumblebees above is almost deafening. Covered with small, round leaves during the growing season, its branches let plenty of light filter through. I underplant fruit and nut trees, bushes, and vines, and trim the locust's lower branches high enough (up to 30 feet [9 m] if necessary) that there is ample

Figure 7.4. Black locust in bloom.

Figure 7.5. Purple Robe locust's attractive blooms can range in color from rose pink to deep purple.

Figure 7.6. Crowned with fluffy white snow, locust's dark seedpods enhance the winter landscape.

light to meet the needs of the plants below. Trimmed branches serve well as trellises for vines and as the base of hügelkultur mounds. In fall, the locust leaves turn a bright yellow. Abundant small, dark brown seedpods hang from the branches throughout the winter.

I integrated black locust in my garden in two ways. I planted these trees 10 feet (3 m) apart, in a row to form a windbreak hedge. I also planted one at each end of a 20-foot-wide (6 m) bed on the northwest side of the garden, to provide dappled shade for shade-loving plants beneath. (This arrangement is illustrated in figure 14.3 on page 274.)

Downsides of the black locust include short, but sharp, thorns and a tendency for the trunk or branches to break in strong winds. My trees' trunks and branches have snapped in windstorms, but the trees are so resilient that they grow back quickly. The black locust also has a suckering habit, whereby new shoots arise from the widely spreading roots, that has led to its categorization as an invasive in New York State, even though it has naturalized almost everywhere in the state. I do not find the suckering hard to control; I mow or clip these volunteer seedlings whenever I see them. They can also be dug up and transplanted elsewhere. I even let one sucker mature into a tree because its chosen location appeared advantageous.

Purple Robe locust is a black locust cultivar that is distinguished by its strikingly colored blooms, its shorter mature height (30–40 feet / 9–12 m), and its lack of thorns. I planted my first Purple Robe locust in an area with a high water table and learned that it does not tolerate wet ground. The tree died two years later during an unusually wet spring. I surmise the saturated soil robbed its roots of oxygen. I turned the dead tree into a trellis for a groundnut vine. I planted subsequent Purple Robe locusts on higher, better-drained ground, between fruit trees in my garden's hillside orchard rows.

Because of black locust's "invasive" categorization, check with your local or regional authority to find out under what circumstances, if any, you are permitted to plant it. In New York State, you are allowed to plant these trees if they grow naturally in the area surrounding your location. There is a natural thicket of black locust on my next-door neighbor's property, so I figure I am in the clear.

Chinese Chestnut

N

Castanea mollissima

Native Range: Korea and China
Height: 30 to 60 feet (9–18 m)
Canopy: 30 to 60 feet
Soil Conditions: Most soil types, likes moisture but is drought-tolerant
Sunlight: Full sun for optimal nut production, but can tolerate partial shade
USDA Hardiness Zones: 4 to 8

Chinese chestnut forms a dense, rounded canopy that casts deep shade below. Its elongated oval leaves, attractive in summer, turn yellow and then brown in fall. The blooms, which look like giant, white-legged spiders, appear in late spring and can grow 5 to 12 inches (12–30 cm) long. Clusters of spiked round husks are also decorative as they mature from green to brown.

Chinese chestnut trees are naturally resistant to chestnut blight, a fungal disease that decimated the American chestnut (*Castanea dentata*) population during the early 1900s. Although they can grow in the understory, they produce a greater volume of nuts when they receive full sun. Despite their relatively slow growth rate of 1 to 2 feet (30–60 cm) per year, they can begin to bear nuts in four to five years when grown from seed. Two different cultivars are required for pollination, which occurs primarily by wind.

I planted four Chinese chestnut seedlings just inside my garden's fenced boundaries. If they reach their maximum size, half of their canopies will extend outside the fence line, leaving more room inside the garden for other trees. Removing the tree's upper branches is another option for managing their size. The trees I planted were pummeled by the cold during several winters. Perhaps some of my trees were acclimated to a climate warmer than mine because they came from a southern nursery. Though all grew again, not one has yet achieved a height over 3 feet (1 m). I am hopeful that our winters will become milder as the climate warms, giving these trees an opportunity to become productive.

English Walnut

Juglans regia

Native Range: Eurasia
Height: 40 to 60 feet (12–18 m)
Canopy: 40 to 60 feet
Soil Conditions: Adaptable, but needs well-drained soil
Sunlight: Full sun
USDA Hardiness Zones: 5 to 9

English walnut, which is also called Carpathian walnut and Persian walnut, yields nuts with the

Figure 7.7. A Chinese chestnut seedling.

Figure 7.8. A walnut seedling regrows after being winter-killed.

large, round, tan shells familiar in holiday mixes. Black walnut (*Juglans nigra*) produces nuts with round, black, furrowed shells that are tough to crack open. Black walnuts are hardy to at least Zone 4, and in fact grow wild so prolifically in our region they are considered nuisance trees. I chose not to plant them because I was not interested in the labor involved to remove the meat from the shells. I chose instead to plant the hardiest cultivars of the Eurasian species I could find, though very few are rated as hardy for conditions as cold as I experience in my northern New York garden.

Finding a suitable location for walnut trees in my garden was a challenge due to their expansive mature canopies and the requirement of two trees for pollination. Moreover, their roots emit juglone, a chemical that deters the growth of some plant species (but does not bother others). According to permaculture expert David Jacke in his book *Edible Forest Gardens, Volume 2,* mulberry, elderberry, pawpaw, and black locust are unaffected by walnuts; black raspberries (but not red raspberries or blackberries) are also compatible, as are American persimmons (but not Asian persimmons). In *Gaia's Garden,* Toby Hemenway suggests that mulberries can be planted as an intentional buffer between walnuts and other plants that are intolerant of juglone, such as apples. I kept this information in mind when deciding what to plant near my walnuts.

I started with two English walnuts, which are rated no colder than Zone 5, sited 40 feet apart inside my garden's northeast fence line. One appeared dead the next year, though when I dug it up to replace it, I discovered too late that the roots were still vital. I replaced it with a new seedling. The next year, I thought the other tree had died, but waited to be sure. It eventually leafed out but barely grew. In the years since, both the original tree and the new seedling were winter-killed above and below the snow twice, but both still survive. Part of the problem might be the trees' location; with little flexibility in terms of space, I had no choice but to plant them in a frost pocket. At this point, I am more hopeful for success with pecans (see the "Pecan" entry on page 111) than with English walnuts.

Ginkgo

Ginkgo biloba

Native Range: China
Height: Up to 100 feet (30 m)
Canopy: 25 to 35 feet (7.5–10.5 m)
Soil Conditions: Tolerates a wide range of soil types, requires well-drained soil with ample moisture
Sunlight: Full sun or partial shade
USDA Hardiness Zones: 3 to 8

Also known as the maidenhair tree, ginkgo is a living fossil—one of the oldest living species of tree on Earth, dating back over 250 million years. Ginkgo trees originated in China, but they are cultivated throughout the world today. In Traditional Chinese Medicine, the leaves are used to treat Alzheimer's disease and other ailments. I first came across ginkgo trees as a child in New York City, where their durability enables them to survive when exposed to stresses such as air pollution, pests, and diseases. Ginkgos can live to be hundreds, even thousands, of years old, often beginning with a slender pyramidal canopy that spreads out as the tree ages. Ginkgo leaves are a decorative fan shape, turning saffron yellow in fall.

Ginkgo trees are either male or female—the males producing pollen-bearing cones, and the females producing ova that, when pollinated, become edible nuts surrounded by a pulp. The pulp exudes a distinctly unpleasant odor, similar to rancid butter, as it decays. You can source male and female cultivars, and will need one of each if you desire nut production. I planted two unsexed seedlings mainly for their aesthetic appeal. If it turns out I have one of each sex, maybe I will have nuts in the future as

Figure 7.9. One of my ginkgo seedlings. These trees take a few years to establish their deep taproots before beginning to gain in height.

Figure 7.10. This nurse honey locust (*background*) is placed to the north of a sweet cherry (*foreground*) and is trimmed up as the cherry tree grows beneath.

well. As these trees mature, I may enlist their trunks as trellises for a food-producing vine.

NF

Honey Locust

Gleditsia triacanthos

Native Range: Central US
Height: 50 to 75 feet (15–23 m)
Canopy: Up to 45 feet (14 m)
Soil Conditions: Adaptable, tolerates wet soil
Sunlight: Prefers full sun
USDA Hardiness Zones: 4 to 9

Honey locust is distinguished from other locusts by its wider and rounder canopy, its long strands of filigreed yellow flowers, its ability to tolerate wetter soil, and its immense, flat seedpods, which can grow up to 2 inches (5 cm) wide and 8 inches (20 cm) long. When they are young and green, the pods contain a sweet, sticky substance that can be scraped off the interior walls and eaten like honey—hence the name "honey" locust. Like black locust, honey locust has four-season aesthetic appeal, with its huge seedpods adding even more drama to the winter landscape. While some people consider the mature brown pods a nuisance when they fall and litter the ground, I view them as nutritious fodder for pigs and ruminants. However, the trees produce these pods only when there is at least one male plant to pollinate the females. If you decide to

Nitrogen-Fixing Trees as Trellises

Overstory and understory trees can double as trellises for vining plants. It is best not to use fruit-producing trees for this purpose, though, because the shade cast by the vines could block sunlight from the ripening fruit. On the other hand, nitrogen-fixing trees are ideal supports for vines. Most grow rapidly, so they can develop into sturdy trellises in as few as three years. Many can be trimmed to the desired height or shape without harming the tree, and shading fruit is not a concern. However, most of these trees need full sun and could be harmed by an aggressive, densely growing vine.

Any of the nitrogen-fixing trees described in this and the following chapter can work as a trellis as long as the vine you choose is suited to the same habitat and the amount of shade cast by the tree. Locusts are a good match for tall vines adapted to dappled shade and dry-to-moist soil. A vine comfortable with saturated soil and partial shade can partner successfully with alder. One adapted to deep shade would do well climbing a redbud.

It is best to wait until your nitrogen-fixing tree is well established, tall, and strong enough to support the vine you choose to grow up it. Your choices are varied: from annual vegetables such as peas or pole beans, to annual flowers like morning glories or climbing nasturtiums, to any number of perennial flowering or food-producing climbers. I suggest several options in chapter 12.

Figure 7.11. An akebia vine I planted at the base of a black locust in the tree's third year. As the tree grew taller, the vine, which is well adapted to the dappled shade, climbed higher.

include honey locust in your garden, I recommend the thornless variety (*Gleditsia triacanthos* var. *inermis*), which does not possess the standard dagger-like protrusions.

There is a dispute in the literature regarding whether or not honey locust is a nitrogen fixer. However, Martin Crawford states in *Trees for Gardens, Orchards, and Permaculture* that "it now appears that it does [fix nitrogen], though not via nodules like other legumes."

I interplant this tree in orchard rows and use it as a "nurse" plant. Nurse plants provide shelter and sustenance to more vulnerable seedlings to ensure their healthy development. My honey locusts are located near and to the north of sweet cherries and American persimmons, providing these more tender trees with nutrients and shelter from cold north winds (see "Designing Orchard Rows," page 274).

N

Korean Nut Pine

Pinus koraiensis

Native Range: Eastern Asia
Height: 50 to 75 feet (15–23 m)
Canopy: Up to 25 feet (7.5 m)
Soil Conditions: Adaptable, but needs well-drained soil
Sunlight: Shade for the first two years
USDA Hardiness Zones: 3 to 7

Many species of pines produce edible "pine nuts" inside their cones. The one I chose to plant, Korean nut pine, grows straight and tall, with needle-covered branches extending out and upward from the trunk forming a tepee-shaped canopy. The needles remain green all winter, adding a touch of color to the winter landscape. As compared with its relative the rapidly growing Eastern white pine (*Pinus strobus*) that populates the rocky ridges surrounding our farm, the Korean nut pine grows slowly at first.

I planted three Korean nut pines at the top of a south-facing slope where they have full sun and good drainage, spacing them 25 feet apart to leave room for their mature canopies. I surrounded each with pallet slats hammered into the ground in a semicircle to provide the shade preferred by the young seedlings.

When I purchased these trees, I learned that it could take 40 years for them to produce nuts. That became my motivation for living a long life: I will have to live past age 100 so I can be sure to experience the fruits of these trees.

Figure 7.12. Slow-growing Korean nut pine seedlings benefit from shade during their first couple of years. For this one, I fashioned sun protection from pallet wood.

Mulberry

FR

Morus species

Native Range: South Asia (*M. alba*), eastern US (*M. rubra*), Middle East (*M. nigra*)
Height: Up to 40 feet (12 m)
Canopy: Up to 40 feet
Soil Conditions: Tolerates a wide range of soil types and moisture
Sunlight: Full sun or partial shade
USDA Hardiness Zones: 4 to 10, depending on species

This fruitful tree has three distinct species: white, red, and black (in Latin, *alba*, *rubra*, and *nigra*). Mulberry trees grow quickly and begin fruiting by the third year. The berries are similar in appearance to blackberries, but narrower, and can be white, lavender, or black. The berries on the cultivars I

Figure 7.13. This mulberry branch is loaded with ripening fruit.

grow gradually turn from red to black as they ripen. Once they turn a reddish black, the berries can be harvested one by one. If they are perfectly ripe, you can spread a drop cloth underneath the tree and shake the tree so the fruits fall onto the cloth. Any unharvested fruit will be cheerfully consumed by birds. My favorite variety is the Illinois Everbearing mulberry, which fruits all summer and into fall, producing the largest and tastiest fruits of the cultivars I grow. The fruits can be eaten fresh, baked into pies and tarts, frozen, and processed into jams, syrups, and even a sweet wine.

Mulberry leaves are also edible and appetizing in salads. One way to encourage mulberry trees to produce a flush of new leaves is to use the pruning technique known as pollarding. One summer I severed the trunk and branches of a mulberry tree at a height of 3 feet (1 m) from the ground (see figure 7.14). The stump soon sprouted many new stems covered with tender leaves.

Of the three mulberry species, only white mulberry and white-red hybrids are hardy to Zone 4. I planted several, including two dwarf varieties: a weeping mulberry for children to play under and a "contorted" mulberry as a specimen plant.

These adaptable trees are disease- and pest-resistant, and some can live for more than 50 years. When choosing where to plant a mulberry tree, keep in mind that they have spreading root systems that can become invasive, so avoid planting them near septic systems or utility lines. The fruit drops can make a mess on the ground, so it is best to site these trees where that will not become a problem. I was careful to avoid placing mine near stone walkways. I sited one so that some of its mature branches will extend over a pond, dropping its berries to feed the pond life.

When I began planning my edible garden, I was not acquainted with these versatile trees beyond the children's song "Round and Round the Mulberry Bush." Since then, I have become so enamored with them that I am gradually planting more to replace other trees that have failed to flourish under my growing conditions. If you have room in your

Figure 7.14. Multiple new shoots are evident in this winter view of a mulberry tree pollarded the previous summer.

landscape to incorporate even one dwarf variety of this productive tree, which could even be small enough to occupy the shrub layer, I am confident you will be glad you did.

N

Pecan

Carya illinoinensis

Native Range: Southern US
Height: Up to 60 feet (18 m)
Canopy: Up to 60 feet
Soil Conditions: Adaptable and drought-resistant
Sunlight: Full sun for optimal nut production
USDA Hardiness Zones: 5 to 9

Mature pecans are large, densely shading trees that can live for over 300 years. If spaced widely and grown in the open, they form oval canopies with branches beginning close to the ground. In forest settings, the trunks are bare until farther up, where they form branches that widen into a compact pyramidal crown at the height of the surrounding

Figure 7.15. This pecan seedling has outgrown its protective upside-down tomato cage, which can now be removed, as the tree has grown higher than a deer can reach.

Figure 7.16. Notice the varying shapes of the leaves on this sassafras seedling.

forest canopy. Pecans are pest- and disease-resistant and adapt to all kinds of soil, where their deep taproots can access water in times of drought. At least two trees are necessary for pollination.

When I began the edible forest garden, I didn't imagine that I could grow pecans, which I knew as native to the southern United States. However, pecan trees are native as far north as northern Illinois, and with some research I came across some that just might be hardy enough for me to grow. Oikos nursery offered a variety they called Michigan that is allegedly hardy to Zone 4. I planted two of these promising trees along the garden's northwest edge—the only space I had available where I could allow for their 60-foot-wide canopies. I had previously observed that trees with deep taproots often take a few years to establish their roots before beginning any significant upward growth. At first, the small pecan seedlings grew very little. Then, around the fourth year, they began to exhibit annual growth spurts. In the sixth year I had a slight problem with small caterpillars nibbling on the growing tips, but I was able to control this predation with a couple of applications of insecticidal soap. As of this writing, both continue to grow well. It may be another five years or more before they begin to produce nuts. I look forward to that day, at which time I hope our growing season is long enough to allow the nuts to ripen.

Sassafras

Sassafras species

Native Range: North America
Height: 30 to 60 feet (9–18 m)
Canopy: 25 to 40 feet (7.5–12 m)
Soil Conditions: Tolerates a wide range of soil types
Sunlight: Full sun or partial shade
USDA Hardiness Zones: 4 to 9

I remember the sassafras tree from my childhood, when I marveled at its three kinds of distinctly shaped leaves: one-, two-, and three-lobed, all on the same tree! Two species of sassafras are native to a broad swath of North America. In the wild, it is found in open woods and fields, tolerating various soil types. Sassafras trees are either male or female, and both are required to produce fruit, which the birds enjoy. Because of its unusual leaves, yellow flowers, and dramatic fall color, sassafras is often used as a specimen plant. In addition to its landscape appeal, all parts of the plant, including its roots, stems, twigs, leaves, bark, flowers, and fruit, are aromatic and used in culinary and medicinal applications. Sassafras root was the main ingredient in traditional root beer, and the ground-up leaves are used to flavor Creole cuisine, including gumbo. The leaves and flowers make a tasty addition to salads. Native Americans used the plant to treat wounds and other ailments, and sassafras extract has been used as an anesthetic in traditional dentistry.

Mainly for visual interest, I planted three sassafras seedlings close together in a border hedge, where they receive alternating sun and shade throughout the day. Only one of the original seedlings took hold. I had to replace the other two more than once, possibly because their deep taproots make these bare-root trees difficult to transplant successfully.

Tamarack

MT

Larix laricina

Native Range: Central Canada to northeastern US
Height: 30 to 60 feet (9–18 m)
Canopy: 15 to 25 feet (4.5–7.5 m)
Soil Conditions: Tolerates a wide range of soil types, prefers wet soil
Sunlight: Full sun
USDA Hardiness Zones: 2 to 6

Tamarack, also known as American larch, is an unusual conifer: It is not evergreen. Each fall, its needles turn golden yellow and then drop off the tree. Tamaracks prefer wet lowland habitats, and while adaptable to many soil types, they do require

Figure 7.17. In early summer the tamarack's lovely rose-colored cones are a standout.

full sun and lots of moisture. They are also extremely cold-hardy, found naturally near the Arctic Circle, but don't tolerate hot climates. The tamarack's short, flat, pale green needles cast a light shade, providing suitable conditions for many plants to grow beneath. Their cones are small, rounded, and a striking rose color when they first appear, turning brown as they age and adding visual interest in the winter. Tamaracks grow quickly, which makes them a good choice for an effective windbreak. Their wood is rot-resistant, excellent for building fence posts and corduroy roads (see "Corduroy Roads," page 61). In fact, the name *tamarack* has its origins in the Algonquian word for "wood used for snowshoes." Native Americans also had several medicinal uses for it.

In my garden, I installed tamarack trees to serve as a windbreak when my Korean nut pines, intended to serve the same purpose, did not grow quickly enough (see "Plans Are Made for Changing," page 45). However, this location turned out to be an unsuccessful one for the tamaracks, which did not have enough moisture to satisfy their needs. I also planted them in the expansion garden's boggy western corner, for both their ornamental appeal and their attractiveness to wildlife.

My Tamarack Travesty

I have a small south-facing slope where I planned to place an American persimmon. The cultivar was allegedly suitable for Zone 4 conditions, but the fruit doesn't ripen until late in the season. The south slope placement was intended to extend the growing season in this tree's microclimate, enough to ripen its fruits. To further enhance this possibility, I thought about the wind, particularly cold winds that occasionally blow from the north. To create a favorable microclimate, I planted three Korean nut pines on 25-foot (7.5 m) centers, on the crest of the slope to provide a windbreak. The pines, however, grew too slowly to create an effective windbreak. (I have since learned that if they had been grafted to white pine rootstocks, they would have grown a lot faster.)

I decided to interplant three tamaracks among the pines. I expected that the tamaracks would grow much faster and thus provide the desired windbreak considerably sooner. Although this was a good motive, I also knew that this high-and-dry location was not the ideal habitat for tamarack. This species of conifer occurs naturally in low-lying bogs where the soil is always wet and often saturated.

A couple of years after the tamaracks were established, there was a significant summer drought. I was preoccupied with other tasks and did not immediately spot the effect of the drought on these trees. I finally noticed their distress, and I was able to save one with repeated watering, but it was too late for the other two. The next year, I incurred the additional labor and expense of replacing the two that died. After that, each spring I made sure to apply a heavy layer of mulch at the base of each tamarack in this location to insulate the ground and preserve the moisture near their roots. Lesson learned, I vowed never to plant tamaracks in a high-and-dry location again.

MT White Birch

Betula papyrifera

Native Range: North America
Height: Up to 60 feet (18 m)
Canopy: Up to 25 feet (7.5 m)
Soil Conditions: Tolerates a wide range of moist soil types, has some drought tolerance
Sunlight: Full sun or partial shade
USDA Hardiness Zones: 2 to 7

The white birch, or paper birch, is so named for its white bark that naturally peels off in paper-like strips, which have historically been used for writing. The bark is a great fire starter and has many traditional uses, including as a water-resistant covering for birch-bark canoes. The tree can be tapped for its sap, which boils down to a sweet, tasty syrup.

Birch trees are pioneer plants—among the first to move in after a disturbance such as a fire. They typically grow in pure stands, giving way after several decades to more long-lived hardwoods such as oaks and maples.

After visiting England and strolling down a path lined with these charming trees, I was

Figure 7.18. A mature white birch stand occupies a rocky ridge on our farm. This stand probably arose shortly after dairy cattle stopped grazing here.

inspired to plant some white birch in my own garden. I envisioned a wedding party gathering at the top of the garden, then promenading down a curving, birch-lined path to the back gate. I planted seven groups of three white birch seedlings 25 feet apart, staggered along both sides of my existing path. My main motivation was their aesthetic appeal—green foliage contrasting with white bark during the growing season, brilliant yellow leaves in fall, and ornamental trunks and branches in winter. Someday, years from now, I might tap them for syrup.

If you have the space and a suitable habitat, I hope I have inspired you to plant one or more of the overstory trees discussed in this chapter. However, if your garden space is limited to smaller trees, there are many options that may suit your taste and growing conditions. In the next chapter, I describe trees growing no more than 30 feet (9 m) tall. These can be planted below the overstory trees, or themselves serve as the top layer of your edible garden.

CHAPTER 8

THE UNDERSTORY

The understory is defined as the vertical space from 10 to 30 feet (3–9 m) above the ground. A few nitrogen fixers, most fruit trees, and some nut trees occupy this level. In this chapter, I describe my experience with those I have grown.

If your garden space does not allow for plants that can reach 20 or 30 feet (6–9 m) tall, there are numerous dwarf and semi-dwarf options for most of the fruit trees I describe. However, these trees are grafted to less vigorous rootstocks than standard trees—hence their less vigorous growth. While dwarf and semi-dwarf trees can do well in warmer climates, standard trees' rootstocks make them more resilient in colder areas. If your climate makes standard trees the better choice, you can often keep them to a manageable size with regular pruning.

If you are partial to cherry fruit, I offer several examples of cherry-bearing trees in this chapter. However, for those whose space is too limited for understory plants, I also describe more diminutive cherry bushes in chapter 9.

As you plan the understory layer of your garden, be sure to space your food-bearing plants to accommodate their mature canopies in order to promote maximum fruit or nut production.

Figure 8.1. Sour cherry and many other types of fruit trees are excellent choices for the understory of an edible forest.

Figure 8.2. Almond's beautiful pink blossoms.

N

Almond

Prunus dulcis

Native Range: Middle East
Height: Up to 20 feet (6 m)
Canopy: Up to 20 feet
Soil Conditions: Well-drained soil
Sunlight: Full sun
USDA Hardiness Zones: 5 to 9

Almond's elongated, pointed leaves and rounded canopy resemble those of the peach tree, to which it is closely related. In spring it sports magnificent large, pink blooms with deep rose centers, which stand out in the landscape. Fall colors can range from yellow to deep red. Like peach, almond is subject to environmental impacts such as injured blossoms from cold spring weather and fungal diseases fostered by humidity. Unlike the short-lived peach, in the right conditions almond trees can live hundreds of years.

It's generally recommended to grow almonds in Zone 7 or warmer to achieve commercially worthwhile yields. Given the apparent unsuitability of my Zone 4 northeastern location, I would not have attempted growing almonds were it not for a few cultivars rated close to my zone (Zone 5 but borderline Zone 4). First, I tried two Hall's Hardy almonds, rated for Zone 5, in the loosened soil of some hügelkultur mounds with southwest exposure. These trees were promising at first, growing rapidly and even flowering beautifully. In the fourth year, not having borne any nuts, they both died. Clearly, they were not hardy enough.

Meanwhile, I thought I would try my luck with two Ukrainian cultivars, Bounty and Oracle, also rated for Zone 5. I planted these on raised mounds in the crescent bed (see "The Crescent Bed," page 290), where they survived no more than two years. I might give almonds one more try if and when I have a vacancy on the extra-warm, well-drained south-facing slope where my Asian pears are currently located. Almond's gorgeous blooms, and the potential scrumptious nuts, are incentive enough to plant these trees. If you have the space, give some extra-hardy cultivars a try if you are in Zone 5 or warmer.

American Plum

FR

Prunus americana

Native Range: Most of the US and Canada
Height: 10 to 20 feet (3–6 m)
Canopy: 10 to 20 feet
Soil Conditions: Tolerates most soil types and wet conditions
Sunlight: Full sun
USDA Hardiness Zones: 3 to 8

This hardy native tree grows as a multi-stemmed thorny shrub or single-stemmed tree with a spreading crown and suckering habit. Due to its dense

Figure 8.3. An American plum in bloom, sited on the bank of a pond.

My Nemesis: The Plum Curculio

The plum curculio is a weevil—a type of beetle—native to the United States east of the Rocky Mountains. It infests the fruitlets of both pome and stone fruits, including plums, apples, pears, peaches, cherries, and quinces. In my garden, it hits the plums and the Asian pears especially hard. Although I have never observed one, the ¼-inch-long (6 mm) mature insects fly in from the surrounding woods each spring. Females lay their eggs in the immature fruit, leaving a telltale crescent-shaped scar no bigger than the nail of your pinkie finger. The infested fruit falls to the ground as the eggs hatch into larvae that eat the fruit from the interior. These larvae then enter the soil, where they develop into adults that reemerge in August and fly off to winter in the woods, returning the next spring to repeat the cycle. This pest has infested my plums every year they've fruited. Kaolin clay (Surround) spray can deter this pest, but since I have sworn off using sprays of any kind, controlling plum curculios is a challenge for me. I could remove the infested fruits by hand, but the baby plums are so small and numerous and mostly out of reach that this approach is not practical. Since most of the fruits of my Asian pears are larger and within reach, each year, when the fruits are about the diameter of a quarter, I remove all the scarred fruits to feed to the pigs and chickens. Of the hundreds of fruits, rarely are there more than half a dozen left on a tree. Sometimes every fruit is infested.

Figure 8.4. The crescent-shaped marks of the plum curculio scar this immature Asian pear.

Early in my edible forest adventure, I learned of a native nematode that the Shields Lab at Cornell University is studying as a biocontrol for an alfalfa pest but which is thought to have potential for fruit pests as well. This nematode inhabits the soil, where it preys on insect larvae. In the hope that this nematode would enjoy eating plum curculios, I contacted Cornell and acquired enough to inoculate my entire acre with the tiny worms. A representative from the Shields Lab visited my farm to take soil samples before and after the nematode application. Whereas none of these nematodes were detected in my soil before the treatment, they were still present two years later, suggesting that the application was a success.

It is too soon to tell if this biocontrol is having its intended effect. For one thing, it takes time for the nematode to reproduce and infiltrate the totality of the soil beneath the susceptible trees. Second, given that the plum curculio's life cycle is not contained within the garden, new weevils can always fly in from the woods to replenish any brethren lost to the nematodes.

Another curculio-control option is to run a flock of chickens through the garden to consume the fallen fruits. I have avoided trying it thus far for two reasons: I use a lot of wood chip mulch that chickens love to scratch away, leaving the soil exposed and generally making a mess. Also, chickens roaming loose make tasty prey for raptors, mink, and raccoons, all of which are prevalent on our farm. Soon, I intend to bring some chickens into the garden housed in a protective cage. In the meantime, I hope that a recent planting of garlic at the base of my plum trees might help deter this troublesome pest.

One upside of the plum curculio, when its numbers are modest, is that it can perform the annual task of thinning fruit for you, so the fruits that are left grow bigger and healthier. I look forward to the day when this pest's numbers are such that they might perform this useful function for me.

branches and extensive roots, American plum is traditionally used in windbreaks and to secure stream banks. Its fruits are no more than 1½ inches (4 cm) in diameter, round, and yellow-to-red in color. The ripe flesh is sweet, while the skin remains quite tart. I eat the plums fresh and have also made a tasty sweet/tart jam by cooking down the whole fruit until I could run it through a food mill, then recooking the pulp with sugar.

I incorporated American plums in my garden along the pond's banks and in overly wet areas. Although it survived in the saturated environments, it has done much better on the pond banks, where the drainage is better. Perhaps because it is located farther from the surrounding woods or blooms later than the Asian/American plum cross-cultivars, the one American plum tree in my garden that has bloomed and fruited has been minimally affected by the plum curculio. Beginning in the sixth year, it has flowered profusely and has borne a multitude of unscathed fruits.

When do I know it's time to harvest my American plums? When there are signs that the local chipmunks have begun to dine on them. Then, if I don't pick them fast, they will soon disappear at the hands of these dexterous rodents.

American plum is a relatively carefree, adaptable native fruit that takes up little space. If you have a taste for stone fruits and are partial to native plants, it may be a fine choice for your garden's understory.

Apple

FR

Malus domestica

Native Range: Central Asia
Height: Up to 35 feet (10.5 m)
Canopy: Up to 35 feet
Soil Conditions: Well-drained soil
Sunlight: Full sun
USDA Hardiness Zones: 3 to 8, varies by cultivar

While apples are native to Central Asia, they have been cultivated worldwide for centuries. Thousands of cultivars exist, selected for the size, shape, and color of their fruits, as well as their suitability for cooking, eating fresh, and making cider. Standard apple trees reach 35 feet tall and wide at maturity, but with regular pruning, they can be maintained at

Figure 8.5. The Freedom apple tree is a natural dwarf, growing no more than 15 feet (4.5 m) wide and tall.

smaller dimensions. In May, apples trees are adorned with abundant pinkish white blossoms very attractive to pollinators. Maturing fruits, which range in color from green to yellow to red depending on the cultivar, are quite decorative from late summer deep into fall. Most apples are sold as grafted trees where the choice of rootstock often determines tree size. If you live in a region that is warmer than Zone 5, you have the option of choosing the smaller dwarf or semi-dwarf apple trees. In colder regions, it is best to choose the hardier standard-sized trees—which have more substantial root development—and, if necessary, prune them to fit your space. A few "natural" dwarf apple varieties are available, which are as hardy as standard-sized trees but naturally more compact in growth habit.

When Dave and I bought our farm, we inherited several mature apple trees planted on a west-facing slope near the farmhouse. During each of the first seven years we lived there, these trees were loaded with fruit. To our disappointment, no more than a handful of the apples were ever suitable for fresh eating; the rest were scarred and disfigured by pests and disease. Given these observations, I doubted I would succeed with apples elsewhere on the property, so I left them out of my original garden design.

As time went on, I needed to replace some other trees in the garden that had failed to survive. I decided to take my chances on a few apple trees, and carefully selected those rated for my hardiness zone and bred for disease resistance. Freedom (a natural dwarf), Liberty, and MacFree are three such cultivars. Choosing the best-drained sites, I planted Freedom and Liberty on slopes with full sun. They both appeared to adjust to these locales, though yellow blotches appeared on the leaves during wet growing seasons, indicating fungal disease. Then one spring I noticed the Liberty tree leaning at an angle. I grabbed the trunk to push it back upright, and the tree came out of the ground in my hand. Apparently, despite a trunk guard, rodents had chewed through each root until the trunk was completely severed just below ground level. So much for that apple.

I planted the MacFree on a small raised mound, once again hoping to raise its roots far enough above the water table. It grew slowly and looked peaked during all but the driest periods. As of this writing it is still alive but clearly struggling. Despite the mound, I believe the soil in the crescent bed is just too poorly drained for its liking.

It can take four to eight years for an apple tree to begin to fruit. Mine have only been in the ground for seven years as of this writing, and have not flowered or borne fruit to date.

If you want to include apple trees in your garden, site them in the best-drained, sunniest spot you have. If you live in the North, choose standard trees or natural dwarfs. Be patient. And unless you are lucky, be prepared to deal with pests and disease.

FR

Apricot

Prunus armeniaca

Native Range: Central Asia
Height: 15 to 20 feet (4.5–6 m)
Canopy: 15 to 20 feet
Soil Conditions: All types, well-drained soil
Sunlight: Full sun
USDA Hardiness Zones: 4 to 8

Apricot is a fast-growing tree with a spreading canopy that looks festive when adorned with ripening orange fruit. It is particularly attractive in spring, when it is one of the earliest fruit trees to bloom, sporting white or pinkish white flowers that surround each branch. But you don't have to wait until spring to enjoy the blossoms. Cut a branch in late winter, bring it inside, and put it in a vase filled with water; in a couple of weeks your branch will bloom in your living room.

Apricot trees are best suited to climates that feature spring temperatures that rise gradually as well as lots of dry weather during the growing season. Thus, most of the commercial production of apricots in the United States is in California. In the Northeast, where spring temperatures can swing erratically, apricot trees usually have no problem surviving but have less success in producing fruit. In spring, early-developing flower buds are often killed by freezing temperatures that follow a warm spell.

Because I enjoy the sweet-tart flavor of their fresh fruit, I planted several different apricot cultivars, siting them all on north-facing slopes to slow bud development in spring. With one exception, they all grew happily. Most flowered beautifully, but so far, none have developed ripe fruit. Since the apricot has a projected life span of 20 to 30 years, there is time to wait.

Figure 8.6. The early-spring blooms of the apricot are a welcome sign of warmer days ahead.

FR

Asian Pear

Pyrus pyrifolia

Native Range: East Asia
Height: 15 to 20 feet (4.5–6 m)
Canopy: 15 to 20 feet
Soil Conditions: Well-drained soil
Sunlight: Full sun
USDA Hardiness Zones: 5 to 9

Though native to East Asia, Asian pears are grown commercially in temperate climates throughout the world, including along the US Pacific Coast. A few cultivars are self-fruitful, but most require proximity to another variety of Asian pear for pollination. Asian pears boast three seasons of appeal: showy white blossoms in spring, large buff or yellow fruits during the summer, and a pastiche of leaf color in fall. The fruits have smooth or russet skin and are round, crisp, sweet, and juicy. Unlike some other fruits that continue to ripen post-harvest, Asian pears need to ripen completely on the tree. You can tell they are ripe when the stem comes loose as the fruit is gently lifted up at an angle (or when a mature fruit or two has fallen to the ground). Once picked, the fruits of some Asian pear varieties can maintain their quality for months when stored in the fridge.

Figure 8.7. Russet-skinned Asian pear fruits are so tempting when they are almost ready to pick.

Two Asian pear trees I planted on a south-facing mound struggled during several harsh winters but bounced back and continue to produce sporadic fruit (see “My Asian Pear Adventures” on page 92.) Undeterred by the challenges with those first trees, I eventually planted six more. I chose several cultivars (including replacements) based on their Zone 4 nursery ratings: Chojuro, Kosui, Hosui, Shinko, Shinseiki (also known as New Century), Shinsui, Korean Giant, and Drippin' Honey. The inspiration for siting more Asian pears occurred on a crisp October afternoon as I stood in the middle of a south-facing slope. The ambient temperature there was a good 10°F (6°C) warmer than the surrounding area. I instantly decided that this was where I would plant four of the new Asian pears. They would benefit from the extra warmth when their fruit was ripening late in the season.

Outwitting my climate conditions was only one challenge; another was animal pests. The section of the garden containing the newest four pears was protected by a T-post fence with electrified wires. This barrier, however, did not deter deer once the surrounding foliage leafed out and obscured it. Like most sensate beings, deer enjoy novelty. They browse the least familiar plants first. As there are no Asian pears growing wild in our woods, these were the first trees the deer stripped of their leaves. Soon after the leaves grew back, the deer stripped them

A Well-Intentioned Remedy

When I decided to plant four Asian pear trees on a south-facing hillside, I was worried about the heavy clay soil that I knew might result in poor drainage for these sensitive plants. I decided to do something about it. Just before planting, I went out to the garden with a broadfork—a tool whose long metal tines create spaces within the soil for water, air, and roots to penetrate, without disturbing the soil to the extent that using a rotary tiller would.

My broadfork has five tines extending down from an 18-inch (45 cm) crossbar, but there are many other models. Long wooden handles extend up from each end of the bar. To use it, you insert the points of the tines vertically into the soil, grasp the handles, and step onto the bar, balancing your weight and pressing down one foot, then the other, until the tines completely penetrate the ground. You then step off and back from the bar, pulling the handles toward you and then pushing them down until they are at a 45-degree angle to the soil surface. Once you see the soil above the tines gently lift, you push the handles back to the vertical, pull the tines up and out of the ground, and start the procedure again 6 to 10 inches (15–25 cm) back from the original spot. Working backward ensures that you don't stand on and compress the ground you just loosened. Continue this procedure until the desired area is completely worked over.

It was grueling to manipulate the broadfork in the dense, heavy clay on my hillside, but I persisted. I was not able to push the tines more than halfway into the ground, and then could only lift the soil slightly. I moved along backward, repeating this partial performance four times. As I inserted the tines and then lifted them for the fifth time, I was suddenly thrown off balance, found myself on the ground, and heard a loud snap. The clay was so resistant that one of the wooden handles had broken off at its junction with the metal bar. With my broadfork smashed and my good intentions dashed, I would just have to plant the Asian pears and let them fend for themselves.

again, and again for a third time, all in the first season. Out of desperation in the wake of this devastation, I devised protective cages to surround these trees going forward (see "Tomato Cages," page 80).

Despite this setback and the distortion of their shape caused by the browsing, the Asian pear trees continued to grow. Then, like my original Asian pears, they suffered winter-kill down to the snow cover. In a display of resiliency, they grew back from above the grafts to produce fruit again two years later. One of the four, planted in the lowest spot on the hill, perished during an unusually wet growing season. The other three, plus two more planted since, are still alive.

The joy and sense of accomplishment I feel at the sight of the beautiful Asian pears ripening on their branches more than compensate for the setbacks and frustration along the way. If you have a south-facing space or a sunny spot in your yard and only have room for one fruit tree, a self-fertile Asian pear could be a good choice.

FR

Cornelian Cherry

Cornus mas

Native Range: Southern Europe and Southwest Asia
Height: 10 to 20 feet (3–6 m)
Canopy: 10 to 20 feet
Soil Conditions: Tolerates most well-drained soil
Sunlight: Full sun or partial shade
USDA Hardiness Zones: 4 to 8

Cornelian cherry is a small tree that is more closely related to the familiar dogwood than to sweet or sour cherry trees. Its bright yellow blossoms, dark red fruits, occasionally vibrant fall leaf color in tones of deep red and burgundy, and prominent winter flower buds coupled with variegated bark provide four seasons of landscape appeal. Its dense branches are suited for a border hedge or windbreak and provide inviting habitat for nesting birds. The fruits, which ripen in late summer, are small and tart. They are traditionally dried, made into jams and sauces, and used medicinally or to flavor vodka.

I planted a pair of Cornelian cherry trees from different nurseries, as two different seedlings or cultivars are recommended for pollination. They both grew slowly but steadily for the first few years. After their leaves fell in the sixth year, I noticed some round flower buds jutting out from the branches of one tree. It flowered the following spring. The next year, both trees flowered, but still no fruit. A scourge of rodents ran rampant under the snow that winter, almost completely girdling one of the trees. Remarkably, it flowered in the spring, but lost vigor and eventually died during the droughty summer that followed. Late that August, I noticed a few bright red orbs on the remaining tree—my first fruit! I gleefully harvested them, ate them on the spot, and saved the seeds to plant another day.

While these trees were growing, I planted several more Cornelian cherries in a windbreak hedge elsewhere in the garden. Since they are more than 250 feet (75 m) from the surviving original tree, they are probably too far away to pollinate it. If I want to see fruit again, I may need to plant another closer by.

Figure 8.8. A unique feature of Cornelian cherry is the tiny bright yellow flowers that cover its branches in early spring, even before forsythia blossoms appear.

One or more of these small, densely branched trees may grow well on an east-facing side of your home or a semi-shaded section of your yard or blend in well as part of a screening hedge along a property line.

FR

Crab Apple

Malus hybrids

Native Range: Central Asia
Height: 15 to 30 feet (4.5–9 m)
Canopy: 15 to 30 feet
Soil Conditions: Well-drained soil
Sunlight: Full sun, but can tolerate some shade
USDA Hardiness Zones: 3 to 8

Crab apples are reliable producers of small fruits best processed for syrup, apple butter, and jellies. They are attractive as specimen trees, can take up relatively little space, and offer four seasons of landscape appeal. The

Figure 8.9. The Firecracker crab apple in bloom.

Firecracker cultivar that I grow is a natural dwarf that boasts showy pink blooms in spring, edible fruits in late summer, and colorful leaves in fall. If not harvested, the deep red fruits of some varieties of crab apple look festive as they remain hanging from the tree's bare branches in winter.

I came across the Firecracker crab apple while thumbing through catalogs in search of a small fruit tree. Its description—pink blossoms and fruit shaped like miniature red delicious apples with edible rose-colored flesh—sounded too good to be true, but I ordered two anyway. These two trees fared much better than the other apples in my garden. Both flowered and bore fruit the second year, which looked and tasted just as described. The only downside was the insect pests that infested and distorted every fruit on one of the trees, leaving the other untouched. This phenomenon has occurred each year that both trees have borne fruit, perhaps because the tree with unscathed fruit is located across a small pond and downwind from the infested tree, making it less accessible to pests.

N Dwarf Korean Chestnut

Castanea crenata

Native Range: Korea and Japan
Height: 10 to 20 feet (3–6 m)
Canopy: 15 to 30 feet (4.5–9 m)
Soil Conditions: Well-drained soil
Sunlight: Full sun
USDA Hardiness Zones: 4 to 8

When I was searching for a small chestnut tree hardy to Zone 4, the owner of Oikos nursery in Michigan recommended the dwarf Korean chestnut. This tree has elongated oval leaves with pointy edges and assumes an irregularly shaped bushy canopy. Its flowers, which appear in July, are large upright catkins—drooping, cylindrical flower clusters—giving rise to spiny orbs containing three to seven nuts that ripen in October. It can begin to flower in 3 to 4 years but does not produce heavy nut yields until it is 5 to 10 years old.

At first, I believed the dwarf Korean chestnut trees I purchased would grow to only 10 feet tall and about 15 feet wide, but I have since learned they can reach double that size. If that turns out to be the case, these trees can be coppiced or pollarded to reduce their size.

I planted three dwarf Korean chestnuts toward the top of a well-drained, south-facing slope in July of the garden's second year. These trees survived despite my accidentally mowing over two of them twice. They appear to tolerate drought as well as our unpredictable winters. To my delight, three husks appeared on the most mature tree (the sole escapee of my mowing) in its seventh year. Unfortunately, chipmunks reaped the harvest before I did.

Based on my experience thus far, I recommend dwarf Korean chestnuts for a Zone 4 location. In addition to its cold hardiness, this tree sometimes holds its leaves all winter, making it an effective screen or windbreak even during the cold months.

Figure 8.10. A dwarf Korean chestnut seedling.

FR

European Plum and American/Asian Crosses

Prunus domestica; *Prunus* hybrids

Native Range: Europe, North America, and Asia
Height: Up to 20 feet (6 m)
Canopy: Up to 20 feet
Soil Conditions: Well-drained soil
Sunlight: Full sun
USDA Hardiness Zones: 3 to 8

Plum trees produce lovely white blossoms in the spring, bear luscious fruits in the summer, and their lush green leaves assume tones of yellow, orange, and red in fall. They are classified into pollination groups based on flowering times, and need a companion from the same group for successful pollination.

Though most of the commercially grown plums in the United States originated in Eastern Europe and Asia, there are many native varieties as well. I grow several non-natives, some crosses between native and Asian varieties, and one native plum (see "American Plum," page 119). I planted five different Asian/American cultivars from an early-blooming pollination group in a thicket on a southeast-facing hillside, and three European plums from a late-flowering group toward the base of the same hillside, spaced closely in a row. They all flowered

The Plum Patch

Plum trees require cross-pollination to develop fruit, but they flower early in the season when the air is still cool and some larger pollinators like honeybees are not yet active. It can be difficult for small pollinators to navigate in cool and windy conditions, especially to fly a large distance between trees. For these reasons, organic orchardist Michael Phillips recommends in his book *The Holistic Orchard* planting plums close together, so close that once they reach mature size, the branches of adjacent trees will cross one another. This enables small pollinators to easily travel among blossoms of neighboring trees and ensures good cross-pollination. Following this advice, I planted five or six plum trees in a close grouping. This happens naturally in wooded areas and is known as a plum patch or thicket.

If you already have plum trees planted at a distance from one another, or if your plan doesn't allow for planting several trees together, there is an alternative. In the spring when the plums are flowering, cut a branch from one tree and place it in a bucket of water set close to another plum tree in flower. Be sure to provide each of your trees with a pollinator branch from another tree, and if the flowers on a cut branch begin to wilt, replace it with another fresh-cut one.

Figure 8.11. Branches from two closely spaced plum trees cross, enabling optimum pollination

Figure 8.12. Plum trees are unmistakably beautiful in bloom.

Figure 8.13. Hazelbert husks are decorative in themselves.

consistently, producing a multitude of immature plums, almost none of which ripened fully. The culprit: the plum curculio (see "My Nemesis: The Plum Curculio," page 120).

N

Hazelbert

Corylus hybrid

Native Range: North America and Europe
Height: 12 to 20 feet (3.5–6 m)
Canopy: 6 to 10 feet (2–3 m)
Soil Conditions: Well-drained soil, drought-tolerant
Sunlight: Full sun or partial shade
USDA Hardiness Zones: 3 to 9

Hazelbert is a cross between two related species in the birch family: the hazelnut (*Corylus americana*), native to the eastern United States and Canada, and the filbert (*Corylus avellana*), native to Eurasia.

Hazelnut is hardy to Zone 3. Filbert is less hardy but bears larger nuts. Combined, they form a rapidly growing, multi-stemmed, suckering shrub or small tree. Hazelberts have interesting features in all four seasons. Their decorative catkins hang from the branches through winter. They turn yellow as they elongate in late winter, bringing some bright interest to the otherwise drab landscape. In summer, hazelbert's multiple stems with dense leaves provide an effective visual screen. Clusters of light green, husk-encased nuts develop on the branches by mid-summer. Later in the season, hazelbert leaves turn eye-catching shades of yellow, pink, orange, and mauve. Within three years of planting, hazelberts begin to produce their small, tasty, nutritious nuts. These can be eaten fresh, made into a paste like Nutella, or left for wildlife. The tree's straight stems can be harvested for use as garden stakes.

If you plan to include hazelberts in your design, you will need at least two trees close to each other for pollination by wind. I interplanted hazelberts with sea buckthorn along a northwest fence line, where they serve as an effective windbreak. Hazelbert nuts will fall to the ground when ripe, but this has rarely happened in my setting—if I don't harvest them before they are completely ripe, the squirrels and chipmunks beat me to it. I wait until the husks open enough to reveal the tips of the nutshells, pick them green, and leave them at room temperature for a couple of weeks to dry down. They mature enough for me to remove the husks, crack them open, and eat. The flavor is improved by simply roasting the nuts in their shells for 20 minutes in the oven.

Medlar

Mespilus germanica

Native Range: Southwest Asia and southeastern Europe
Height: Up to 15 feet (4.5 m)
Canopy: Up to 15 feet
Soil Conditions: Well-drained soil
Sunlight: Full sun or partial shade
USDA Hardiness Zones: 5 to 9

The fruit of the medlar, which ripens in late fall, was a favorite Christmas treat in Europe in medieval times. It may have been one of the only fresh fruits available at that time of year (in the absence of modern refrigeration). Medlar fruit is also unusual because it needs to be bletted, or stored at room temperature, until it ripens and softens. The softened fruit has a flavor like cinnamon apple butter. Medlar is a short, stocky tree that takes up little space, can tolerate quite a bit of shade, and has three seasons of visual interest. It has lovely large (2½-inch / 6 cm) white blooms in spring, dense green foliage sprinkled with developing fruit in summer, and branches adorned with yellow-brown fruit and leaves in fall.

Even though medlar is rated hardy only to Zone 5, I decided to try planting one—the Breda Giant cultivar—in a spot on a southwest-facing slope where I had an opening of just the right size for its compact mature height and canopy. This medlar tree flowered and formed two fruits the second year. I waited to harvest them until the fruits seemed as ripe as they were going to get on the tree, then placed them on a shelf in the house to blet. I checked them regularly, but they never did soften.

Figure 8.14. Immature fruits of the Breda Giant medlar tree.

Instead, they turned completely black inside, as if infested with some kind of mold. I put them in the compost. The same thing happened in two subsequent years. Perhaps I waited too long before I picked them, or didn't wait long enough. Regardless, I was reaching the conclusion that I would probably never be able to taste the fruit of this tree. The next year, the tree was winter-killed down to the snow. I didn't even check to see if it was alive above the graft. I became so frustrated with its failure to yield edible fruit that I dug it up and replaced it with a much hardier crab apple that very spring.

Despite my poor luck with this tree, if you live in Hardiness Zone 5 or warmer and have a semi-shaded spot suitable for only one small fruit tree, the unusual, self-fertile medlar might just be the one for you.

Pawpaw

Asimina triloba

Native Range: Eastern US and Canada
Height: 15 to 20 feet (4.5–6 m)
Canopy: 15 to 20 feet
Soil Conditions: Prefers moist soil
Sunlight: Requires some shade when young
USDA Hardiness Zones: 5 to 8

Pawpaw produces the largest fruit of any tree native to the United States. Its large, oval-shaped leaves cast dense shade and remind me of trees seen in the tropics. Pawpaw flowers are unique—large, purplish maroon, downward-facing bells that are pollinated by flies. Two genetically different trees—either two different cultivars, two seedlings, or one named cultivar and one seedling—are necessary for pollination. The fruit is mango-sized and -shaped, contains large black seeds (that are easy to germinate to propagate new trees), and tastes like banana or vanilla custard. Depending on the hardiness zone, seasonal weather, and the tree's genetics, the fruit may start to ripen as early as August or as late as October. A tree's worth of fruit can ripen over a few days or take as long as a month. For optimal flavor, it is best to gather the fruit when it softens or falls off the tree. Alternatively, it can be picked slightly unripe and left to ripen at room temperature. Pawpaw fruit is rarely, if ever, seen in stores because it has a short shelf life, but it can be frozen for future use in smoothies, pies, and puddings.

There are numerous grafted varieties of pawpaw chosen for the quality of their fruit. Though some are rated for Zone 4, none of the grafted varieties rated for my zone that I have planted survived our winters. I have only had success with seedlings.

My initial pawpaw planting was a staggered row of four trees—three grafted varieties and one seedling. In nature, pawpaws begin their lives in the understory, shaded by other trees. Thus, young seedlings require shade for their first two or three years. I created shade by hammering wooden pallet slats into the ground in a semicircle, with the tallest slats facing south to better block the sunlight (see figure 7.12, page 109, for an example of this technique). Two of the grafted trees, named NC-1 and Prolific, did not survive the first winter. I replaced them with two native seedlings from different nurseries. The third grafted tree, Pennsylvania Golden, held out until the third winter. After its demise, I replaced it with a seedling from yet another nursery. Ungrafted seedlings are unpredictable in the size and flavor of their fruit, so I figured if I planted seedlings from different sources, at least one would produce palatable fruit.

The three oldest seedlings produce flowers, but no fruit so far. I was hopeful the second season when I observed a fly in one of the blooms. One seedling has "given birth," sprouting new shoots (called suckers) from the plant's root system, which I transplanted that fall. None survived. The next year, it sprouted two. I waited until the following spring to temporarily pot these up.

In another section of the garden, I planted three pawpaw seedlings at the damp base of a slope to the north of some existing tall trees that offered partial shade. One was winter-killed above the ground, but reemerged from the base of its stem. All three trees are growing slowly in this location. Like the medlar, it may be that even if fruit appears, pawpaw does not have a long enough growing season in my environment to enable ripening. Only time will tell.

Figure 8.15. Even if they never produce palatable fruit, pawpaw trees add interest to the landscape with their lush tropical-looking foliage and striking yellow/brown fall color.

FR

Peach

Prunus persica

Native Range: Northwestern China
Height: Up to 20 feet (6 m)
Canopy: Up to 20 feet
Soil Conditions: Well-drained soil
Sunlight: Full sun
USDA Hardiness Zones: 4 to 9

The peach made its way to North America via the 16th-century Spanish explorers. Today it is grown commercially in many US states, including Washington, Michigan, Pennsylvania, and New York. In addition to the colorful contrast of the deep green leaves with the ripening red-orange fruits, these attractive trees sport lovely pink blossoms in spring and fall colors in shades of bright orange to burnt sienna. Peach trees can be kept much smaller than their maximum size through rigorous pruning, and do best in low humidity. Unfortunately, peaches have a life span of no more than 7 to 15 years.

Few peaches are rated for Zone 4. The first cold-hardy cultivar I came across, named Reliance, sounded promising, so I planted one in a sunny spot on a well-drained hügelkultur mound. It adjusted well and grew nicely the first year and even faster the second. The third year, it failed to flower or leaf. Clearly, this unreliable tree was misnamed. To this day, its death is inexplicable to me.

Undeterred by one failure, I searched further and discovered a second promising cultivar, Contender. I planted two at either end of the crescent bed (see "The Crescent Bed" on page 290). They both flowered beautifully the second year and bore a few

Figure 8.16. Perfectly ripe peaches cling to a branch of the Contender cultivar.

fruits. I checked these peaches regularly as they matured, waiting for them to become perfectly ripe. I knew that day arrived when an indentation the shape of a bird's beak tip appeared in one of the orbs. I picked the peach and took a bite. Sweet juice gushed from the succulent flesh and dribbled down my chin. Birds are very selective. Perfectly ripe it was. I ate two more on the spot, then harvested the remaining three to save for another day.

Encouraged by my early success, I purchased three more peach trees, including a third Contender, another Reliance, and a new variety called Intrepid. The only cultivar that grew reliably for me was the Contender, better adapted to wetter soil than the others. Although they survived and grew, my Contenders have not fruited consistently. Some years, they failed to bloom. During others, no fruits appeared even after they flowered. As with apricots, peach blossoms are vulnerable to damage by spring frosts. I believe this is the reason they have not always borne fruit. I am still hopeful that I will reap another harvest or two from these trees. Unless this occurs, I likely won't replace the peaches at the end of their life spans. Instead, I have planted seedlings of longer-lived fruit trees such as mulberries and American persimmons nearby, expecting them to grow to fill the space left vacant upon the peaches' demise.

FR Pear

Pyrus communis

Native Range: Temperate regions of Europe, Asia, and North America
Height: Up to 30 feet (9 m)
Canopy: Up to 20 feet (6 m)
Soil Conditions: Tolerates a wide range of soil types
Sunlight: Full sun
USDA Hardiness Zones: 3 to 9, varies by cultivar

I first became acquainted with pear trees when we bought our farm, which, in addition to several mature apple trees, featured two mature pears in the side yard. In mid-spring, they were covered with masses of white flowers that attracted all kinds of bees. By late August, a multitude of small pears hung from the branches. We probably harvested 100 pounds (45 kg) of fruit from the two trees that first year, without even reaching those on the top branches. Some of the fruits were distorted by indentations on their surfaces. I have since learned that such dimple-like puckers can be the result of incomplete pollination. Other than this flaw, I observed only occasional evidence of pest or disease damage on or in the fruit. I definitely wanted to plant some pears of my own.

Most pear trees are self-infertile, which means they require another tree of a different cultivar for pollination. Depending on the cultivar, the fruit can ripen from summer well into fall. Based on my nursery's information about ripening times, I chose cultivars that would ripen at two-week intervals beginning in late summer, so I could spread my harvest out over a couple of months. I planted them on sunny sites to the north of other, shorter trees so

Figure 8.17. The characteristic appearance of fire blight.

they would not cast shade over those trees. If your space is limited, pear trees can also be sourced in dwarf and semi-dwarf sizes.

My pears appear to tolerate the heavy clay soil and have exhibited no disease problems except for fire blight, a bacterial infection becoming more problematic with climate change because it thrives during hot, wet weather. An infection is evident when leaves inexplicably turn black, as if singed by fire. The remedy is to cut off the infected branch at least 8 inches (20 cm) below the infection. Be sure to sterilize the pruning tool by wiping it with isopropyl alcohol before and after use to avoid spreading the bacteria to other susceptible plants. Burn, or otherwise dispose of, the cutting. If caught early, this disease can be controlled. If not, it can completely infest and kill a tree. Luckily, I caught it early when it presented on one of my pear trees, and was able to prune the infected limbs. If you decide to plant pear trees in your edible landscape, it makes sense to choose cultivars with resistance to fire blight.

It takes five to seven years for a standard pear tree to begin to produce fruit. Some of my trees have just begun to bear, and their fruits have been small. Recently, a friend suggested that I prune my trees during the dormant season to reduce the number of branches, which should increase the size of the fruits. It is also advisable to thin the fruits as they begin to develop so those remaining grow bigger.

Pears ripen from the inside out. They should be harvested slightly unripe. Leave those you want to eat right away at room temperature for a few days to allow them to ripen fully. Refrigerate those you wish to store when slightly unripe: Later, when you are ready to use them, remove them from the fridge to allow them to ripen. Do not wait too long to harvest or eat your pears, or you may find them rotten at the core, as I did before I learned about their inside-out ways.

If you have enough space for them in your garden, pear trees could give you relatively trouble-free harvests for decades to come.

Figure 8.18. Fruits ripening on a pear tree branch.

FR Quince

Cydonia oblonga

Native Range: West Asia
Height: 10 to 15 feet (3–4.5 m)
Canopy: 9 to 12 feet (2.75–3.5 m)
Soil Conditions: Well-drained and slightly moist soil
Sunlight: Full sun or partial shade
USDA Hardiness Zones: 4 to 9, depending on cultivar

Quince was another tree I had never seen when I started to plan the edible forest, though I had heard of its fruit. When I investigated these "exotic" plants, I learned that they have been cultivated for centuries and in fact are speculated to be the famous fruit in the tale of the Garden of Eden. I learned that most quince fruits are unpalatable unless processed; their flavor is tart and astringent and their raw flesh coarse and grainy. However, once cooked and sweetened, they reportedly transform into an aromatic delight. Traditionally, quinces are used to make marmalade, pies, jams, and a thick paste that is sliced and served with cheese.

I was not particularly interested in a fruit I had to process to enjoy, but when I read that there was a Russian cultivar hardy to my zone that produced fruit sweet enough to eat fresh, I wanted to plant one. This cultivar, the self-fertile Aromatnaya quince, has large white blooms in spring, giving rise to showy yellow fruits that resemble large, bumpy apples and ripen in fall. Its multi-stemmed trunk, which becomes gnarled with age, adds visual interest in the leafless season. Quinces are reputed to have no major disease or pest problems.

I sited my quince on a south-facing mound, surrounded by a ground cover of assorted herbs that I thought would do well in the warm, well-drained conditions, and encircled by a spiral staircase of large limestone slabs. Remarkably, it soon bore fruits that looked and tasted just as described in the catalog, with a mild, pineapple-like flavor rendering them sweet enough to eat unprocessed. Subsequent years were not as fruitful. Soon after this stellar year, the tree was winter-killed down to the snow. It quickly regrew from above the graft, but failed to fruit before it was winter-killed again. This resilient,

Figure 8.19. The Aromatnaya quince with ripening fruit.

small tree recovered once more with vigor, and at the time of this writing has recently flowered and fruited once more. Motivated by this success, I have since planted another Aromatnaya quince.

NF

Redbud

Cercis canadensis

Native Range: Eastern US
Height: 20 to 30 feet (6–9 m)
Canopy: 20 to 30 feet
Soil Conditions: Adaptable, but does not like saturated ground
Sunlight: Full sun or partial shade
USDA Hardiness Zones: 4 to 9

Also known as eastern or American redbud, this understory nitrogen-fixing tree is a member of the legume family with four-season aesthetic appeal. Before its leaves appear in spring, the entire tree is covered with lovely rose-pink, edible blossoms, which give way to edible seedpods. The heart-shaped leaves have a shiny red cast when young, turning to a medium green during the summer and shades of yellow in fall. In winter, redbud's bare dark bark and graceful trunk and limbs are attractive in the landscape.

Redbud grows naturally in the forest understory, where it can tolerate a good deal of shade. Like many plants adapted to shade, its leaves grow larger in shadier locations, maximizing surface area for the absorption of sunlight. In a designed landscape, redbud can be planted alone as a specimen tree, bunched in groups for dramatic effect, or planted as a component in a hedge. It does not tolerate wind well, so best practice is to avoid that kind of exposure. Redbud's large leaves form a dense canopy, providing optimal shade for plants below.

Redbud is a fairly new addition to my garden. I didn't even learn about this lovely tree's existence until five years after I began my forest garden project. Although it is a borderline Zone 4 plant, its

Figure 8.20. A mature redbud in bloom. You can see the graceful vase-like form of the trunk and branches, which will continue to provide aesthetic interest even in the winter months.

four-season attractiveness and nitrogen-fixing ability were so appealing that I decided to plant a couple. I situated both in a long hedgerow where they would be sheltered from the wind and shaded for part of each day. Both trees survived the first year and began to grow. After about three years, I noticed that the tree on higher ground was thriving, having grown about twice as big as the other, which appeared to be struggling. *Why is this?* I wondered. Was the smaller tree genetically different? Was it receiving less light?

The next spring, I decided to transplant the smaller tree to what I hoped would be a more favorable location. When I dug up the dormant tree, water seeped into the hole, filling it halfway. "That was the problem!" I exclaimed to myself. The ground in that location was too saturated for this

A Propagation Story

One spring shortly after I had become acquainted with redbud, I noticed two huge trees in bloom in front of a house as I drove toward town. These were the only redbuds I'd ever seen in my area. I decided to stop and talk to the owners. The trees were magnificent in full flower and still had a few of the previous year's seedpods hanging from their branches. The owners offered me about a dozen of the pods, which I gladly accepted.

Oftentimes, the seeds of temperate-climate plants require a period of *stratification,* meaning exposure to cold, in order to germinate. Since these pods had remained on the trees all winter, I knew they had already been stratified. Most legume seeds also require *scarification*—some form of physical disturbance that makes their hard outer covering porous, allowing water to penetrate and begin the germination process. You can scarify seeds by soaking them in boiled water for up to 24 hours, or by abrading them. After removing the redbud seeds from the pods, I chose to place them between two pieces of sandpaper, gently rubbing them until I felt enough of the covering had been removed. Then I planted the seeds in potting soil in a small flat and left them in a solar-heated greenhouse to germinate.

Figure 8.21. The redbud I started from seed, in its second year of growth atop a shaded mound. Notice the large, heart-shaped leaves and spreading canopy.

Four of the seeds germinated and began to grow. In mid-summer, once they had a few leaves, I transplanted them into individual 8-ounce yogurt cups with holes poked in the bottoms for drainage. I placed these cups in a semi-sheltered, partly shaded spot outdoors. That fall, I planted the biggest, strongest seedling atop a hügelkultur mound that was naturally shaded by existing tall trees. The seedling was only a couple of inches (5 cm) tall at the time. The next spring, I saw no sign of the tiny tree where I had planted it, and assumed it had been consumed by a rodent during the winter. Then, to my delight, in mid-spring it began to grow. Two beautiful branches emerged from beneath the soil surface, where the root system must have remained intact, and grew to create a 2-foot-wide (60 cm) canopy. Each year since it has grown taller and wider. I am particularly hopeful about this tree because its parent has survived and grown to maturity in the climate conditions of my area.

tree. I moved it to higher ground, where it is doing much better. Since then, I have interplanted two more small redbud seedlings in rows of fruit trees, grouped several together in a border hedge, and topped several hügelkultur mounds in the woods with these shade-tolerant, nitrogen-fixing gems. My pride and joy, however, is the one I grew from seed (see "A Propagation Story").

FR Shipova

Sorbopyrus auricularis

Native Range: Yugoslavia
Height: 15 to 20 feet (4.5–6 m)
Canopy: Up to 15 feet
Soil Conditions: Fertile, well-drained soil
Sunlight: Full sun
USDA Hardiness Zones: 3 to 9

Shipova, an unusual pome fruit from Yugoslavia, is a cross between pear and European mountain ash. The catalog description intrigued me: a medium-sized tree with gray-green leaves that produces small, sweet, seedless pears with a rose-like aroma. Somewhat self-fertile, shipova is best pollinated by a European pear or mountain ash, the latter beautiful with its filigreed leaves and edible orange berries. The shipova I encountered was rated as hardy to Zone 4, so I decided to plant one, accompanied by an ash for pollination.

I sited both on a northeast-facing slope. The mountain ash quickly grew straight and tall, flowering and yielding tart, nutritious berries by its fifth year. The shipova appeared happy but grew much more slowly. Finally, in the spring of the eighth year, the tip of every branch sported multiple flower buds. I was so eager for the buds to open! I waited and waited, but the flowers never appeared. The buds remained closed, as though petrified, for the rest of the growing season. Not one leaf emerged on the tree. This particular spring had seen several successive hard frosts at night during a single week in May. In hindsight, I hypothesized that these repeated assaults on the tender flower buds were too much. Moreover, since most of the tree's energy was tied up in the abundant nascent blossoms, it may have had few resources left to promote further growth in the form of leaves or branches. Or maybe, unobserved by me, the trunk had been girdled below the surface by a rodent over the winter.

In any case, I did not give up. Periodically, I checked the branches for life by gently scratching off a small section of bark. They remained bright green throughout the spring and summer, sustaining my hope. It wasn't until later that fall that my scratching uncovered only dull brown, the telltale sign of death. Disheartened but not discouraged, I ordered two more (to hedge my bets) of these unique trees to plant the following spring.

If you have the patience—these trees can take 10 years or more before bearing fruit—this uncommon tree may be perfect in a foundation planting or as a specimen surrounded by edible flowers and herbs in your yard.

Figure 8.22. A young shipova tree.

Figure 8.23. A sour cherry tree loaded with fruit.

FR

Sour Cherry

Prunus cerasus

Native Range: Europe and Southwest Asia
Height: 10 to 20 feet (3–6 m)
Canopy: 10 to 20 feet
Soil Conditions: Well-drained soil
Sunlight: Full sun or partial shade
USDA Hardiness Zones: 4 to 9

The year we started farming our property, I planted several sour cherry seedlings from a local nursery, labeled "natural dwarfs" since they grow no more than 10 feet (3 m) wide and tall. Although I don't know the variety of these sour cherries, they could be the Evans cultivar, which is described as a hardy natural dwarf. They flower beautifully with picturesque white blossoms each spring and bear reliable crops of bright red, tart cherries each July.

These trees have a suckering habit, and mine generate suckers each year. The year I began planning the edible forest, I dug up and potted nine of these suckers to install the next year in a staggered row on a small hilltop in the garden.

I have learned a lot about harvesting these versatile fruits over the years. At first, I waited until most were ripe before beginning to harvest. I noticed that birds enjoyed eating the ripe fruits, sometimes removing the flesh but leaving the pit hanging from the stem. Other times they would just peck a hole in the fruit, causing the cherry to slowly rot. One year, I decided to net the trees to prevent this carnage, but only got around to enclosing four. It made little difference: The netted trees still furnished rotting cherries, as did the unnetted. Something wasn't right. I thought about the expression "don't cherry-pick." I had previously interpreted this to mean, "wait until all the cherries are ripe before you pick them." But the opposite is conceivable, too: that you need to pick cherries as they ripen and leave the rest for another day. The next season, I implemented this insight. I picked the cherries daily, beginning when the first ones were just about ripe. I noticed far less bird predation and fewer fruits rotting on the trees. I have followed this procedure ever since with consistently good results.

The only problems I have experienced with these cherries are black spots and puckered shapes in some of the fruit. If they are not too damaged, I use them in processed treats like spreads and sauces, which I prepare by cooking the fruit with the seeds until it softens, then straining it through a food mill. I eat the unblemished cherries fresh and use them in pies. One of my favorite desserts is a cherry mousse made by mixing cherry spread with freshly whipped cream. Yummy! If you enjoy tart cherries, these hardy, small trees can brighten your landscape with reliable spring blooms and summer fruits for 10 to 20 years.

NF Speckled Alder

Alnus incana

Native Range: Northeastern US and southeast Canada
Height: 15 to 25 feet (4.5–7.5 m)
Canopy: 15 to 25 feet
Soil Conditions: Tolerates poorly drained soil, needs ample moisture
Sunlight: Full sun or minimal shade
USDA Hardiness Zones: 3 to 6

Also known as green alder, speckled alder is a fast-growing, native nitrogen fixer. Like other nitrogen fixers, it is a pioneer plant that prepares the soil for more demanding plants that follow. The soil is often quite acidic and oversaturated around swamp edges, where speckled alders are found in nature. Alder's roots penetrate and open up the soil to create better drainage and air pockets, while the leaves that fall in autumn add organic matter that encourages soil life like earthworms whose activities improve drainage. All of these functions make the soil more hospitable for plants less tolerant of soil saturation. A member of the birch family, speckled alder has pointed, oval, deeply veined leaves and sports decorative catkins and cones that add interest to the winter landscape. It has shallow roots and a clumping growth habit, whereby its multiple trunks grow close together.

Well adapted to saturated soil conditions, this alder is ideal to incorporate as your understory nitrogen fixer in a wet habitat. It grows naturally in the boggy sections of our farm, but it does not do well in dry soil, which I learned the hard way. I planted seven alders as part of a set of plant groupings that stretched from high to low points of a hillside (as described in "A Planting Dilemma" on page 281). The only one that survived was at the bottom of the hillside, where the soil remained moist throughout the season. In the higher locations, the soil became too dry for the alders in the summer. I also planted some alders in the saturated southwest quadrant of the garden, where they have done well. If you are looking for a medium-sized nitrogen-fixing tree for a wet, sunny location, this alder may fill the bill.

Figure 8.24. In this late-winter scene, the alder catkins are already coming to life, turning their characteristic golden color.

FR

Sweet Cherry

Prunus avium

Native Range: Eurasia
Height: Up to 20 feet (6 m)
Canopy: Up to 20 feet
Soil Conditions: Well-drained but moist soil
Sunlight: Full sun
USDA Hardiness Zones: 5 to 9

Who doesn't relish the flavor of fresh sweet cherries when they appear in the produce aisle in early summer? I find them irresistible, which is why I wanted to plant some, even though the odds were against me with my heavy clay soil and Zone 4 climate. Their lovely white blooms in spring and spreading canopies replete with ripe fruit in summer were also appealing. I discovered two cultivars that were rated as nearly compatible with my zone: White Gold, a yellow sweet cherry; and Black Gold, a dark red one. Perfect! I bought the pair and planted them 20 feet apart on a southeast-facing slope.

The first years were promising: My sweet cherry trees grew quickly to form vase-shaped canopies, and they flowered and fruited by the third year. Unfortunately, due to my lack of attention, birds harvested that first crop. The following year, I netted the trees before the cherries ripened and enjoyed some for myself. I was encouraged by this success, and over time, I planted more of these cultivars.

A couple of years later, the fall season was unusually temperate, staying warm with few frosts well into December. Suddenly, in mid-January, winter arrived with a vengeance, bringing a week of temperatures down to –20°F (–29°C). That spring, my two sweet cherries began to leaf out and flower normally, but then ceased their growth. Their leaves turned brown, and when I scratched the bark from the trunk I discovered no life there, either. What had happened?

Figure 8.25. My first harvest of White Gold sweet cherries. They tasted so much sweeter than cherries from the grocery store!

Stone fruits like sweet cherries and peaches "harden off" (prepare for the cold season) from the outside in. First the slender branches, then the larger limbs, and finally the trunks prepare for the colder weather to come. I surmise that the sudden change in weather caught these trees by surprise, before they had completed their preparation, killing both trees at the trunk. Since their slender branches had time to prepare and the buds had formed before the cold snap hit, the flowers and new leaves appeared as expected during the spring, but couldn't develop further due to the lack of nutritional flow through the dead trunks. Of the later sweet cherries I planted, the youngest two survived, while one other mature tree perished. I imagine that the younger trees had time to prepare for winter while the mature one, with more substantial branches and a thicker trunk, did not.

If your climate is a bit warmer than mine (Zone 5 or better) and you have the space, one or more sweet cherry trees can provide beautiful spring blooms and delectable fresh fruits for 20 to 30 years.

MT

Witch Hazel

Hamamelis species

Native Range: Eastern US
Height: 15 to 25 feet (4.5–7.5 m)
Canopy: 15 to 25 feet
Soil Conditions: Prefers moist soil
Sunlight: Full sun or partial shade
USDA Hardiness Zones: 3 to 8

Although it doesn't bear edible fruit or nuts, witch hazel is a unique addition to the understory layer. Native to wooded areas in the eastern United States, its fragrant yellow flowers bloom in late fall to early winter, providing food for pollinators that are out and about on warm days late in the season. It can adapt to both part shade and moist soil and will grow in Zones 3 through 9. Witch hazel has medicinal applications as well. Native Americans use the leaves and bark to treat inflammation, while a commercial solution derived from the stems is used to soothe skin irritations.

I planted one of these trees in the wet, shady end of a border mound. It struggled for a few years before it finally died due to the combination of oversaturated soil and too much shade. I recently installed a pair of these small trees on the same mound, but in a higher location, where the soil is

Figure 8.26. A mature witch hazel tree in the state park near our farm, recovering from a caterpillar attack the same year that mine were hit.

moist but not saturated and there is partial sun. Here they are off to a slow start and have been set back by caterpillar predation, but hopefully will recover and do better in the future.

If you have room in your plot for plants that occupy the understory, I hope this chapter has inspired you to include one or more of these familiar or unusual trees in your garden plan. If your space can only accommodate smaller shrubs, the next chapter offers numerous options from which to choose.

CHAPTER 9

SHRUBS

If your plot will not accommodate the taller plants described in the "Overstory" and "Understory" chapters, you will discover that there are many appealing food-bearing bushes to choose from in the shrub layer, which comprises woody plants that grow up to 12 feet (3.5 m) tall. It is an encouraging layer to incorporate because many shrubs bear fruit as early as the first year after planting. If you plan on including understory or overstory trees, many shrubs tolerate partial shade and so can be nestled under these taller plants. Or they can be placed in areas of your landscape that do not receive full sun, such as a northwest- or northeast-facing side of your house, or even in the shade of an existing tree in your yard. Whether your habitat is wet, dry, sunny, or partially shaded, there are a number of attractive options, even for cold climates.

As you will see from my examples, I don't always have positive results; I have made mistakes as well as successful choices along the way. When you place your bushes in suitable habitats and give them proper care, most of them will be fine. Discover your planting options, and don't be afraid to try and try again until you find plants that will thrive under your unique conditions. Your local cooperative extension and horticulture groups in your area can be helpful in this regard.

Figure 9.1. Ripe red raspberries provide a splash of color and delicious fruit in the shrub layer.

B Aronia

Aronia species

Native Range: Eastern North America
Height: 4 to 8 feet (1.25–2.5 m)
Canopy: 3 to 4 feet (1–1.25 m)
Soil Conditions: Tolerates saturated soil and drought
Sunlight: Full sun or partial shade
USDA Hardiness Zones: 3 to 8

Commonly known as chokeberry, aronia is often found in wet woods and swamps, where it tolerates both shade and saturated soils. This adaptable plant can also flourish in sunny, drier habitats. In mid-spring, these decorative plants are covered with umbrella-shaped bunches of pink buds, or umbels, that open into attractive white flowers. The resulting fruit appears in clusters and turns a deep purplish black as it ripens in late summer. When clasped by hand, the clumps of berries pull right off the stems, making harvest easy and efficient. Their dark color signifies an abundance of flavonoids and antioxidants, making this berry popular among the health-conscious.

The fresh fruit is astringent, causing the mouth to dry and pucker with aversion, hence the moniker *chokeberry*. The berry loses its astringency when cooked. I once brought these fresh berries to two chefs for sampling. After only one taste, they rejected them as too tannic for use. Later, I brought them some sauce that I had made by cooking the berries and adding some sugar. The chefs loved the texture and complex flavor of these processed fruits. This sauce is great as an addition to yogurt or smoothies. I have also mixed aronia berries, which lack pectin, with red currants, which contain it in abundance, to create a jam that jells well and has excellent flavor. I imagine aronia would form a tasty jam paired with any other fruit that has high pectin content. It can be used in herbal teas and converted to juice, which is used as a coloring agent in commercially produced beverages and yogurts. These berries can be stored for weeks in the fridge or frozen for later use.

In the fall, aronia leaves turn dramatic tones of burnt orange and scarlet, completing this plant's three seasons of landscape appeal.

Several aronia cultivars are bred for improved fruit size and production. Aronia have a suckering habit, so with time they will naturally increase their width. I planted these bushes at the low, moist end of an edible hedge; as an understory component in a partially shaded orchard-type row; and in both sunny and shady locations with saturated soil. They survived in all four habitats but produced more fruit in those with better light. They require no particular care, although they may need to be staked to prevent their heavy fruit-laden branches from reaching the ground. I haven't observed any pest or disease problems other than deer, who enjoy nipping the leaves and growing tips. Aronia can be propagated from seed, dormant cuttings, or by transplanting suckers.

Figure 9.2. Aronia branches laden with ripening berries.

Figure 9.3. An aronia bush in flower.

NF

B

Autumn Olive

Elaeagnus umbellata

Native Range: Eastern Asia
Height: Up to 12 feet (3.5 m)
Canopy: Up to 12 feet
Soil Conditions: Adaptable
Sunlight: Full sun or partial shade
USDA Hardiness Zones: 3 to 8

Autumn olive, a large nitrogen-fixing shrub, has attractive oval, gray-green leaves with silver undersides. The leaves appear early in spring and are among the last to drop off in the autumn. This feature, coupled with the shrub's dense branches and rapid growth habit, makes it ideal for inclusion in a windbreak hedge. In spring, autumn olive branches are covered with clusters of pale yellow bell-shaped flowers—very attractive to pollinators and sweet smelling to humans. Small, round berries, containing little flecks that sparkle like silver glitter, develop in clusters along the thorny branches and ripen to red or orange in mid-fall. Their flavor is pleasantly sweet and tart. Autumn olive berries contain the antioxidant lycopene and can be eaten fresh, frozen whole, or processed into jams and fruit leathers. I enjoy the fresh or frozen berries mixed with breakfast cereals and yogurt and in baked goods like scones and cookies.

Figure 9.4. The sweet-smelling blooms of autumn olive completely cover the plant.

Autumn olive can be maintained at your preferred size with frequent pruning, which does not hurt the plant. I prune mine often to keep them from intruding on nearby berry bushes and fruit trees. I let the prunings drop to the ground, where, as they decay, their high nitrogen content nurtures the surrounding plants.

Autumn olive was introduced to the United States in the early 1800s. Like most nitrogen fixers, autumn olives are "pioneer" plants. By the end of the 20th century, autumn olives were invading areas where the natural habitat was disturbed by burning or logging. Today it is considered invasive in the central and eastern United States in Zone 5 and warmer. If you live in an area where natural habitats are disturbed by logging, development, or fire, it is best not to incorporate this plant in your garden. Birds will spread the species by eating the berries and depositing the seeds elsewhere through their excrement. If you live in Zone 5 or warmer and choose to plant autumn olive, protect your nearby environment from its spread by netting the bush before the berries ripen. This will keep the birds from eating the fruit.

I have incorporated autumn olive in my garden in designed beds, as a windbreak component, and as a nurse plant. During a couple of harsh winters, the branches of some of my bushes were winter-killed above the snow. However, they regrew vigorously from their roots the next spring.

FR

Beach Plum

Prunus maritima

Native Range: Eastern US
Height: 4 to 8 feet (1.25–2.5 m)
Canopy: 4 to 8 feet
Soil Conditions: Well-drained but moist soil
Sunlight: Full sun
USDA Hardiness Zones: 3 to 6

Beach plum is commonly found growing in sandy soils along the ocean shores of the eastern United States. I remember bringing home a jar of beach plum jelly as a souvenir from a childhood vacation in Maine, but I had never actually seen the plant or its fruit before I read about it in a nursery catalog. Motivated by my pleasant memory, I decided to plant some.

Although beach plum is generally described as a shrub, some grow tall enough to be considered trees. This plant is eye-catching in late spring, with abundant white blooms covering branches that grow from a central trunk like widespread arms embracing the sky. The ½-inch-wide (13 mm) spherical fruits grow in clusters that hang from the branches on short stems. When ripe in late summer, these small decorative plums range in color from yellow to maroon to dark blue or purple. Beach plum leaves turn deep red and orange before they fall.

I use beach plum fruits to make a delectable jelly with a viscous texture and unique sweet and tart flavor. Rather than pitting each small fruit individually, I stew them with a little water until the flesh becomes soft, then run them through a sieve to remove the pits and skin. I then reheat the pulpy juice with some sugar and wait for the natural pectin to work its magic.

Since beach plum's natural habitat is sandy, I was unsure how this plant would fare in my poorly drained clay soil, but it surprised me by doing well. If your soil is on the sandy side, they will likely do well for you. Since I have yet to net these bushes, I share my harvest with visiting birds. Like many shrubs, these bushes benefit from regular pruning to open their centers for better air circulation and exposure to light. They are best propagated from seed, as with most stone fruits.

Figure 9.5. Beach plum fruits.

B

Blackberry

Rubus species

Native Range: Different species found in temperate climates throughout the world
Height: 4 to 10 feet (1.25–3 m)
Canopy: 3 to 4 feet (1–1.25 m)
Soil Conditions: Fertile, well-drained soil
Sunlight: Full sun
USDA Hardiness Zones: 5 to 10

When adorned with bunches of ripening berries, a blackberry patch is a delight to observe. Later in the season, the occasionally dazzling colors of blackberry leaves enliven the fall landscape. Blackberries can be sourced with upright or trailing growth habits, and with or without thorns. The stems and leaves of upright varieties look similar to those of red raspberries, while the trailing varieties resemble black raspberries in growth habit. However, blackberry flowers are much more prominent than those of raspberries—an inch (2.5 cm) in diameter with five showy, white or pinkish white petals. Blackberries can be eaten fresh or frozen, or made into jams, pies, and other pastries. The plant's leaves have medicinal uses and can be included in herbal teas.

Blackberry thorns can be a turn-off. I was familiar with native blackberries, whose sharp prickles can draw blood and rip clothes. I planned to invite U-pickers into my garden and felt they might find thorny plants aversive. The thornless cultivars now available seemed more desirable, and I recommend them to you if they are hardy to your

Figure 9.6. Ripening blackberries punctuate the late-summer landscape.

zone. I didn't plant any blackberries during the early years of my edible forest because my research suggested that most thornless varieties are rated for Zone 6 or warmer, so I assumed they wouldn't survive in my Zone 4 location. Then I went to England and was struck by the sight of thick blackberry canes loaded with ripe fruit arching out of the hedgerows. This experience convinced me to try growing them.

Like raspberries, blackberry canes live for two years. In nature, they only bear fruit on the second-year stems, or floricanes, which die after fruiting. Modern breeding has produced primocane varieties that fruit late in their first year, as well as thornless cultivars that I imagined would be more appealing to future U-pickers. In my initial research, the only thornless blackberry cultivar that came close to being hardy in Zone 4 was a primocane called Prime Ark Freedom, an upright variety rated for Zone 5. I planted them in a hedgerow that mimics a forest edge—one of the locations where blackberries are found in nature. This location had sun for part of the day. The plants survived but never thrived, nor did they produce any fruit. I noticed them gradually suckering toward an area that got more sun. I concluded there was not enough light in the spot I had chosen and too short a growing season in my region for this variety to be productive. I was disappointed, but did not give up.

Two years went by as I continued to pore over catalogs, looking for another thornless variety I could try. Finally, I found a catalog that rated the Triple Crown variety as hardy in Zone 5, though it was rated as Zone 6 elsewhere. This is a semi-trailing or semi-upright variety, meaning that its canes arch outward, with the tips draping down to the ground surface. I planted them on a well-drained south-facing hillside, a favorable microclimate with full sun that mimics a Zone 5 or 6 location during the growing season. Due to their trailing nature, I knew they would need to be trellised. I placed each bramble on the south side of a fruit tree, draping the sprawling canes through the branches for support. That summer, the plants grew strong. To my surprise, they survived the first winter. In August the following year, I was astonished to see big fat berries ripening on the tips of some second-year canes. When the berries were completely black and easy to slip off the canes, I tasted them. Delicious! Encouraged by this success, I ordered more of the same variety from the same nursery. I planted them to trail over some bushes in another well-drained, sunny location. The following year, my original Triple Crowns fruited once more. Moreover, where some of the arching primocanes touched the ground, they rooted to form new plants! I marked these babies with flag-taped stakes so I could locate them the next spring to transplant them to a new location.

Blackberry care is similar to that of raspberries with respect to removing spent floricanes and the weakest primocanes. Air circulation and sanitation are important as well. The primocanes of the Triple Crown variety should be tipped in summer to your desired height, usually 4 to 6 feet (1.25–2 m). This stimulates the growth of side shoots that should also be trimmed once the canes go dormant in late fall or winter. It is on these side shoots that the fruit will form the next year. A reputable nursery usually provides specific pruning advice for the plants they sell.

While the canes can be trellised upright, I do this only during the growing season, allowing them to lie closer to the ground over the winter. There, they are less exposed to cold winds and may become covered with an insulating layer of snow. Each fall, I apply a thick mulch of wood chips around their base to add extra protection to the roots from the winter cold and to conserve moisture during the following growing season.

If you have a spot on your property that is sunny, is well drained, and can be kept moist when the berries are growing, with the proper care blackberry plants should yield delicious fruit for up to 15 years.

Figure 9.7. Black currant berries ripen in bunches along the stems.

B Black Currant

Ribes nigrum

Native Range: Central and Northern Europe and Northern Asia
Height: 4 to 6 feet (1.25–2 m)
Canopy: 4 to 6 feet
Soil Conditions: Fertile, moist soil
Sunlight: Partial shade, but tolerant of full sun
USDA Hardiness Zones: 3 to 8

Currants are delightful fruits that range in color from white or pink to red or black. Most widely available varieties originated in Europe and were brought to North America by early settlers. One or two red currant bushes used to be a common sight near the foundations of traditional farmhouses. These berry plants were banned in the early 20th century, when it was determined that they were carriers of the invasive white pine blister rust fungus that can devastate white pine trees. Now there are cultivars resistant to this fungus, but some municipalities still restrict these plants. Check with your local cooperative extension office to find out if they are allowed in your area.

Black currant fruit is savory, pungent, and complex—an acquired taste for some palates. Chefs treasure this unique flavor for dessert confections and sauces for meat and fowl. I had never eaten black currants before I harvested my first fruit but developed a taste for them before long. High in vitamins and other nutrients, the juice is a staple source of vitamin C in Europe, similar to orange juice in the United States. The fruits can be made into wine or liquor, and are a popular flavoring for vodka. Black currant leaves can be used to make a fruity herbal tea. They contain beneficial nutrients and have medicinal properties as well.

When they are perfectly ripe, I enjoy eating the berries fresh and adding them to cereals and rice pudding. For processing into jams or pies, it is best to pick ripe berries mixed with underripe berries, which have the greater pectin content. Black currant jam is one of my favorites: thick and dark, with an intense flavor.

Black currants of European origin are available in a multitude of varieties—Raintree Nursery in Washington State lists over 25 cultivars in its catalog. I grow three: Ben Sarek, Consort, and Titania. The first two grow up to 4 feet tall and wide, while Titania can reach 6 feet in both dimensions. The year after I planted them, all of these bushes were loaded with ripening fruit beginning in mid-July. Harvesting of black currants is very similar to that of clove currants ("see Clove Currant," page 156). Once established, these shrubs require little attention other than harvesting and annual pruning, as described in "Pruning Currants and Gooseberries" on page 174. They tolerate more sun than red and pink currants and do well in two full-sun locations in my garden.

B Black Raspberry

Rubus occidentalis

Native Range: Eastern US and Canada
Height: 6 to 9 feet (2–2.75 m), unpruned
Canopy: 2 to 3 feet (60–100 cm)
Soil Conditions: Well-drained but moist soil
Sunlight: Full sun to partial shade
USDA Hardiness Zones: 4 to 9

Black raspberries grow on long, arching canes that form new plants where they touch the ground—a process known as tip layering. Their canes are a lovely pale purple but are extremely thorny. Small flowers give rise to deep purple, almost black, berries that ripen in July. They resemble red raspberries in shape, but have a unique flavor. Like blackberry and red raspberry canes, black raspberry canes are spent after two years, at which point they should be pruned out to make room for new canes emerging from the crown. It is best to prune the tips of the primocanes, called tipping, when they reach the length you desire to encourage side shoots, which will bear fruit the next season. Tipping the canes also prevents them from touching the ground; if allowed to do so, they will soon form an expanding and impenetrable tangle of thorny brambles. Tipped canes benefit from support to keep them upright. I tie them together with a string a couple of feet above the crown to keep them erect, or support them by draping them over the branches of a neighboring bush.

I planted black raspberries among elderberries and in a border hedgerow, with disappointing results in both locations. I suspect my failure to prevent weed encroachment in the first location, and the combination of heavy clay, too much shade, and wet ground in the second did them in. Based on my experience, I recommend planting black raspberries away from competing plants, where they will receive at least some sun, and where the ground is moist but not saturated.

Figure 9.8. Side shoots, which will produce flowers and fruit the following year, grow from a tipped black raspberry cane.

NF B

Buffaloberry

Shepherdia species

Native Range: Central and western North America
Height: 6 to 12 feet (2–3.5 m)
Canopy: 6 to 12 feet
Soil Conditions: Tolerates most well-drained soil, drought-tolerant
Sunlight: Full sun or partial shade
USDA Hardiness Zones: 3 to 9

If you are partial to native plants and looking for a cold-hardy nitrogen fixer similar in size and use to autumn olive and goumi (see "Autumn Olive," page 148, and "Goumi," page 163), buffaloberry may be a good choice. This thorny shrub is slender at the base and wider at the crown. In very early spring, before its small silver-gray leaves appear, buffaloberry produces tiny pale yellow flowers that provide food for intrepid pollinators. Small red berries ripen in summer. Although they are quite tart and relatively unappealing when eaten fresh, buffaloberries can be processed into jams, pies, syrups, and a cranberry-like sauce. These shrubs can be included in a windbreak or used as a specimen plant. I planted three buffaloberry bushes in a row, creating a feature that often attracts the interest of guests.

Figure 9.9. Even in winter, buffaloberry's intricate thorny branches and vase-like shape draw attention.

FR

Bush Cherry

Prunus japonica × *Prunus jacquemontii*

Native Range: This hybridized plant has no native range
Height: Up to 4 feet (1.25 m)
Canopy: Up to 4 feet
Soil Conditions: Well-drained, loamy soil
Sunlight: Full sun
USDA Hardiness Zones: 3 to 8

These hybrid shrubs are also called Meader cherries, after the plant breeder Elwyn Meader. There are three cultivars: Jan and Joel, which need each

What about Blueberries?

Blueberries (*Vaccinium* spp.) are a popular fruit, well known for their nutritional value, and widely grown commercially. They can range in height and width from 2 to 12 feet (0.6–3.5 m) and are native to North America, Europe, and Asia. The lowbush variety, which produces tiny berries, grows wild on the rocky ridges of our farm, and I have picked huge berries from highbush blueberries growing wild along the coastline in the Northeast. Depending on the variety, they can grow in Hardiness Zones 3 to 9.

Although I have grown blueberry bushes in an isolated plot that I acidified annually with sulfur, I did not plant them in my edible forest precisely because they require acid soil. In keeping with permaculture principles, I chose not to amend the soil there. However, if your plot happens to be acidic, there are numerous varieties and cultivars from which to choose. Blueberries need full sun, have shallow root systems that do not tolerate drought, require annual pruning, and can be prone to pests and disease.

Figure 9.10. The dazzling fall colors of bush cherry.

other for pollination, and the self-pollinating Joy. These were the smallest cherry bushes I came across, and I thought I had the perfect spot for them with full sun and fairly good drainage. I fell in love with these plants the first year. They bloomed beautifully in spring, produced a small crop of tart cherries later that summer, and turned magnificent colors in the fall. I had hit upon a gem! As time went on, however, my ardor faded.

During the second year, when my bush cherries were loaded with unripe fruit, I netted them to protect them from hungry birds. However, I did not notice that my resident chipmunks were systematically stripping the cherries off the branches as they ripened. When their discouraging pilfering finally caught my attention, I was able to salvage just enough of the fruit to make a tasty topping for a cheesecake. My enchantment was fading.

Another year went by, during which the chipmunks again burrowed under the bird netting to help themselves. Moreover, I noticed that the three-year-old branches were dying, suggesting that they should have been pruned out that spring. I was no longer infatuated.

Then, as the snow melted the following spring, I was appalled to observe that nearly every bush was completely girdled at its base by rodents burrowing under the snow. That growing season was an unusually wet one. The combination of the girdling and the saturated ground destroyed what was left of these bushes, and of my enthusiasm for them.

My experience with these cherries taught me a valuable lesson: that not every plant that sounds wonderful from catalog descriptions and does well at first will work out in the long run, through no fault of the gardener. Bush cherries may be a good choice for you under the right conditions: a sunny location with well-drained loamy soil, an absence of rodents, and a willingness to prune them annually.

B Clove Currant

Ribes aureum

Native Range: Most of US, parts of Canada and Mexico
Height: 4 to 8 feet (1.25–2.5 m)
Canopy: 4 to 8 feet
Soil Conditions: Adaptable and drought-tolerant
Sunlight: Full sun or partial shade
USDA Hardiness Zones: 4 to 8

Clove currant is one of my favorite fruiting shrubs. Also known as golden currant, this berry bush was traditionally used as a source of food and medicine by Native Americans. In May, its attractive yellow flowers exude a spicy aroma that hints of cloves or vanilla. Clove currant fruits are up to ½ inch (13 cm) in diameter and begin to ripen in mid-summer. In fall, the intricately shaped leaves turn dramatic shades of red. Thus, this bush has landscape appeal for at least three seasons.

I enjoy clove currant fruits straight off the bush when they are very ripe. Their flavor is not as pungent as the fruits of European black currants. They will keep refrigerated for at least two weeks. I freeze them to add to cereals, pastries, and yogurt in the winter months. And I make clove currant jams.

I am familiar with the Crandall cultivar, which grows to about 4 feet by 4 feet and has a sprawling habit, lending it to placement on a hillside. I first planted two clove currant bushes on slopes at the ends of a 50-foot-long (15 m) mound. The following spring, while chatting with my neighbor, standing at least 150 feet (45 m) from the nearest clove currant bush, I detected a pleasant aroma in the air. What was the source of this delightful odor? Following my nose, I wandered through the garden and landed near one of the clove currant bushes in full bloom. I understood at once how the name *clove currant* came about. Each spring, the scent from these bushes permeates the entire garden.

Figure 9.11. Clove currant blooms, whose distinct aroma can be detected from hundreds of feet away.

You can expect to harvest fruit from this plant during the second year, or possibly the first if you are lucky. The berries are quite tart when they first turn black, but become sweeter the longer they remain on the bush. As with most berries, their quality improves the longer you wait, providing the

Figure 9.12. These clove currants begin to ripen in late July or early August. Notice the first signs of fall color in the delicately shaped leaves.

weather is dry. When there is a lot of rain, the ripe berries absorb too much water and rapidly lose their quality and begin to decay. Since these currants ripen unevenly, they need to be "cherry-picked" by hand for fresh eating. For processing, however, partly ripe and fully ripe berries can be picked all at once, as the less mature berries contain more pectin to help jams jell. I like to pick them all before a predicted rain to catch them when their quality is at its peak. To pick the bush clean, I use a berry-picking comb. This inexpensive tool consists of a plastic box with a handle on top and a metallic comb attached under the open end. To harvest, slide the teeth of the comb along a branch, capturing the berries as you go.

I liked my first two Crandall clove currants so much that I planted an additional dozen. Then, as I continued to develop the garden, I added another 10. I plan to incorporate another handful to replace the bush cherries (see "Bush Cherries," page 154) that failed to thrive in my garden. These shrubs are quite adaptable. I have observed no pest or disease problems, and they require minimal pruning.

Clove currants are most easily propagated by layering (as described in "Propagating Berry Bushes," page 82). I have not been successful propagating them from cuttings taken in early spring. I have not attempted to propagate them from seed.

EF

B

Elderberry

Sambucus species

Native Range: Europe, temperate Asia and North America
Height: Up to 12 feet (3.5 m)
Canopy: Up to 12 feet
Soil Conditions: Moist but not saturated soil
Sunlight: Full sun to some shade
USDA Hardiness Zones: 3 to 10

These large, many-stemmed shrubs with broad canopies can grow to their full height in one year. The tiny ivory-white blossoms appear in large, flat, rounded umbels in early summer and can continue blooming into August. Elderberry's aromatic blooms are attractive to pollinators and beneficial insects such as parasitic wasps. The small, round red or black berries ripen from early August onward and are a favorite of birds. Of the species native to North America, I am only familiar with American black elderberry (*Sambucus canadensis*), which grows wild on our farm.

Named cultivars of American black elderberry produce larger fruit than the wild type, but after growing several cultivars, I decided that I prefer the berries from wild shrubs. Although the berries are small, to my taste they have a more complex and intense flavor. All parts of the plant can be toxic if consumed raw, but are safe once cooked. The flowers can be dipped in a batter and fried up as fritters. In England, I came across a syrup made from the blossoms that, mixed with sparkling water or soda, makes a refreshing summer beverage. The French use the blooms to make a liqueur. The berries, bursting with vitamin C and other beneficial nutrients, are traditionally used to make wine as well as medicinal syrups and tinctures to boost immunity and provide other health benefits. I like to use the fruits to make elderberry pies and a preserve that is delicious mixed with yogurt or used as a topping for ice cream or cheesecake.

I planted my first elderberries at the edge of the garden, along a roadside fence where I knew the soil would stay naturally moist for most of the growing season. Though not my original intention, the ripe elderberries serve as a trap crop for birds that eat them in preference to nearby grapes that ripen at the same time. Other than being a favorite browse for deer, these plants have not displayed any significant pest or disease problems in my experience. Elderberries are easily grown from cuttings, transplanted suckers, or seed. They fruit best on one-to-three-year-old stems that should be pruned to the ground before their fourth year.

Figure 9.13. The aromatic flowers and dark purple fruits of the elderberry.

NF False Indigo

Amorpha fruticose

Native Range: US east of the Rocky Mountains
Height: 4 to 12 feet (1.25–3.5 m)
Canopy: 6 to 18 feet (2–5 m)
Soil Conditions: Tolerates moist soil and drought
Sunlight: Full sun or partial shade
USDA Hardiness Zones: 3 to 10

There are well-known herbaceous perennials and dye plants that also go by the common name of false indigo, but the false indigo I describe here is a deciduous woody plant, *Amorpha fruticose*. It has several bare stems crowned by bushy masses of leafy branches wider than the plant is high. False indigo flowers, arranged along slender spikes, are purple with yellow-orange anthers. These fragrant blooms are attractive to butterflies and other pollinators and can be used to make dye. The seeds are easy to propagate, once stratified and scarified. In fact, this plant has naturally self-propagated in my garden, with new seedlings growing around the base of each shrub. To control their spread, I either weed them out or transplant them elsewhere.

In contrast to most other woody legumes, false indigo can tolerate some shade and wet conditions, making it a desirable understory plant in low-lying habitats. I include it as a nitrogen fixer beneath fruit trees in orchard rows, atop hügelkultur mounds, and in the crescent bed in my garden.

Figure 9.14. A false indigo shrub in bloom.

B Goji Berry

Lycium species

Native Range: Asia
Height: Up to 10 feet (3 m)
Canopy: Up to 4 feet (1.25 m)
Soil Conditions: Adaptable, but requires well-drained soil
Sunlight: Full sun or partial shade
USDA Hardiness Zones: 4 to 9

Goji berry, also known as wolfberry, is used both as food and medicine in China and surrounding countries, where it originates. However, the berries have become quite popular as "superfoods" in the West as well. Their health benefits include significant quantities of iron, vitamins A and C, and antioxidants. This shrub has thin arching branches, sparsely covered with thorns and narrow gray-green leaves. The branches sprawl outward from the base and benefit from trellising.

Goji belongs to the same plant family as tomatoes, and its small, bell-shaped flowers resemble those of the tomato but can appear in a range of pastel colors, from pale yellow to pink to lavender. The young shoots and leaves can be cooked and eaten. When conditions are suitable, these plants bloom from mid-summer well into

Figure 9.15. The delicate goji berry flowers are lovely viewed close up.

fall with continuously ripening fruit. The berries are extremely resistant to damage by subfreezing temperatures and remain viable on the bush well into December in my garden. The ripe red-orange, torpedo-shaped berries hang from the branches on short stems. On my plants, they reach at most 1⁄16 inch (2 mm) in diameter and three times that in length. However, I have seen images of commercially grown plants with berries that are much plumper, about as big as a medium-sized blueberry. When eaten fresh, goji berries are a pleasant, juicy explosion of sweet and tart flavor. The berries are traditionally dried for preservation, but I find that they freeze as well as any other type of berry. I enjoy adding them to cereals and salads where, due to their intense flavor, a small quantity goes a long way.

In my garden, I notice that these resilient shrubs become dormant when the soil is either too wet or dry, but resume their growth once conditions improve. I grow them successfully in both full sun and moderate shade, and propagate them easily by layering.

Gooseberry

Ribes species and hybrids

Native Range: Europe, western Asia, and northern North America
Height: 4 to 5 feet (1.25–1.5 m)
Canopy: 4 to 5 feet
Soil Conditions: Well-drained but moist soil
Sunlight: Partial shade
USDA Hardiness Zones: 3 to 8

Gooseberry is the larger, sweeter, thorny cousin of red and black currants. American gooseberry (*Ribes hirtellum*) is native to the northern reaches of the American continent, but I have only planted cultivars that are hybrids of American gooseberry and European gooseberry (*R. uva-crispa*). These tend to have larger berries, some reaching an inch (2.5 cm) or more in diameter. In contrast with currants, which grow in clusters, gooseberries hang off the sprawling stems in staggered rows, like small Christmas tree ornaments.

Gooseberry varieties differ slightly from one another in size, growth habit, time of ripening, and

Figure 9.16. Gooseberry fruits, similar to but larger than their cousins, currants.

berry color, which ranges from yellow, red, or maroon to pink or purple. I find virtue in each of the cultivars I grow. The inconspicuous flowers open early in spring, attracting hungry pollinators, especially the early foraging bumblebees. The fruits are ready to harvest beginning in late June or early July. Fully ripe berries give a bit when squeezed and can be quite sweet. I find gooseberries easy to harvest using a berry picker comb (as described in "Clove Currant," page 156) or by hand. In a properly pruned plant, if I lift each branch by the tip, I can pluck or skim off the hanging berries and avoid being pricked by the sharp thorns. The tarter, unripe berries are preferred for confections such as pies and jams because they contain more pectin and can be sweetened to taste. I make a thick, delicious preserve with fruits of varying ripeness. When the berries begin to drop from the bush, it is time to harvest them all. The ripe ones can be eaten fresh or mixed with the others to preserve through freezing or cooking.

Care of gooseberries is similar to currants. They prefer a cool and moist but well-drained location and can take a good deal of shade. An eastern-facing side of your house or another location that receives afternoon shade is ideal. These plants are susceptible to fungal diseases, a trait that makes good air circulation, soil drainage, and annual pruning (see "Pruning Currants and Gooseberries" on page 174) important preventive measures. I once observed my plants entirely stripped of leaves overnight by what I later learned are the caterpillar-like larvae of the gooseberry sawfly. The plants survived their defoliation, but sawfly-induced damage can affect the fruit yield. In the case of these pests, I believe the best defense is a good offense: Create a habitat friendly to insect-eating birds and ground beetles,

How I Cooked My Goose(berries)

The first year my gooseberry bushes produced fruit, I noticed that a robin nesting nearby made a daily round to these plants, helping herself to the ripest berries. I didn't want that to happen again. The next year, with a bumper crop of gooseberries beginning to ripen and no bird netting handy, I decided to lay a wide piece of floating row cover over the bushes to keep birds away from the soon-to-be-ripe berries. This lightweight synthetic fabric is designed to protect annual vegetables from insect pests, and it holds in heat to some degree, enough to provide a few degrees of frost protection early or late in a growing season. Three unusually hot and humid early-summer days passed before I lifted the cover to begin my harvest. When I did, I knew immediately that something was wrong. All of the berries on the upper branches had lost their shine and bright color. The pale orbs felt mushy to the touch. Had I waited too long, letting the berries become overripe on the vine? I investigated further and found that near the ground, the remaining berries looked and felt fine. I realized that I had inadvertently "cooked" most of that year's bounty under the insulating blanket, which held in warm air and allowed it to intensify under the sun's rays. Meanwhile, the fruit near the bottom of the bush next to the cool soil surface escaped this unanticipated fate. Lesson learned: I made a mental note to purchase proper bird netting in advance of the next season so I would not cook my gooseberries again.

and you will minimize sawfly problems. If you see an infestation, view the pests as food for your beneficials (such as wrens and chickadees), which will soon discover and devour them.

NF

B

Goumi

Elaeagnus multiflora

Native Range: Asia
Height: Up to 6 feet (2 m)
Canopy: Up to 6 feet
Soil Conditions: Tolerates all well-drained soil
Sunlight: Shade-tolerant
USDA Hardiness Zones: 4 to 8

Goumi is similar in appearance and growth habit to autumn olive, but the Sweet Scarlet cultivar, which I grow, is considerably smaller. In mid- to late summer, red-orange oval or round edible fruits hang from its branches on cherry-like stems. The juicy berries have a pleasant, sweet flavor when completely ripe and are rich in flavonoids and vitamins A, C, and E.

Figure 9.17. Goumi branches loaded with ripening berries.

Although this self-fertile shrub is borderline Zone 4, its small size and shade tolerance persuaded me to take a chance in my garden. I've placed it in several locations: between small nut-producing plants, under the shade of a plum thicket, and as a nurse plant to Cornelian cherry trees. So far, it has survived in all of these places and recently produced an abundance of fruit.

B

Highbush Cranberry

Viburnum trilobum

Native Range: Northeast Canada through north-central US
Height: 8 to 12 feet (2.5–3.5 m)
Canopy: 8 to 12 feet
Soil Conditions: Wet, boggy soil
Sunlight: Full sun or partial shade
USDA Hardiness Zones: 2 to 7

Before I embarked on my edible forest adventure, I was aware that cranberries grow on low, trailing vines, but had no idea that they grow on bushes as well. Highbush cranberry is a cold-hardy native shrub that grows in wet woods, along streams, and on moist, wooded hillsides. The berries these plants produce are not true cranberries (*Vaccinium macrocarpon*), but they look and taste similar to the "real thing." Highbush cranberries produce the most flowers, fruit, and autumn color when grown in sun, but are shade-tolerant as well.

Highbush cranberry shrubs have a narrow, multi-stemmed base and wide canopy. In the spring, they are covered by bunches of white flowers that attract pollinators. In the fall, clusters of ⅜-inch (1 cm) tart, edible berries turn a deep red. If left unharvested, these berries can remain on the bushes all winter, providing food for birds in early spring. The leaves turn eye-catching shades of crimson in the fall. Highbush cranberries are firm and acidic until hit by a few frosts, after which they soften and sweeten slightly. You can process them into sauce or add them to baked goods just as you would regular cranberries. The bark of this shrub has applications in traditional medicine.

I installed the Wentworth cultivar of this shrub in my garden. Several of these ornamental plants are growing well and providing four-season landscape appeal in a sunny section of the garden where the water table often rises above the soil surface. Another Wentworth is thriving in a low spot that is shaded after mid-morning by a clump of trees. I am delighted to find that such a decorative shrub can thrive in these challenging habitats. From my observations, these plants have no pest or disease problems and require no particular care. They can be propagated through layering, hardwood cuttings, and seed. If you have a sunny or shady spot on your property with wet ground, these large, attractive bushes may serve you well, too.

Figure 9.18. Highbush cranberry covered with ripening fruit.

B Honeyberry

Lonicera caerulea

Native Range: Northern latitudes of the Northern Hemisphere
Height: Up to 4 feet (1.25 m)
Canopy: Up to 4 feet
Soil Conditions: Adaptable
Sunlight: Full sun or partial shade
USDA Hardiness Zones: 2 to 8

Honeyberry, an edible honeysuckle, produces oblong berries that have the same color and blush as blueberries, but it doesn't need acidic soil. I decided not to grow blueberries in my permaculture-inspired garden because they would require annual applications of acidifying amendments. If your soil is more acidic, blueberries may do well for you (see "What about Blueberries?" on page 155). For me, honeyberry is the perfect substitute. Also known as haskaps, honeyberries are cultivated in Japan, Russia, China, Canada, and, more recently, the United States. These shrubs tolerate colder climates than blueberries—overwintering even in the harsh winters of Zone 2.

Honeyberries are among the first berry bushes to bloom in the spring and to bear ripe fruit, which they do as early as mid-June in my region. At this time of year, the bright blue fruit is irresistible to birds with nests full of hungry chicks, so if you want any for yourself, be sure to net your bushes just before the berries begin to turn blue. Three weeks later, the berries are ready to harvest. The fruits expand in width as they gain in sweetness, appearing to puff out when they are perfectly ripe and best for fresh eating. They will be tart if picked

Figure 9.19. Honeyberries grow as individuals and in small clusters attached to the undersides of the plant's dense branches, which can make them hard to reach and time consuming to pick.

earlier, but good for processing into jam or jelly or baking into pies and crumbles. I like to freeze them and add them to cereals and yogurt throughout the later summer months. Some cultivars readily drop their berries when they are ripe. To harvest these, simply spread a drop cloth underneath the plants to catch the berries as you shake the bush. (This is how they are harvested commercially.)

I grow more than a dozen honeyberry varieties in my garden. Plant at least two different bushes in close proximity for good pollination. The cultivar Berry Blue is known as a universal pollinator because it blooms over a six-week period beginning in April, overlapping with the blooming times of most other varieties. In terms of fruit size and quantity, evenness of ripening, and ease of harvest, I favor a cultivar called Tundra.

Honeyberries benefit from annual pruning when they are dormant, to open the centers and reduce congestion among the densely growing branches. This facilitates harvest and can also result in larger berries. They can be propagated from dormant stem cuttings. I observed that our local deer have little interest in them, perhaps because the woodland creatures are sated from browsing the related invasive honeysuckle that overpopulate our woods. Remarkably, honeyberries don't appeal to tent caterpillars, either—they remained untouched during one season when these pests attempted to devour every other berry plant in my garden. A blueberry alternative that can adapt to a variety of soil types, has minimal pest and disease problems, and ripens a month earlier even than early blueberry varieties do—what's not to like?

Two Counterintuitive Discoveries

When I decided to plant a windbreak hedge along a southwest-facing fence, I realized I would need to move four honeyberry plants because they were growing in a location that I wanted to turn into a wide path next to the hedge. Honeyberry plants have fine, fibrous roots that grow close to the soil surface (as do blueberries). From my past experience, I knew that berry bushes with this type of roots need a steady supply of moisture, or they will die.

I was pondering this change to my garden in the fall, just after I had attended a talk by Stefan Sobkowiak of Miracle Farms in southern Québec. To my surprise, Stefan mentioned in passing that fall was a good time to transplant bushes and trees. As the plants' leaves wither and dry, he explained, the nutrients they hold are reabsorbed by the roots. The roots then experience a growth spurt in response, which helps them get established in a new location.* This was counter to everything I had learned, namely that in my zone, spring is the best time to plant or transplant shrubs and that fall planting should be avoided. Miracle Farms' hardiness zone is comparable to mine, so I decided to try transplanting the honeyberry bushes that same fall.

In mid-August that year, a group of students who were taking part in a pre-college team-building experience (through New York State's Hamilton College) had visited our farm for a weekend. I gave three of them the task of building a raised bed between mulberry trees using dry oak leaves. As the students applied the leaves in a 10-inch-thick (25 cm) layer, I asked them to water the bed thoroughly. These students completed what they thought was a thorough watering and then moved on.

This bed seemed like the perfect place to relocate the honeyberries early that November. During the intervening weeks, the "thoroughly" watered leaves should have begun to decompose. Unfortunately, saturating a pile of dry leaves takes far more time than one would expect, and the students must have been too impatient. When I dug into the bed of leaves to make room for the root balls of the honeyberries, I discovered that the conditions inside the bed were still crisp and dry. I worried that the honeyberries would die surrounded by this dry medium, but went ahead and planted them anyway, adding a little additional soil from a recently delivered pile of topsoil to fill in the narrow space between the dry leaves and the root balls.

The following spring and summer were dry, and the oak leaves surrounding the honeyberry roots remained crisp. Any composting that took place was negligible. Nonetheless, all four plants flowered, leafed out, and bore ripe fruit. I marveled at how such fine-rooted plants could thrive in these harsh conditions and came to the counterintuitive conclusion that the honeyberries were in fact drought-tolerant.

The following winter, as is my habit, I ordered many more plants than I had specific planting plans for. Among the excess

* Sobkowiak, "Growing a Permaculture Orchard with Stefan Sobkowiak—Kingston Workshop."

purchases were 10 Tundra honeyberries. Faced with the dilemma of where to place them, I suddenly remembered how drought-tolerant these plants proved to be, and I thought of a trench full of crushed stone that had resulted following a mistake made by contractors (as described in "A Cautionary Tale" on page 59). I expected that this strip filled with composting wood chips with crushed stone beneath would quickly dry out in the absence of frequent rain, but perhaps the honeyberries could prosper in this drought-prone niche. Sure enough, to date they remain alive and well in this challenging location. I had verified two counterintuitive phenomena: that it is fine to transplant woody plants in mid-fall, and that some plants with fine, fibrous roots can be drought-tolerant.

Figure 9.20. Occupying the shrub layer, beneath a row of plum trees, this row of honeyberry bushes is healthy and happy despite growing a mere 6 inches (15 cm) above a thick layer of compacted crushed stone.

B

Jostaberry

Ribes hybrid

Native Range: Hybrid of plants from North America and Europe
Height: Up to 6 feet (2 m)
Canopy: Up to 6 feet
Soil Conditions: Moist fertile soil
Sunlight: Full sun or partial shade
USDA Hardiness Zones: 3 to 8

Jostaberry is a three-way cross between a European black currant and American and European gooseberries, retaining the sprawling growth habit of the latter but lacking its thorns. Its flowers, like those of its parents, are not particularly noticeable. Jostaberry fruit is sweeter and less pungent than black currants and smaller than gooseberries. You can eat them right off the bush or process them into jams, pies, and other concoctions.

These are not my favorite bushes, as they take up a lot of space (and their suckering habit leads to them taking up even more) and have thick branches that die by the third year. Because of their girth, the branches take a lot of strength to prune out. The fruit grows in small, spaced-out clusters that can be tedious to harvest. However, if you have the garden space with the right conditions, strong biceps and pectorals, and time to spare, these shrubs will give you a sizable, thornless harvest of sweet, fresh berries for years to come.

B

Juneberry

Amelanchier species and hybrids

Native Range: Northern Hemisphere
Height: 4 to 25 feet (1.25–7.5 m)
Canopy: 4 to 20 feet (1.25–6 m)
Soil Conditions: Adaptable, but does not tolerate soggy soil
Sunlight: Full sun or partial shade
USDA Hardiness Zones: 4 to 9

Juneberry, also known as shadbush, saskatoon, and serviceberry, is a diverse genus of plants, with species ranging in size from a few inches (8 cm) to

Figure 9.21. These jostaberry fruits will gradually ripen to black, resembling large black currants.

Figure 9.22. Princess Diana juneberry bears showy early-spring flowers.

60 feet (18 m) tall. Juneberries grow in the wild in every state except Hawaii, and in the northern prairie region of Canada, where they are also grown commercially. Related to the apple, they share many of the same pests and diseases. Based on my observation, they are the first plants that deer choose to browse and are a favorite of caterpillars. In my heavy clay soil, they are prone to fungal disease; I saw fungal spores emerging from the berries during several moist growing seasons.

I am familiar with two small trees, the *Amelanchier × grandiflora* hybrids Princess Diana and Autumn Brilliance. They can grow to 20-plus feet tall, and their graceful, vase-like shape and smooth, dark gray bark make them attractive specimen plants. Their small leaves cast a dappled shade, so the trees can be underplanted with shrubs and ground covers that grow well in partial sun. Their berries, when they have some, are small and mostly out of reach but accessible to birds, who savor them. I have planted several shrub-like cultivars—notably Regent, a cultivar of *A. alnifolia* that grows to about 6 feet (2 m) tall and was bred for superior berry production.

Both the shrubs and the small trees are multi-stemmed and have a suckering habit that lends them to propagation by division. These plants can also be propagated by seed and dormant hardwood cuttings.

Juneberries are one of the earliest plants to flower in spring, ripen in late June, and look and taste similar to blueberries. Like blueberries, they can be eaten fresh or processed. Native Americans traditionally mix dried juneberries with dried meat and fat in the dense, nutritious food called *pemmican*.

Unfortunately, I have not had a successful harvest from any of my juneberry shrubs over the 10 years that I have grown them. Although they manage to flower, these plants seem ill suited to my deep, dense clay soil and have not produced palatable fruit in any significant quantity. Either the fruit never seems to form, or it becomes diseased, or the birds remove it before it is ripe enough for me to harvest. The wild trees found in the woods on our farm prefer the shallow, well-drained soil and crevices found on rocky ridges. Their scattered presence stands out from afar when they are bursting with spring blooms. If you have well-drained, loamy soil in a sunny or partially shaded location, juneberry shrubs may do well for you. For me, the large tree-sized cultivars' four-season visual interest, including brilliant fall colors, compensates for their lack of food production.

FR

Nanking Cherry

Prunus tomentosa

Native Range: China, Japan, Himalayas
Height: Up to 8 feet (2.5 m)
Canopy: Up to 8 feet
Soil Conditions: Adaptable, but requires well drained soil, drought-tolerant
Sunlight: Full sun
USDA Hardiness Zones: 3 to 7

Nanking cherry is a hardy East Asian bush cherry whose dense branches and quick growth make it an ideal choice for an edible hedge or specimen plant. The first cherry to bloom in spring, the Nanking

Figure 9.23. Nanking cherry fruit.

cherry comes alive with pink buds that open to white flowers just as its bright green, deeply veined leaves emerge. With two of these bushes to ensure pollination, you can have fruit in the first year. The pleasant sweet-tart, bright red fruits, about ½ inch (13 cm) in diameter, ripen in July and August. However, if you don't net them, the birds will beat you to the harvest. The leaves turn a soothing yellow in the fall. In winter, the bush's shiny reddish brown bark peels to reveal a bright orange underlayer, giving it four seasons of landscape appeal.

Early on, I planted four Nanking cherries at the base of a small hill near the entrance to my edible forest garden, where I thought their lovely spring blooms would be inviting. That summer was relatively dry, and the plants did fine. The next year was wetter, and I noticed some of their leaves turning a mottled yellow and brown. I determined the location was too wet for them and made a mental note to transplant them to higher ground. The following spring, I incorporated the three survivors farther up the hill in a diverse edible hedge on the southwest side of the garden. Here, their new leaves remained a vibrant green all summer, so I could tell my decision to transplant them to higher ground was the right one. If placed in a suitable habitat, and with occasional pruning, these durable fruit-bearing bushes can grace your garden landscape for up to 30 years.

New Jersey Tea

NF

Ceanothus americanus

Native Range: Eastern North America
Height: 3 to 4 feet (1–1.25 m)
Canopy: 3 to 5 feet (1–1.5 m)
Soil Conditions: Tolerates poor, dry soils
Sunlight: Full sun or partial shade
USDA Hardiness Zones: 4 to 8

New Jersey tea is the most diminutive of the nitrogen-fixing woody shrubs I have encountered. Its name dates back to the American Revolution, when its leaves were used as a substitute for true tea leaves. In summer, its thin stems are topped with white flowers attractive to butterflies. Its dense, fibrous roots also have medicinal uses. I have paired these small shrubs with fruit tree seedlings on the well-drained crest of a mound, where they provide an extra boost of nitrogen to the developing trees.

Figure 9.24. New Jersey tea bears frilly white blooms in July.

NF H B

Northern Bayberry

Myrica pensylvanica

Native Range: Coastal eastern North America
Height: Up to 10 feet (3 m)
Canopy: Up to 10 feet
Soil Conditions: Tolerates a wide range of soil types, lots of moisture, and salt
Sunlight: Prefers full sun
USDA Hardiness Zones: 3 to 7

I was surprised to discover the many virtues of this semi-evergreen, densely branched shrub. Its leaves are bright green in spring, turning to dark green or brown in winter. If the winter is not too severe, the leaves may stay affixed to the plant adding color and dimension to the winter landscape, until they drop off in spring to allow for new growth. Northern bayberry's aromatic leaves are used to flavor soups and stews and have become a favorite of chefs in my area. Its waxy berries can be transformed into the iconic bayberry candles.

Figure 9.25. A young northern bayberry shrub.

While northern bayberry can grow to 10 by 10 feet and spread by suckering, it grows slowly. If necessary, it is easy to control its size and spread through pruning. I have planted many of these useful plants in saturated areas of my garden, often employing them as nurse plants. Unfortunately, they are also appealing to rodents, who have circumvented my trunk guards and killed several of my newly installed bushes by completely severing the trunks just beneath the soil surface.

B

Red and Pink Currants

Ribes rubrum

Native Range: Central and Northern Europe and northern Asia
Height: 4 to 6 feet (1.25–2 m)
Canopy: 4 to 6 feet
Soil Conditions: Moist soil rich in organic matter
Sunlight: Shade-tolerant
USDA Hardiness Zones: 3 to 8

I have experience with three varieties of red currants—Jonkheer Van Tets, Red Lake, and Rovada—and one pink currant, Pink Champagne. In spring, they all sport tiny yellow blossoms massed along strings cascading down from the branches. By early July, the flowers give way to strings of sparkling red and pink berries, about ¼ to ⅜ inch (6–10 mm) in diameter. Whether loaded with flowers or fruit, these bushes enhance any landscape.

Currants produce an abundance of fruit beginning the second year, with culinary as well as medicinal uses. Harvest is easy to do by hand, but using a berry picker comb (described in "Clove Currant," page 156) can be more efficient.

Red currant fruits can be quite tart, but I find if they are left on the plant for a couple of weeks after turning color, they become considerably sweeter. When perfectly ripe, the gorgeous Pink Champagne fruits—which look like translucent pink pearls—are

Figure 9.26. These pink currants are ripening up nicely (*top*), and the red currants are ready to pick (*bottom*).

Pruning Currants and Gooseberries

Annual pruning improves the vigor and fruit production of most *Ribes* species. Another goal is to open up the center of the plants so the sun can penetrate to minimize moisture, which helps protect against fungal disease and also helps speed fruit ripening. A pair of pruning shears sterilized with isopropyl alcohol should suffice, though you may need a pair of loppers for older, thicker branches.

Red and pink currants and gooseberries bear the most fruit on their two- to three-year-old stems, while black currants produce the best on one- and two-year-old stems. For red and pink currants and gooseberries, the goal in pruning is to select and retain four to six well-spaced stems of each of the three ages: one-year-old, two-year-old, and three-year-old stems. For black currants, the goal is to select and retain four to six well-spaced one-year-old stems and four to six two-year-old stems; you may also choose to retain some three-year-old stems that bear lots of one-year-old side shoots.

It is best to prune during late winter or early spring, when the plants are still completely dormant. First, prune off the low-hanging branches. If left on the shrub, these branches will be pulled to the ground by the weight of fruit, increasing the likelihood that the fruit will rot. Next, remove any dead or damaged wood.

Step back and observe the plant to determine the age of the remaining branches. The age can be surmised from the width and color of the branch as well as how many side branches have grown from it. The lightest brown, narrowest, and straightest shoots that lack side branches are one year old. The somewhat darker, thicker branches that have some side shoots are two years old. The darkest ones with a larger number of side shoots are three years old or older.

After this assessment, (one that will become easier as you gain experience), prune out all the branches that are more than three years old. Then observe the plant again and decide on your target—will you aim to leave four branches of each age level, or six? With that goal in mind, beginning with the one-year-old stems, remove the thinnest and those that are touching another branch. Continue until the desired number remain. Do the same with the two- and then three-year-old stems.

To increase the productivity of red and pink currants and gooseberries, you may take the additional step of trimming each branch back by about a third of its length, cutting just above an outward-facing bud (red and pink currants) or upward-facing bud (gooseberries) so the resulting branch will grow outward or upward respectively.

For black currants, do not trim back the branches. Do make a second assessment of the three-year-old stems. Choose two to four of them that have the most one-year-old side shoots. Retain those, and cut off the rest of the three-year-old stems at the base.

As a final step in pruning all types of currants and gooseberries, trim out any cluttering side shoots that are blocking the light or impinging on other branches.

When you are finished, an equal number of well-spaced stems from each year should remain—8 to 15 in total depending on the vigor and growth habit of the plant. For black currants, I leave the maximum number of branches on Titania because it is a large, vigorous cultivar overall, and the minimum on the Ben Sarek variety, which tends to be smaller and less vigorous.

If this procedure sounds complicated and overwhelming, you are not alone. At first, I was reluctant to remove so much of these plants, fearing I would harm them. I also felt insecure about my ability to make the proper judgments about which stems to remove and which to leave. Nonetheless, following the advice of experts whom I respected, like Lee Reich, I held my breath and went at it. I find, as you will, that I gain in confidence each year as I improve my technique with practice and observe the plants' positive responses. The improvement in the resulting harvest is validating.

palatable for fresh eating. Traditionally, these currants are used to make a clear jelly and as an ingredient in baked goods. I like to eat them fresh mixed with yogurt or cold cereal in summer and freeze them to add to hot cereals during the colder months. The chefs I know love to use these sparkling currants as a garnish for appetizers, entrées, and desserts.

Currants belong to the gooseberry family, and all of the members of this plant family are cold-hardy and shade-tolerant. Although they benefit from moist soil rich in organic matter, currants grow well in several locations in my garden where the clay soil has been amended over time with leaves and wood chips. I have planted currants on the north side of my 50-foot-long (15 m) hügelkultur mounds, in the shaded part of the crescent bed, in a partially shaded hedgerow, and on a well-drained north-facing slope. In my experience, they have no pest or disease problems. All can be propagated through cuttings, layering, and seed. For optimal production, these plants require annual pruning (see "Pruning Currants and Gooseberries"). You may also need to net these berries to keep birds at bay.

Red Raspberry

B

Rubus idaeus

Native Range: Eastern US and Canada
Height: 3 to 5 feet (1–1.5 m)
Canopy: 2 to 3 feet (60–100 cm)
Soil Conditions: Well-drained soil
Sunlight: Full sun or partial shade
USDA Hardiness Zones: 4 to 8

Red raspberries are a popular bramble and a must-have in most home perennial fruit gardens. The berries develop from small flowers at the tops of stalks called *canes*, which are covered with a layer of soft prickles that are rarely troublesome. These delectable berries can be eaten fresh, converted into jellies and jams, used in baked goods, or frozen. The best way to freeze them so they don't stick together in clumps is to spread them out in a single layer on a cookie sheet. Once they are frozen, you can transfer them to a freezer bag or another airtight container for long-term storage. Raspberry leaves can be used to flavor teas.

There are two types of raspberry—primocane varieties and floricane varieties. Primocane varieties produce the majority of their fruit during their first year of growth, generally in late summer or fall. Floricane varieties produce their major crop the second year, during the summer months. The former are known as fall varieties and the latter as summer. No matter the variety, raspberry canes live for only two years and then die back and should be removed.

In nature, raspberries are continually on the move, gradually sending up new suckers in successive patches of adjacent ground. They are often found at forest edges or in bright areas inside the woods. It is ideal to plant them in a location where you can allow them to move in this way. Their suckering habit also makes raspberries easy to propagate: Simply dig up a new sucker with its roots and transplant. You can expect a patch of raspberries to live for about 10 years before they begin to succumb to fungal disease, at which point it is wise to obtain fresh disease-free stock to plant in a new bed. It is best to plant where other members of its plant family, Rosaceae—including roses, strawberries, and pome/stone fruits—have not been grown before.

Figure 9.27. Raspberries ripen at the tips of canes.

To flourish, raspberries need well-drained soil with plenty of moisture while the fruit is forming. One year, I did not irrigate one of my raspberry patches during a summer drought. The flavor of the berries was intense, but the fruit was stunted and there was barely any juicy pulp surrounding the seeds. The following spring, I installed a drip line so I could irrigate these plants if necessary.

For protection from pests (such as the spotted wing drosophila) and disease, good air circulation, timely and thorough pruning, and excellent sanitation are the best preventive measures. By "sanitation," I mean regularly picking your berry patch completely, so no overripe, pest-infested, or rotting fruit is left on the stem or the ground. Good air circulation is important to prevent disease and pest infestations in raspberries. Site them in a location where there is natural air movement, and thin the canes seasonally. The spent second-year canes of floricane raspberries should be removed as soon as possible after their berries have been harvested to allow room for the emerging shoots that will produce fruit the following year. If your growing season is long enough, these new shoots (the primocanes) might bear a small fall crop. The earliest summer variety I grow, Prelude, gives me two crops a year—the first beginning in early July on the floricanes and the second starting in late August on the primocanes. My second summer variety, Nova, gives me a reliable summer crop beginning in mid-July. It always produces a second round of flowers on the new primocanes, but our fruiting season for raspberries ends when we have our first hard frost, sometimes as early as late September or early October, almost always before the second crop of Nova ripens. If your growing season is longer, you might get a reliable fall crop from this cultivar as well. A third summer cultivar, Encore, begins to

The Raspberry Cane Borer

Early on in my edible forest gardening adventure, I planted two varieties of fall raspberries in the northeast section of a hedgerow. They yielded a small crop each fall until one year, after a dry summer, when I noticed that the tips of several of the plants had wilted and there were no fall berries to be had. *They must have needed water*, I thought, and installed a drip line the following year.

That next year, despite ample water, the canes' condition did not improve. Worse, more tips began to wilt. I contacted my local cooperative extension's horticulture specialist, who immediately diagnosed the problem: The raspberry cane borer was attacking my raspberries. This small beetle girdles the stems of primocanes in two bands, spaced about an inch (2.5 cm) apart, 6 inches (15 cm) from the tips. She then lays her eggs between the girdles. When the eggs hatch, the larvae feed on the stem between the bands. The girdling causes the tops of the canes to wilt. The next year, the larvae burrow down the entire stem and enter the ground, later emerging in adult form and repeating the cycle.

To manage this pest, the horticulture specialist told me to cut the cane below the lower girdle and burn or otherwise dispose of it. I began to do this whenever I saw a drooping tip, but it was too late. The beetles had already completed a life cycle, completely killing the year-old canes and reemerging to infect new ones. Before I could stop it, this planting of fall raspberries became so infested with the borer (and possibly overwatered as well) that it was all but destroyed.

I noticed drooping tips in two other places where I had planted raspberries in my garden. I caught the infestation in the summer raspberry (Prelude) planting sooner and pruned all the brambles to the ground. The following year, though I lost the entire summer Prelude crop, most of the new primocanes remained healthy and provided a decent fall crop. Now I know to nip the borer in the bud (literally) by methodically pruning out any wilted tips as soon as I observe them.

My raspberry cane borer encounter wasn't a total loss. As soon as I recognized and accepted the demise of my fall raspberries, I replaced them with gooseberry and currant bushes that bear their fruit much earlier in the season.

fruit in late July or early August. With these three floricane varieties, plus the late-summer crop from the Prelude, I have a continuous supply of raspberries for most of the summer into early fall.

In late winter or early spring, when they are still dormant, floricanes should be thinned to three or four per square foot, leaving the tallest, straightest, and strongest, and pruning the rest flush with the ground. Any dead stalks missed the previous year should be removed as well. Then snip the tops of the floricanes just below the level where the past year's fruits or flowers grew. Your summer berry crop will appear toward the tops of these tipped, thinned-out canes.

To maximize fruit yield, primocane varieties (those that fruit in fall) should be mowed or pruned to the ground after their fruiting cycle is complete and they go dormant in late fall. Or you can wait and mow or prune as soon as the snow melts the following year. New primocanes will emerge in the spring, flower in late summer, and produce fruit in the fall. Each year during the dormant season, I add 2 to 4 inches (5–10 cm) of wood chips on the ground below all my raspberries to deter weeds, maintain moisture, and add fresh nutrients. Even though most commercial producers trellis the canes to keep them upright, I haven't found this necessary.

When raspberries are ripe, they should be picked clean, meaning all ripe fruit should be removed daily, or on alternate days, to maximize the harvest and prevent the fruit from becoming overripe and attracting fungal disease and insect pests. You can tell that raspberries are ripe when they turn a true red color, appear plump and shiny, and slide off the plant with ease when grasped between your thumb and forefinger. For best storage life, harvest raspberries when they are dry. If picked when wet with dew or rain, fungal molds may grow on the berries, reducing their shelf life. The harvested fruit should be refrigerated as soon as possible. On hot summer days, I precool my harvest by placing pints of raspberries inside a bin containing a liter bottle filled with frozen water.

One wet summer, I neglected to pick my Nova patch before a heavy rain that saturated many of the berries and caused them to drop. I was heading into the garden the next morning to clean them up when I saw a wild mama turkey enter the area, followed by her brood of chicks. I watched her stroll along the row of berries as her chicks ran alongside her, scrambling under the brambles where they devoured every fallen fruit. I silently thanked that avian family for providing this valuable sanitation service.

In addition to the above preventive practices, there are two strategies you can use to reduce the impact of one particular pest—the invasive spotted wing drosophila (see "A Looming Threat," page 37)—if it is a problem in your area. First, choose early-fruiting summer varieties because the population of these pests expands as the season progresses. Second, you can apply fine netting over your raspberries to prevent fruit flies from accessing the fruit as it forms. Knock on wood, I have yet to observe this invasive pest in my garden.

Rugosa Rose

FR

Rosa rugosa

Native Range: Eastern Asia
Height: 3 to 6 feet (1–2 m)
Canopy: 3 to 6 feet
Soil Conditions: Adaptable and drought-tolerant
Sunlight: Full sun or partial shade
USDA Hardiness Zones: 2 to 7

Rugosa roses are thorny, suckering bushes that I have learned have both virtues and vices. Their flowers, with a single layer of pink or white petals, are edible. If spent blossoms are removed and the plants receive adequate moisture, these roses will continue to bloom from early summer well into fall. If left on the bush, the spent blossoms become bright red edible fruits 1 inch (2.5 cm) in diameter and high in vitamin C. They can be eaten fresh, made into "rose hip" jelly, or dried for a nutritive tea. Incidentally, I've observed that rose hips serve as a trap crop for chipmunks, who prefer them to other fruits nearby.

When I learned that rugosa rose was a maintenance-free, adaptable, long-blooming rose that can thrive in my hardiness zone (few roses can), of course I wanted to include it in my planting plan. Who wouldn't? What would be better than to have these romantic flowers as a border for the limestone access route near the back entrance, where my imagined wedding party would promenade past them and into the garden? I planted 10 bushes 6 feet apart along this major pathway.

All was well for the first couple of years. The plants grew rapidly, flowered profusely, and even served as trap crops for Japanese beetles. I could pluck the pests from the flowers' blooms and drown them in a cup of soapy water.

Then the trouble began. I noticed that the bushes, which have an aggressive suckering habit, were beginning to sprout new shoots near the trunks of two Asian pear trees close by. This would not do. The last thing I needed was roses competing with the roots of these valuable and vulnerable trees. Meanwhile, I tasted the rose hips and did not find them appealing eaten fresh. They were mushy and insipid in flavor. I decided that equally attractive shrubs bearing tastier fruit, such as bush cherries, would be a better choice in the space occupied by these roses.

I began to remove the bushes. This was quite the chore. In just three years, the thick, prickly stems of each bush had pushed outward underground to form a shrub that extended 6 feet from the center in all directions. The underground runners often crossed over those from a neighboring rosebush. I started with an outermost emergent stem from one bush and worked my way inward, slowly digging out the sucker until it was freed to the point where it attached to the central trunk. Then I attacked the next stem. I continued around the entire circumference until I uprooted all the runners, 15 to 20 in all. Then I took a pointed shovel and began to dig out the thick, deep taproot at the center, trying to avoid the thorns on the surrounding uprooted stems. I dug and dug, but could not extricate the root. Finally, I brought in a golf cart, wrapped a heavy chain around the base of the bush, attached it to the cart, and drove outward, yanking the root out of the ground. One down, nine to go.

After repeating this laborious process five times, I decided I needed a break. I hired a team of two female landscapers 10 years younger than me to complete the work. They struggled for an hour and abruptly quit, complaining that the job was too difficult for them. I had no choice but to complete the excavation myself.

During several subsequent seasons, small pieces of runner that I'd missed sent up new stems. Rugosa seedlings popped up, too. I needed to remove all of these vestiges of the roses or the nightmare would repeat.

On a positive note, when I planted the 10 rugosa roses along the stone access route, I also planted some Dart's Dash hybrid rugosa roses in two additional locations: between the limestone route and the upper fence, and also in a spot bordered by

Figure 9.28. The multi-petaled blooms of the Dart's Dash cultivar.

a limestone patio and a pond. Where these built-in barriers prevent the roses from impinging on other vulnerable plantings, they continue to bloom, providing beauty without threatening the landscape.

Sand Cherry

Prunus pumila

Native Range: Eastern and central North America
Height: 2 to 6 feet (60–200 cm)
Canopy: 2 to 6 feet
Soil Conditions: Well-drained soil, drought-tolerant
Sunlight: Full sun or partial shade
USDA Hardiness Zones: 2 to 7

If you are partial to native plants, sand cherry may be the shrubby cherry for you. It gets its name from its tendency to grow on sandy dunes and shorelines, and it is quite drought-tolerant. In spring, sand cherry sports clusters of pink flowers, which develop into fruit that hangs from the branches on long stems and turns a deep purple as it ripens. Once it is completely ripe, the pulp is sweet and succulent, in pleasing contrast with the tangy skin. The bush's fall leaf color is an attractive deep red. *Prunus × cistena* is an ornamental sand cherry hybrid that has purple leaves.

I planted this cherry in an edible hedge, and later in a sheet-mulched bed where I could net it more effectively to prevent birds from eating the fruit. A downside of the sand cherry is its vulnerability to infestation with black knot, a fungal disease prevalent in the wild cherry trees in our woods and characterized by enlarged blackened growths around the twigs and branches. This disease is manageable by cutting back the diseased branches, making the pruning cut at least 6 inches (15 cm) below the knot, and burning or discarding the cuttings.

This cherry would do well in the shrub layer of a south- or southwest-facing foundation planting, or as an element in a sunny perennial bed with sandy or otherwise well-drained soil.

Figure 9.29. Sand cherry fruit hanging among its dark green oblong leaves, which have a leathery texture.

Sea Buckthorn

Hippophae rhamnoides

Native Range: Mongolia, Russia, and Northern Europe
Height: Up to 20 feet (6 m)
Canopy: Up to 20 feet
Soil Conditions: Tolerates dry soil
Sunlight: Full sun
USDA Hardiness Zones: 3 to 6

Sea buckthorn, also known as seaberry, is a thorny, deciduous, cold-hardy nitrogen-fixing shrub. It has a suckering growth habit, but is a different plant than the invasive "buckthorn" (*Rhamnus cathartica*). Its narrow, silver-gray leaves are packed along densely growing branches that make for excellent bird nesting habitat.

Its round, orange, edible berries ripen in late summer and can stay on the bush throughout the

Figure 9.30. The narrow leaves and sharp thorns of the sea buckthorn.

Figure 9.31. Bright yellow blooms adorn Siberian peashrub in spring.

winter, providing food for wintering birds. Rich in vitamin C and carotenoids, the berries are quite tart; their flavor reminds me of a vitamin C tablet. They can be processed into pastries, jams, juices, and teas, and also have medicinal applications. The berries on the interior of the plant can be hazardous to pick among its sharp thorns. An alternative is to cut off the whole branch, freeze it, and then whack it against a clean surface to remove the berries.

Sea buckthorn is well suited to a hedge at the seaside, or near a road that is salted in winter, because it is resistant to salt. I interplant this shrub with hazelbert in a windbreak hedge near a road. Unlike many fruiting bushes, sea buckthorn has separate male and female plants, and of course only females produce fruit. For good fruit development, plant at least one male close by up to five females.

Siberian Peashrub

Caragana arborescens

Native Range: Northeastern Asia
Height: 6 to 12 feet (2–3.5 m)
Canopy: 6 to 12 feet
Soil Conditions: Drought-tolerant, does not tolerate saturated soil
Sunlight: Full sun
USDA Hardiness Zones: 2 to 8

Siberian peashrub is beautiful in all four seasons. In May, it is covered with visually striking, fragrant yellow blossoms that also appeal to pollinators. The flowers morph into attractive small pods that contain tiny pea-like seeds and serve as excellent chicken fodder. In fall, the shrub's small round leaves turn a bright yellow. After leaf fall, vestiges of the seedpods remain on the branches, adding intricate interest to

Figure 9.32. This Siberian peashrub (*right*) serves as a nurse plant for a young weeping mulberry (*left*).

the winter landscape. The flowers, young pods, and seeds (both fresh and dried) are all edible. Siberian peashrub is an excellent choice for a windbreak and provides optimal bird habitat.

Siberian peashrub is considered invasive in parts of the United States, so be sure to investigate its status in your region before you choose to plant. In my garden, I have incorporated this shrub as a stand-alone specimen plant, a nitrogen fixer in a diversified hedge, and a nurse plant. I positioned one to the north of a young weeping mulberry to protect it from north winds and to provide nitrogen, as shown in figure 9.32. The peashrub accomplished its mission so well that the mulberry eventually grew to shade it out. Overall, this attractive shrub has done well in my garden when sited in sun where the soil is well drained.

From the descriptions of food-bearing shrubs in this chapter, perhaps you have been inspired to try your hand at growing a favorite fruit, or one new to you that sounds appealing. In the next chapter, you will discover a variety of herbaceous plants that yield edible leaves, flowers, stems, and more to add beauty to your garden and delight your taste buds.

CHAPTER 10

HERBACEOUS PLANTS

The herbaceous layer consists of plants from 1 to 10 feet (30 cm–3 m) tall that die back to the ground at the end of each growing season and emerge again the following spring. Plants in this layer may fix nitrogen; provide edible or medicinal leaves, flowers, stems, or roots; or contribute decorative blooms to cut flower arrangements. Many provide more than one of these benefits. Incorporating a variety of these plants beneath and around your trees and shrubs will add beauty, utility, and tasty and nutritious food to your garden. This chapter presents a range of herbaceous nitrogen fixers, perennial vegetables, and herbs and flowers.

Differentiating the plants of the herbaceous layer from those of the ground cover layer is somewhat arbitrary, as many plants overlap this human-imposed distinction. Several plants described below can also serve as effective ground covers, and those entries are marked with the ground cover icon. For example, Russian comfrey is an herbaceous plant that can grow up to 4 feet (1.25 m) tall, but its mature stems will lay down and effectively cover a 6-foot-wide (2 m) swath of ground as well.

If you have cultivated a garden before, you are probably familiar with annual vegetables such as squash, tomatoes, and peas. For most of my life I cultivated annual vegetables, so I was surprised to discover several perennial vegetables I had no idea existed. Once established, these labor-saving plants—many of which have flavors similar to familiar annuals—grow anew each year, yielding nutritious leaves, stems, and roots with little effort from the gardener other than harvesting. The perennial vegetables in this chapter often serve other useful functions as well, including providing habitat for beneficials and effectively covering the ground.

Flowering plants with very small blooms attract tiny parasitic or parasitoid wasps that deposit their eggs on or in the soft bodies of the egg, larval, and mature forms of insect pests. These beneficial wasps especially enjoy the nectar of herb flowers, including those of anise hyssop, apple mint, chamomile, oregano, sweet cicely, valerian, wild fennel, and yarrow, all of which are described here. When you pepper your garden with these beneficial insect-attracting plants, many of your potential insect pest problems will be naturally controlled.

H EF GC CF

Anise Hyssop

Agastache foeniculum

Native Range: North American prairie region
Height: 2 to 4 feet (60–125 cm)
Width: 1 to 2 feet (30–60 cm)
Soil Conditions: Well-drained soil, drought-tolerant
Sunlight: Full sun or partial shade
USDA Hardiness Zones: 4 to 8

Anise hyssop is an herbaceous perennial that spreads through rhizomes underground and self-seeds as well. Its spade-shaped leaves grow in bunches around stalks topped with spikes of deep lavender flowers, attractive to pollinators and hummingbirds. The leaves and flowers are edible, with a distinctly sweet and mild anise flavor. They can be added to summer salads, desserts, and beverages or dried for later use.

Anise hyssop blooms make long-lasting aromatic cut flowers. Cutting back spent blooms will force a second flush of growth and flowering. I planted anise hyssop on the southwest-facing slope of a hügelkultur mound, where the plants have maintained a lush stand for 10 years. I have observed these plants continuing to grow and bloom well into October. In late fall, wrens and finches flit to and from the dried flower heads, devouring the seeds. Easily propagated from seed or by division, and pest- and disease-free (deer and rabbits don't even like it), this herb is a decorative, tasty, and carefree addition to any landscape.

Figure 10.1. This dense stand of anise hyssop in flower forms an excellent ground cover in its sunny, well-drained location.

H GC

Apple Mint

Mentha suaveolens

Native Range: Europe
Height: 1½ to 4 feet (45–125 cm)
Width: 1 to 2 feet (30–60 cm)
Soil Conditions: Adaptable, but prefers good moisture
Sunlight: Full sun or partial shade
USDA Hardiness Zones: 4 to 9

Figure 10.2. Apple mint stems emerge in spring and will grow to create a formidable groundcover.

Versatile apple mint has larger, softer leaves than the more familiar spearmint or peppermint as well as a milder, sweeter flavor. In mid-summer, the stems are topped with curved spikes of pale lavender blooms that attract pollinators and beneficial insects. This aromatic mint is delightful added to summer salads, as a garnish in cold beverages, and even as a component in an edible bouquet. It is the flavor component in "apple mint jelly," a condiment I remember from childhood as a complement to lamb. Apple mint has medicinal applications as well, and like most herbs it dries well for storage.

Apple mint thrives in wet soil with full sun or part shade. Like other mints, it spreads aggressively by stolons, a quality that makes it either an excellent ground cover or a hard-to-control nuisance, depending on the situation. The dense growth habit can be a problem later in the season when powdery mildew, appearing as a white powder over the leaves, can infect the plant. Locating this plant where there is excellent light and air circulation can reduce the likelihood of this fungus becoming a problem.

I've found apple mint an ideal choice to cover ground surrounding tall shrubs and trees in a moist habitat. Years ago, I planted one purchased plant in my nursery, where it spread to cover several square yards of ground. To propagate it in spring, I use a pointed shovel to lift up portions of the young stems along with chunks of their dense roots. I interplanted these divisions between established woody plants in low-lying sheet-mulched beds. Over the next two or three years, they spread to form a dense, aromatic ground cover. I recommend waiting to install mint as a ground cover until nearby shrubs and trees are established; otherwise the mint may outcompete young woody seedlings for soil space or sunlight. As with other mints, if you don't want this plant to take over your garden, plant it in a deep pot and bury the pot, leaving an inch (2.5 cm) or so of the rim above the soil surface to prevent the stolons from snaking out and rooting into the surrounding soil. This technique will contain the plant and still provide ample cuttings for home use.

Figure 10.3. Asparagus ferns form the border of a garden bed.

Asparagus

PV

Asparagus officinalis

Native Range: Europe, Asia, and Africa
Height: Up to 4 feet (1.25 m)
Width: 1 inch to 2 feet (2.5–60 cm)
Soil Conditions: Well-drained soil
Sunlight: Full sun
USDA Hardiness Zones: 2 to 11

This familiar vegetable is the shoot of an herbaceous perennial plant with a root system similar to that of a tree. It takes three years to establish these plants, at which point you can harvest shoots as they emerge over a six-week period in the spring. This period begins in early May in my garden, but may start earlier in warmer climates. Harvest by using clippers or a sharp knife to cut shoots at ground level when they are 6 to 10 inches

(15–25 cm) tall. After you cease harvesting, the plants will continue to send up shoots. Leave these to grow; they will stretch up to 4 feet tall and branch out to form a decorative display that resembles the houseplants called asparagus ferns (which are a related species). You can take cuttings from these delicate ferns to include in bouquets. Leave most of them intact, however, to continue photosynthesizing through the growing season. The food they produce replenishes the roots and ensures the health of next year's crop. You can cut off the spent fronds once they turn brown and dry in the fall, or wait to cut them off the following spring.

Asparagus naturally produces male flowers and female flowers on separate plants. You can source male hybrid varieties, which produce better yields and don't self-seed (because all or almost all of the plants are male). Some varieties produce green shoots; others produce purple shoots (the purple color fades to green when shoots are cooked). If you order asparagus plants from a mail-order supplier, they will arrive as dormant roots called *crowns*. To prepare an asparagus bed, dig a trench 6 to 10 inches deep (shallowest in clay) and mix the loosened soil at the bottom of the trench with compost or manure. Lay the crowns in the trench, spaced 12 to 14 inches (30–35 cm) apart. Cover with 2 inches (5 cm) of soil. As the shoots begin to emerge, add 2 more inches of soil, and continue adding soil in 2-inch increments as the shoots grow until the trench is filled. Keep them well watered and weed-free. You can begin harvesting a few shoots in the second year, and complete a full harvest from the third year on. If maintained with weeding, mulching, or a perennial ground cover, your asparagus bed can provide a delectable spring vegetable for over 20 years.

I planted asparagus in my edible garden along the curved border of a bed containing fruit trees and berry bushes. I followed the procedure outlined above and later added some strawberry plants among them as a non-competitive ground cover.

Blue Lupine

Lupinus perennis

Native Range: Eastern North America
Height: 1 to 2 feet (30–60 cm)
Width: 1 to 1½ feet (30–45 cm)
Soil Conditions: Adaptable, but needs well-drained soil
Sunlight: Full sun
USDA Hardiness Zones: 3 to 8

This nitrogen-fixing perennial, which is also called wild blue lupine, has unusual long, thin, oval leaves that radiate from a central point and are attractive even when the plant is not in bloom. The deep blue or purple flowers open gradually along vertical stalks. Lupine blooms attract butterflies and other pollinators and make for long-lasting cut flowers. However, all parts of the plant are toxic to humans and animals.

In the early days of my edible forest, I scattered these seeds on top of a mound spanning the entire length of the garden. Normally lupine seeds, like those of most other legumes, require scarification, but these seeds germinated well without any special treatment. Ten years later, many of the resultant plants reappear each spring. Lupines have also self-sown throughout the garden, adding nutrients and visual interest.

Borage

Borago officinalis

Native Range: Mediterranean region
Height: 1 to 3 feet (30–100 cm)
Width: 1 foot or more
Soil Conditions: Adaptable
Sunlight: Full sun to partial shade
USDA Hardiness Zones: 3 to 10

Borage is a self-seeding annual and one of the easiest plants to grow. It forms lush mounds of light green, furry leaves topped with stems full of lovely lavender flowers that taste very much like cucumbers. One customer of mine used these

Figure 10.4. This vibrant mass of deep blue lupine adds drama to the landscape while providing ground cover.

Figure 10.5. This hardy borage continues to flower well into the fall when few other annuals are still in bloom, providing food for pollinators like these bumblebees.

delicate flowers to decorate a wedding cake. The leaves are edible as well, the seeds can be pressed for oil, and the plant has multiple medicinal uses.

Since borage self-seeds, I only had to plant it once. Every year, new plants appear in unexpected places. In addition to its other useful qualities, borage serves as a mulch plant, producing copious quantities of organic matter that enrich the soil as these herbaceous plants die and decay each fall.

Chamomile

H

Matricaria chamomilla

Native Range: Western Europe, India, and western Asia
Height: 8 to 24 inches (20–60 cm)
Width: 8 to 12 inches (20–30 cm)
Soil Conditions: Well-drained soil, drought-tolerant
Sunlight: Full sun
USDA Hardiness Zones: 4 to 9

You are probably familiar with chamomile as a component in a relaxing, sleep-inducing tea. This

Figure 10.6. This chamomile has self-seeded into the stone dust between limestone slabs in a garden pathway.

Figure 10.7. Chives in bloom interplanted with daylilies.

annual herb has attractive flowers that look like tiny daisies; they bloom for an extended period in summer and provide food for parasitic wasps. Or you can harvest the flowers to make your own tea. The leaves are wispy fine filigrees and are also aromatic. Another benefit of including chamomile in a garden is its role as a nutrient accumulator; these plants concentrate phosphorus, potassium, and calcium in their leaves and flowers.

These self-seeding plants are easy to grow and take up little space. The seeds are tiny and need light to germinate. They can be started indoors six weeks before last frost by pressing them gently into the surface of a lightweight seed-starting mix, or outdoors in fall, by scattering them over the ground.

H EF GC CF

Chives

Allium schoenoprasum

Native Range: Europe, Asia, and North America
Height: 1 to 2 feet (30–60 cm)
Width: 1 foot
Soil Conditions: Prefers moist soil, but drought-tolerant
Sunlight: Full to partial sun
USDA Hardiness Zones: 3 to 11

Chives are a well-known garnish for baked potatoes with sour cream, but chives are also an easy-to-grow herb that can form an excellent ground cover. Chives grow in expanding circular clumps of long, thin, hollow edible leaves. In late spring, stems topped with edible lavender flowers arise from the mounds of leaves. Both leaves and blossoms have a mild onion-like flavor. The leaves can be used as garnish, in salads, or added to soups, chilis, and stews. Unlike the leaves the flower stems are too tough to eat raw, but the attractive blooms comprise many tiny blossoms that can be separated and sprinkled on salads and other dishes for a pretty and tasty accent. To preserve leaves for later use, freeze them rather than drying them. Dried leaves don't retain their flavor well.

This cold-hardy herb begins to grow in early spring and withstands fall frosts, so it has a harvest window of several months, providing it receives ample moisture. In a small garden, one mature plant—which can be recut several times—can provide enough herb for family use. If your plot is larger, this herb is an attractive and edible choice for the ground cover layer. Chives are easy to propagate through three methods: division, seeding, and self-seeding. I planted chives singly near fruit trees and shrubs, as a border alongside a path, and in masses interplanted with daylilies as a serviceable ground cover.

Herbaceous Wildflowers

Figure 10.8. Wild asters and goldenrod growing between orchard rows bloom in September.

Figure 10.9. Native plants volunteer along the shore of a pond. From left to right are blue vervain, bulrush, cattail, and goldenrod.

For much of my gardening life, I viewed all weeds as the enemy, assiduously removing them at first sight. When I began growing annual vegetables as a business, I made sure to cultivate, hand-weed, and mulch to prevent and remove any weed incursions. With the initiation of the edible forest, my time and attention were split between this new perennial planting and the annual vegetable operation. I could not keep up with weeding, cultivating, and mulching both venues. Weeds quickly moved in.

In time, I came to accept and value some of the "wild" ground covers that appeared in the Enchanted Edible Forest. I also came to embrace other wild, herbaceous plants. Those that attract beneficial insects are relevant here.

When I visited Stefan Sobkowiak's permaculture orchard in southern Québec, he emphasized the importance of wild plants as habitat for pollinators and predators of pests. Not only do these helpful insects feed on pollen and nectar from wildflowers, they also nest and hibernate in these plants' dense mulch, hollow stems, and galls. In order to preserve this habitat, he mows the sod between his orchard rows in a six-week rotation. This allows beneficial insects to move to an adjacent unmowed stand of wild plants when their immediate habitat is cut. The six-week interval is enough time for the first mowed section to grow back enough to provide new habitat. If the entire garden were leveled at once, the surviving insects would need to abandon the garden altogether.*

Following Stefan's example, I designed space for wild plants between orchard rows in my garden that I mow alternately and only once or twice a season. I also make sure to leave these plants standing through the winter months to afford cold-season shelter.

In addition to allowing helpful "weeds" including milkweed, Queen Anne's lace, wild aster, and goldenrod to grow between the orchard rows, I am becoming more comfortable, even appreciative, when I see them volunteering among the trees and shrubs that populate the garden beds and mounds.

* Stefan Sobkowiak (Miracle Farms proprietor), in conversation with the author, May 2016.

Figure 10.10. Coneflowers thriving beneath a row of locust trees, where they have persisted for 10 years.

H GC CF

Coneflower

Echinacea species

Native Range: Eastern and central North America
Height: Up to 4 feet (1.25 m)
Width: 2 to 3 feet (60–100 cm)
Soil Conditions: Tolerates a wide range of soil, but not saturated clay, drought-tolerant
Sunlight: Full sun to light shade
USDA Hardiness Zones: 4 to 10

Also known as echinacea, coneflowers grow wild in woods and fields. The most familiar and durable species is the purple coneflower (*Echinacea purpurea*), distinguished by its many purple petals and spiny orange centers that becomes more pronounced as the flowers age. I like these showy plants because they bloom continuously from mid-July well into the fall, especially if spent blooms are removed, and they require little care. Their long-lasting flowers remain vibrant on the plant for several weeks and for at least one week once cut. In addition to their aesthetic appeal, coneflowers attract beneficial insects. Echinacea is also a popular medicinal herb; extracts from the flowers and roots are used to provide immune system support.

I first planted coneflowers by scattering their seeds on top of a 200-foot-long (61 m) raised bed alongside a fence. Due to their longevity and propensity to self-seed, they continue to bring outstanding color to the garden more than 10 years after this planting. I leave the seed heads standing all winter in the garden to provide food for birds. One December day, I watched a downy woodpecker feeding on a coneflower. His technique was to clasp the stem with his feet and repeatedly peck at the dried flower head to extract the seeds. I use my feet on these plants, too, but for a different purpose—in early spring I stamp down the dead coneflower stalks until they lie flat on the ground, where they compost in place as new growth emerges.

EF

Daylily

Hemerocallis species and hybrids

Native Range: Eastern Asia
Height: 1 to 3 feet (30–100 cm)
Width: Up to 2 feet (60 cm)
Soil Conditions: Tolerates many soil types, drought, and frost
Sunlight: Tolerates full sun and quite a bit of shade
USDA Hardiness Zones: 4 to 9

There are tens of thousands of hybridized daylily cultivars, ranging through the entire color spectrum. Most of these cultivars are derived from the orange, and less commonly yellow, daylilies that are naturalized along roadsides throughout much of North America. True lily species grow from a bulb, have a single stem, and can be toxic to humans. Daylilies are not true lilies, however, and they do not

Figure 10.11. These bright yellow Stella d'Oro daylilies rebloom two or more times during the growing season. Though each lasts just a day, multiple buds on each stem that open sequentially make them worthy cut flowers.

produce bulbs; they have branching roots that often contain enlarged, edible tuberous sections where they store food. Long, thin, densely packed clumps of leaves arch out from a daylily crown in a circular formation. A group of daylilies creates a thick, mounded ground cover. Flower stalks arise from the crown center. The multiple flower buds on each stalk open sequentially, each living about a single day, as the plant's name suggests. The flowers have a sweet, succulent crunch and are delightful to munch right off the plant. Some flowers have thin petals, but I prefer the thicker blooms, which are crisper in texture. I used the dry buds of daylilies in hot-and-sour soup for years without realizing their origin from these lovely plants.

Daylilies are hardy and adaptable, with no pest or disease problems in my experience. I have interplanted them with chives, in a single undulating border row, as ground cover underneath shrubs, on shaded hügelkultur mounds, and massed close together so that their blooms provided a beautiful show above the dense leaves. In fall, the leaves die down to mulch the ground during the winter. Daylilies multiply by seed and division: You can dig up a clump and tease apart the roots of the individual plants to install elsewhere.

Egyptian Onion

Allium × proliferum

Native Range: Indian subcontinent
Height: 2 to 3 feet (60–100 cm)
Width: 6 to 12 inches (15–30 cm)
Soil Conditions: Moist fertile soil
Sunlight: Full sun
USDA Hardiness Zones: 3 to 9

Egyptian onion is a delicious perennial vegetable that can be harvested and used just like its annual counterpart. It can also be interplanted among other sun-loving edibles to serve as an aromatic pest confuser. This plant begins its growth cycle in early spring by sending up a cluster of hollow leaves similar to a bunch of scallions. Next, a thicker, taller stem emerges to form a bundle of small bulbs at its tip. The weight of these bulblets causes the stem to bend to the ground, where the bulblets take root to form new plants. Another common name for this plant, walking onion, reflects this growth habit.

I have planted Egyptian onions at the base of several raised planting mounds, where the soil is naturally moist, and in sunny beds among other plants. They do best when surrounded by mulched ground with no competing plants. This allows their bulblets to take root unobstructed. In spring and fall I harvest the tender scallion-like leaves by pulling them up by the root, which is like a slender onion. In summer when the bulblets form, I simply break them off the stem top and use them as I

Figure 10.12. Tasty Egyptian onions have both a macabre appearance and an unusual way of propagating.

would small onions, or drop them on the ground where I want to propagate new plants.

H EF GC CF

Garlic Chives

Allium tuberosum

Native Range: China
Height: 1 to 2 feet (30–60 cm)
Width: 1 to 2 feet
Soil Conditions: Drought-tolerant
Sunlight: Full sun to full shade
USDA Hardiness Zones: 3 to 9

With a growth habit similar to chives (see "Chives" on page 188), garlic chives' leaves, flowers, and seeds have a distinctive garlic flavor. Also known as Chinese leek, this perennial herb has naturalized throughout the world. Its narrow, flat leaves grow in clumps that, when planted closely, spread into one another to provide an adequate ground cover. In late summer, stalks topped with an umbel of white blooms sprout above the leafy mounds, feeding pollinators and beneficials. These blossoms also make lovely cut and dried flowers. In time, small, round, green seeds form at the base of each blossom. These edible seeds fascinate a chef I know; he sprinkles them over meat and fish entrées. Left unharvested, the seeds gradually turn to black in the fall.

I planted garlic chive sets I started from seed in borders around my beds, on shaded hügelkultur mounds, and below fruit trees so I could enjoy them for both their culinary and their visual appeal. Like conventional chives, garlic chives can be propagated through division or seed and also tend to self-seed.

Figure 10.13. Garlic chives provide four seasons of visual interest. In this winter view, the upright tan stalks and dried flower heads dotted with black seeds contribute interesting form and color contrast against the snow.

PV

Giant Solomon's Seal

Polygonatum biflorum

Native Range: North American woodlands
Height: 1 to 3 feet (30–100 cm)
Width: 1 to 3 feet
Soil Conditions: Evenly moist soil
Sunlight: Partial to full shade
USDA Hardiness Zones: 3 to 7

Giant Solomon's seal is a beautiful plant with arched leafy stems from which its white flowers hang like bells (see figure 10.19). In the fall, the leaves turn vibrant shades of yellow. Once this plant is established, you can harvest its edible first shoots in spring and prepare them like asparagus.

Giant Solomon's seal forms a colony via underground stolons that allow it to send up new suckers, making it an effective ground cover as it spreads. This plant can be propagated by division. I planted giant Solomon's seal in two locations where there is relatively deep shade: on the northwest slope of a bed shaded by two large black locust trees (see "A Shady Black Locust Grouping," page 273) and in the shade of the hedgerow bordering my neighbor's property.

Figure 10.14. Edible giant Solomon's seal shoots emerge in spring.

CF

Gloriosa Daisies

Rudbeckia hirta

Native Range: Developed in US from native black-eyed Susans
Height: 2 to 3 feet (60–100 cm)
Width: 1 to 2 feet (30–60 cm)
Soil Conditions: Adaptable, but needs well-drained soil
Sunlight: Full sun or partial shade
USDA Hardiness Zones: 2 to 11

Closely related to the wild black-eyed Susan and to purple coneflowers, gloriosa daisies are distinguished by their "furry" leaves and variegated flower heads that range from pure yellow to reddish brown and various combinations thereof.

These showy, short-lived perennials begin blooming in July and continue throughout the summer, attracting pollinators and ladybugs. They make excellent cut flowers, lasting at least a week when placed in water. I scattered seeds on open ground, where they germinated readily. The plants continue to self-seed, producing beautiful blooms year after year with no additional effort.

Figure 10.15. These stunning gloriosa daisies self-seeded around one of my ponds, where they add vibrant color to the summer landscape.

Figure 10.16. The flower shoots arising from the dense leaves of good King Henry are edible, too!

Good King Henry

Chenopodium bonus-henricus

Native Range: Central and Southern Europe
Height: 20 to 30 inches (50–75 cm)
Width: 12 to 18 inches (30–45 cm)
Soil Conditions: Well-drained but moist soil
Sunlight: Full sun or partial shade
USDA Hardiness Zones: 3 to 7 (or 9 if planted in cooler microclimates)

Good King Henry is a hardy perennial whose dark green, spade-shaped leaves look and taste similar to spinach, although with a sharper flavor that mellows with cooking. Perennial vegetables generally have a more intense flavor than their annual counterparts—a quality that signifies greater nutritional value ("the bitterer the better"). This is because their extensive and long-lived root systems can access nutrients from a greater swath of soil. Good King Henry is rich in vitamins B_1 and C, iron, and calcium. Its tight floret of leaves covers the ground, giving rise to flower stems that can be cut and cooked like asparagus when they first emerge as shoots in spring. There is no limit to how many flower stems can be harvested, as their presence indicates that the plant is already well established. The leaves can be harvested throughout the growing season.

When planted close together or allowed to self-seed, good King Henry plants form a densely shading ground cover. I first planted a dozen purchased plants in a bed with a high water table, where they became sickly. Their leaves yellowed, and they grew slowly. Not wanting to forfeit my investment, I transplanted them to a better-drained location where they have since done well.

Hollyhock

Alcea rosea

Native Range: Eurasia
Height: 5 to 8 feet (1.5–2.5 m)
Width: Up to 12 inches (30 cm)
Soil Conditions: Well-drained but moist soil, drought-tolerant once established
Sunlight: Full sun
USDA Hardiness Zones: 3 to 9

Showy, self-seeding hollyhocks are biennial, living for two years and flowering in their second year. The flowers form on stalks that can reach 5 or more feet tall. They can be sourced in a range of colors, including black. Their flowers are edible, with the texture and taste (or absence thereof) of lettuce.

I grew hollyhocks from seed in flats indoors, and then planted out the seedlings as a border along the north side of a bed where they would not shade shorter plants. Unfortunately, at this point I had done a sloppy job of researching this plant, and it did not survive in this bed due to saturated soil for a good part of the year. Undeterred, the next year I started more hollyhocks from seed. This time, I planted the seedlings in a more suitable habitat, on the north edge of a bed on a south-facing slope with full sun *and* well-drained soil. Here they grew well the first year, returning to flower profusely over an extended period the second year. The following year, several seedlings appeared, and the cycle repeated with no effort on my part.

Figure 10.17. Hollyhock blooms are edible and make good cut flowers.

Horseradish

Armoracia rusticana

Native Range: Central Europe
Height: 2 to 3 feet (60–100 cm)
Width: Will expand outward indefinitely
Soil Conditions: Moist, fertile soil
Sunlight: Full sun
USDA Hardiness Zones: 4 to 7

The familiar condiment of the same name is made from the grated root of horseradish plants. These plants grow in expanding clumps with large, coarse leaves aboveground and an extensive, thick root system below. A lesser-known fact is that the young leaves of this herbaceous perennial are also edible if harvested when they first emerge in spring. Horseradish hails from the brassica family, also known as the crucifers, which includes kale and arugula. Eaten raw or cooked, horseradish's long, oval, coarse leaves have a sharp, bitter, peppery flavor. The plant also has medicinal uses.

Horseradish is an undemanding plant to grow. Make a planting hole and push in a section of root at a 45-degree angle to the bottom of the hole. Then cover with soil so the higher end is no more than 2 inches (5 cm) below the soil surface. Water it and

Figure 10.18. Horseradish leaves when they first emerge.

Figure 10.19. Clumps of hosta nestle under fronds of blooming giant Solomon's seal.

wait for a few weeks, at which point green leaves will begin to emerge. Meanwhile, the root will extend down and form horizontal rootlets that can be harvested and grated for sauce or used to propagate a new patch. Wait 18 months before harvesting any rootlets. The following spring, you can harvest a few of the first leaves, then more each subsequent year. A clump of horseradish tends to expand outward, so plant it where this will not be a problem or keep the roots harvested to control the spread. The root is best when harvested in early spring before new growth begins or after the plants go dormant in fall. I planted my horseradish on the sunny edge of a naturally wet bed, where it has room to spread outward without competing with other plants, and where harvest does not disrupt others, either.

Hosta

Hosta species and hybrids

Native Range: Northeastern Asia
Height: 6 inches to 4 feet (15–125 cm)
Width: 10 inches to 6 feet (25–200 cm)
Soil Conditions: Moist soil
Sunlight: Partial to full shade
USDA Hardiness Zones: 3 to 9

Hostas are beautiful, elegant clumping plants with distinctive leaves in a variety of sizes and colors, topped by tall flower stalks. These plants spread slowly by rhizomes, providing clumping ground cover over time.

I was familiar with this large-leaved, decorative plant as a component of a shaded perennial flower garden. It was surprising to learn that hosta shoots can be eaten as a vegetable, too: Each individual plant begins its yearly growth with a curled leaf cluster that can be harvested and prepared like asparagus. It is best to wait a few years after planting before harvesting, and then only the first few shoots, allowing the rest to grow out. Each remaining shoot will develop into a clump of broad leaves that hug the ground. In mid-summer, a tall stem topped with a cluster of white, pink, or lavender, trumpet-shaped flowers emerges from the clumped leaves to decorate your landscape.

PV H GC

Houttuynia

Houttuynia cordata

Native Range: Southeast Asia
Height: 1 to 1½ feet (30–45 cm)
Width: 1 foot and greater
Soil Conditions: Moist soil
Sunlight: Shade, but can take some sun
USDA Hardiness Zones: 4 to 10

Houttuynia (pronounced *how-TYE-nee-uh*), also known as hot tuna plant, is a hardy perennial that can form a dense ground cover. Its pretty, heart-shaped leaves are either dark green or a variegated white, green, yellow, and pink. Both types are edible and medicinal. The leaves can be eaten raw or cooked and dried for tea. In the summer, houttuynia stems are topped with small greenish white flowers. Widely eaten as a staple food in East Asia, this plant's intense flavor is pungent and peppery.

Houttuynia spreads by horizontal aboveground stems and underground rhizomes that give rise to new plants. Because of its expanding nature, houttuynia can invade space where it is not wanted, so take care to locate it where you can control its spread. I incorporated this plant as a ground cover in "A Shady Black Locust Grouping" described on page 273.

H CF

Lavender

Lavandula angustifolia

Native Range: Europe and Asia
Height: 1 to 1½ feet (30–45 cm)
Width: 1 to 1½ feet
Soil Conditions: Dry soil
Sunlight: Full sun
USDA Hardiness Zones: 5 to 8

Lavender is a perennial herb known for its fragrant and aesthetically appealing pale purple flowers. This compact plant has small, succulent, gray leaves and tiny flowers that cluster at the tops of spiky stems. Its strong scent is distracting to insect pests in the garden. Lavender leaves and flowers are used to flavor main dishes, desserts, and teas. The plant's

Figure 10.20. The pretty heart-shaped leaves and dainty flowers of the houttuynia plant make for an attractive ground cover where shade and moisture are abundant.

Figure 10.21. Munstead lavender in bloom.

oil is commonly used in perfumes and cosmetics, and its antifungal properties make it a beneficial ingredient in traditional medicines.

Most lavender varieties are not rated as hardy to my Zone 4, but I successfully grow Munstead lavender, rated for Zone 4 or 5. I started some from seed in late fall and planted the seedlings out in late spring on two separate south-facing hillsides. The seedlings that were partially shaded by neighboring plants were stunted and have not yet flowered, while those with unobstructed sunlight grew and bloomed.

H GC

Lemon Balm

Melissa officinalis

Native Range: South-central Europe, Mediterranean Basin, and Central Asia
Height: 1 to 3 feet (30–100 cm)
Width: 1 to 3 feet
Soil Conditions: Adaptable, but prefers moist soil
Sunlight: Full sun to light shade
USDA Hardiness Zones: 3 to 12

This herbaceous perennial in the mint family has a lemon scent and flavor and grows in bushy clumps with deep-veined, bright green, scalloped leaves. In summer, small, nectar-rich, white flowers bloom along each stem, attracting masses of pollinators. The leaves can be used as a culinary flavoring, in teas, and in traditional medicinal preparations. I like to snip a few leaves to add to salads and to garnish cold drinks in summer.

I planted one lemon balm plant along with other herbs on a hügelkultur mound in the garden's first year. I soon observed its self-seeding tendency, finding new plants springing up throughout the garden the following year. I resolved to cut these volunteers back before they set seed. I performed this task over several seasons until I reasoned that lemon balm was just self-seeding where there was open ground uncovered by other plants, so it was actually providing aromatic ground cover that needed no effort on my part. Once I arrived at this insight, I no longer felt it necessary to deadhead the ever-increasing abundance of these plants.

Figure 10.22. This clump of lemon balm self-seeded in a sunny niche beneath a Dart's Dash rugosa rosebush.

NF H

Licorice

Glycyrrhiza glabra

Native Range: Southern Europe and western Asia
Height: 2 to 4 feet (60–125 cm)
Width: Up to 6 inches (15 cm)
Soil Conditions: Adaptable
Sunlight: Full sun or partial shade
USDA Hardiness Zones: 6 to 11

Licorice is a suckering herbaceous perennial. Its short horizontal stems lined with small oval leaves extend from narrow stalks that shoot up from its wide-ranging roots. These roots can be harvested in the plant's second or third year and have both medicinal and culinary uses. An extract made from the root is used as the anise flavoring in candies.

I like to include licorice in densely planted beds and hügelkultur mounds because its erect, narrow stems take up little space and cast minimal shade while fixing nitrogen for the surrounding plants. Even though it is rated to be no hardier than Zone 6, it has thrived in my garden. I started licorice from seed and planted some seedlings during my garden's first year. Every year since, its shoots have surprised me by appearing in unexpected locations; once, a new shoot appeared on the opposite side of a stone staircase from the original planting! It had sent out roots even under the heavy stonework.

Figure 10.23. Two licorice shoots extend vertically among other herbaceous plants.

H EF GC

Marshmallow

Althaea officinalis

Native Range: Europe, West Asia, and North Africa
Height: Up to 8 feet (2.5 m)
Width: Up to 6 feet (2 m)
Soil Conditions: Wet soil
Sunlight: Full sun to shade
USDA Hardiness Zones: 3 to 9

The perennial marshmallow is generally found growing in marshes, flourishing in wet soil. Its light

Figure 10.24. In addition to its edible and medicinal parts, marshmallow also serves as a mulch plant due to the copious quantities of organic matter it produces each season.

Figure 10.25. Maximilian sunflower blooms.

green leaves, pale lavender flowers, and roots are edible, and all are used in traditional medicine. The leaves and flowers can be sprinkled in salads, the leaves can be lightly steamed, and you can actually use the roots to make marshmallow, the popular confection. Marshmallow is the first mallow I planted. I started it from seed and placed a seedling low on the east-facing slope of a hügelkultur mound near the bank of a pond, where its roots could easily reach the water table.

In this favorable habitat, the plant grew huge—6 feet tall and wide, its many tall stems eventually pulled down to the ground by their own weight. That year, I conducted a garden tour with a family with a six-year-old girl. We walked along the narrow path at the pond's edge, eventually encountering the immense marshmallow I had planted there, whose stems had cascaded over the path. As we made our way around and through the mass of foliage, the young girl exclaimed, "This is just like Costa Rica!"

The marshmallow's pastel flowers, spaced along each stem, soon became seedpods. The next year, baby plants were sprouting up everywhere in the vicinity of the mother plant. Lesson learned, the following year I cut back the flowering stems before the seeds were set and used the cuttings for mulch. I have since started more from seed and planted them out in the saturated soil of one of my newer edible forest beds, where few other plants would grow but where the marshmallows are performing well. These plants take up a lot of space, but if you have the room and suitable habitat, their lush growth will reward you with quantities of nutritious leaves and flowers in addition to giving your garden a wild, lush appearance akin to a tropical jungle.

Maximilian Sunflower

Helianthus maximiliani

Native Range: North American Great Plains
Height: 4 to 10 feet (1.25–3 m)
Width: Up to 3 feet (1 m)
Soil Conditions: Well-drained soil
Sunlight: Full sun
USDA Hardiness Zones: 3 to 9

Maximilian sunflowers are perennials whose bright yellow blooms grow 2 to 3 inches (5–8 cm) wide, attract pollinators, and make excellent cut flowers. They bloom continuously from early August into fall, and their abundant seeds feed songbirds. Native Americans used these plants for dye, oil, thread, and food; the thick root can be eaten like sunchokes (see "Sunchoke," page 209).

I started these sunflowers from seed and transplanted them to the north side of a low-lying bed where the soil remained wet. They did not survive. The next year, I again started them from seed, this time planting the seedlings on the uphill edge of a sunny, well-drained hillside bed, watering them occasionally as they became established. Here, the plants grow larger each year, producing a bounty of flowers.

Try planting one—or several—of these beautiful, carefree, multifunctional flowering plants. Their flowers will bring you pleasure in the field and in a vase on your table for years to come.

NF H GC

Milk Vetch

Astragalus membranaceus

Native Range: Mongolia and northern China
Height: 6 to 12 inches (15–30 cm)
Width: Up to 6 feet (2 m)
Soil Conditions: Well-drained but moist soil
Sunlight: Prefers full sun, can tolerate afternoon shade
USDA Hardiness Zones: 4 to 9

Milk vetch, also known as astragalus, is one of my favorites among the herbaceous legumes because it creates an effective ground cover, spanning a diameter approaching 6 feet! Although there are many species in the genus *Astragalus*, I grow only *A. membranaceus*, which has delicate leaf-ensconced stems that radiate out horizontally from the central root like spokes on a wheel. White, pea-like flowers appear at the end of each stem and attract pollinators. The roots, which can be harvested beginning in the fourth year, have numerous medicinal uses.

Figure 10.26. This milk vetch forms an effective ground cover and provides nitrogen for surrounding plants.

I started my original plants with seeds from Fedco Seeds, but I have also propagated these plants from seed that I saved, stratified, and scarified. I even transplanted a few that self-seeded in my garden. With the roots' medicinal properties in mind, I planted a bed with milk vetch plants spaced 2 to 3 feet (60–100 cm) apart, anticipating that I would harvest every other plant to sell to herbalist customers. Over the years, the remaining plants will grow to cover the spaces where others were removed. (See figure 14.13 on page 284 for an illustration of this bed.)

Nasturtium

Tropaeolum majus

Native Range: Andes Mountains
Height: Clumping types, 2 feet (60 cm); vining types can climb to 10 feet (3 m)
Width: Clumping types, 3 feet (1 m); vining types can spread to 10 feet
Soil Conditions: Moist soil, tolerates poor soil
Sunlight: Full sun
USDA Hardiness Zones: Annuals, 4 to 8; perennials, 9 to 11

Even though they are annuals in my hardiness zone, I recommend nasturtiums because of their showy flowers, unique peppery taste, and lush growth. Both the flowers and the disk-shaped leaves are edible, commonly used in salads and as garnish for drinks. A friend of mine reports that as an adolescent, she used to suck nectar out of the bottom of nasturtium flowers on her daily walk home from high school in Berkeley, California, where these flowers were "all over gardens."

Nasturtium plants have two distinct growth habits: clumping and vining. Both are tender plants and quite frost-sensitive. Both can serve as ground covers, with the vining type running over the ground if there is nowhere for them to climb.

Start nasturtium from seeds indoors and then plant the seedlings outdoors after danger of frost has passed. You can also sow seeds directly in garden beds once the soil has warmed. If desired, you can purchase a seed mix of varieties that have several different bloom colors—yellow, orange, and red. Although they are annuals, nasturtiums have the ability to self-seed under the right conditions. In my garden, they self-seeded one year in a microclimate particularly favorable due to its heat retention: a patch of ground surrounded by thick limestone slabs and a driveway paved with crusher run.

This flowering annual is not native to North America, but nasturtiums have naturalized in several US states and are even classified as invasive in Hawaii.

Oregano

Origanum species

H GC

Native Range: Temperate Eurasia and the Mediterranean region
Height: 2 feet (60 cm)
Width: Up to 3 feet (1 m)
Soil Conditions: Well-drained, moderately moist soil
Sunlight: Full sun to light shade
USDA Hardiness Zones: 4 to 9

Oregano is a quintessential ingredient in Italian and Greek cooking. Half-inch-wide (13 mm) leaves grow along sprawling woody stems. Clusters of small purple or white flowers bloom along the stem

Figure 10.27. Although you may have to replant them every year, nasturtium's masses of bright leaves and blooms add a dramatic punch to any landscape or culinary creation.

Figure 10.28. Oregano plants form an effective ground cover.

tips beginning in June and continuing throughout the summer. The leaves can be used as a component of herbal teas as well as in cooking. Leaves are used fresh or dried, which intensifies the flavor. To remove fresh or dried leaves from the stems, hold the stem by the tip and run your thumb and forefinger down its length. Essential oil of oregano has medicinal applications.

I massed oregano as a ground cover under honeyberries and installed it on mounds along with other culinary herbs. The plants can be propagated by division, by layering, or from seed. In the garden, oregano spreads naturally by layering and also self-seeds, expediting its expansion into a dense ground cover. Cutting back the flower heads will force fresh leaf growth and prevent the plant from self-seeding. The leaf-covered stems can be harvested throughout the season, with lighter harvest in late summer and fall so as not to remove too much of the plant as it enters dormancy. I have heard that oregano plants grown from seed can vary greatly in the quality and intensity of their flavor. This won't matter if you are growing oregano primarily as an attractive ground cover, but if you plan to use it fresh or dried for seasoning, it might be best to taste the plant before you harvest it to ensure that the flavor is acceptable.

Peony

Paeonia species and hybrids

Native Range: Europe, Asia, and western North America
Height: 1 to 3 feet (30–100 cm)
Width: 1 to 3 feet
Soil Conditions: Tolerates clay soil but not saturated soil
Sunlight: Full sun
USDA Hardiness Zones: 3 to 9

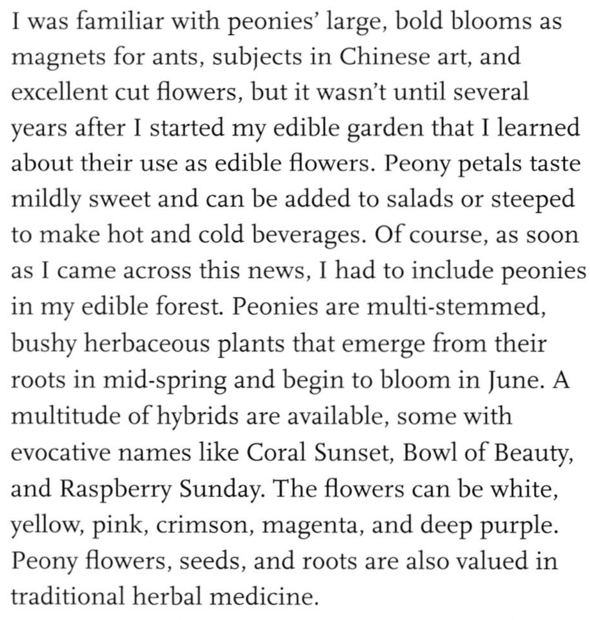

I was familiar with peonies' large, bold blooms as magnets for ants, subjects in Chinese art, and excellent cut flowers, but it wasn't until several years after I started my edible garden that I learned about their use as edible flowers. Peony petals taste mildly sweet and can be added to salads or steeped to make hot and cold beverages. Of course, as soon as I came across this news, I had to include peonies in my edible forest. Peonies are multi-stemmed, bushy herbaceous plants that emerge from their roots in mid-spring and begin to bloom in June. A multitude of hybrids are available, some with evocative names like Coral Sunset, Bowl of Beauty, and Raspberry Sunday. The flowers can be white, yellow, pink, crimson, magenta, and deep purple. Peony flowers, seeds, and roots are also valued in traditional herbal medicine.

It's best to plant peony tubers in the fall; the tubers require a period of subfreezing weather before they will sprout and produce flowers. Several peonies I planted in unmounded parts of a large crescent-shaped bed perished during their first year, no doubt due to the high water table in that location.

Others planted near the base of a gentle, well-drained slope did well. In this more favorable habitat, I planted them as a border and also massed them as a strikingly beautiful ground cover in a large bed.

Once peonies become established, which may take a year or two, these plants produce a larger mound of leaves and more prolific blooms during each subsequent season. They are best planted 3 or 4 feet (1–1.25 m) apart to give them space to expand and at least 4 feet from existing shrubs or trees to reduce root competition. Little maintenance is needed other than supporting the stems topped with heavy blooms to keep them upright and applying a few inches of mulch in the fall. The mulch keeps the roots warmer and less subject to the freeze/thaw cycle during the colder months.

Once established, these magnificent plants can live upward of 25 years. They are a must-have addition to any perennial landscape.

Rhubarb

GC

Rheum × hybridum

Native Range: Northern and Central Asia
Height: 1 to 3 feet (30–100 cm)
Width: 2 to 4 feet (60–125 cm)
Soil Conditions: Evenly moist soil
Sunlight: Full sun to light shade
USDA Hardiness Zones: 3 to 8

Familiar to many home gardeners, rhubarb's mass of broad leaves and stems can effectively shade out weeds and conserve moisture. Early each spring, the leaves emerge from the crown, their stems reaching harvestable length by mid-May. In late spring, the plants send up 3- to 4-foot-tall flower heads covered with fringe-like white blooms that, if not trimmed back, may self-seed, producing possibly unwanted rhubarb "weeds." Rhubarb's tart

Figure 10.29. Bulbous peony blooms are a favorite of many people, and they are also irresistible to pollinators.

green, pink, or red stems are usually cooked and sweetened for use in pies, crumbles, and compotes. They can also be eaten raw with their tips dipped in sugar. The large leaves contain poisonous quantities of oxalic acid and should not be eaten.

I grow rhubarb at the base of sunny mounds and the northern edges of beds where the soil stays uniformly moist. To harvest rhubarb, detach the stem from its base by holding it firmly while gently moving it back and forth and pulling it toward you. It's best to refrain from harvesting more than half of the growth, so the remaining leaves can return sustenance to the roots. In my climate, the stems become woody and inedible in summer, but sometimes have a growth spurt in fall, when the new growth can be lightly harvested again. Once the plants go dormant, replenishing the surrounding mulch helps keep them healthy and conserves moisture in the underlying soil. Rhubarb can be propagated by seed or by dividing and transplanting part of the thick, pulpy root.

Figure 10.30. A young rhubarb plant with deep red edible stems is attractive in the landscape.

H GC

Russian Comfrey

Symphytum × uplandicum

Native Range: Hybrid combo of plants from England and Russia
Height: Up to 4 feet (1.25 m)
Width: Up to 6 feet (2 m)
Soil Conditions: Tolerates most soil types
Sunlight: Full sun to partial shade
USDA Hardiness Zones: 3 to 9

Russian comfrey has large, pointed, elongated leaves arranged on succulent stems that form a dense circular mound. In early summer, sprays of purple, bell-shaped flowers bloom along the tips of the stems. The weight of the stems laden with leaves and flowers causes them to lie down on the ground, forming a thick mulch that snuffs out any underlying plants. As the mature stems sink down, new growth emerges from the center of the plant. When conditions are ideal, this cycle repeats three or four times in a growing season, producing volumes of organic matter that make it a superior mulch plant.

Comfrey is also a nutrient accumulator par excellence. It gains this status due to an extremely large, deep taproot that scavenges the subsoil for phosphorous, potassium, calcium, copper, iron, and magnesium. Its leaves decay rapidly once they touch the ground, releasing these elements for use by surrounding plants. You can also cut the stems and drop them at the base of plants that need a nutritional boost in other parts of the garden. Comfrey is valued for its medicinal uses, especially as a salve or poultice to help heal broken bones and wounds.

I planted one Russian comfrey in my nursery the year I began planning the edible forest. The

Figure 10.31. Russian comfrey stems fall to the ground, where they soon decay, adding nutrients to the soil.

diameter of its taproot at the soil surface expands each year, now measuring over 2 feet (60 cm) across. Imagine how deep the root must extend with such huge surface dimensions! In its ideal nursery location, with full sun and constantly moist soil, it puts forth prodigious new growth four times each year. This is my "mother" plant, which I use to propagate new seedlings.

Most varieties of comfrey produce seed and self-seed aggressively. This can be a problem because once a comfrey plant takes root, it can outcompete most other herbaceous plants and is nearly impossible to eradicate. Russian comfrey is particularly desirable because its flowers are sterile, and even though pollinators frequent them, they do not produce viable seed. Rather, root cuttings propagate this variety. In spring, when leaf buds start to appear on the root surface of my mother plant, I use a pointed shovel to chop off a chunk of the surface root containing many small nascent shoots. Then I divide the chunk into 1- or 2-inch (2.5–5 cm) pieces, each containing a bud. These small chunks can be planted out directly or potted to transplant later. One caution: Any piece of root will self-propagate, so be careful not to scatter pieces of root around by rototilling or accidentally dropping a cutting.

I waited four years before incorporating comfrey in the edible forest. Realizing that they would be nearly impossible to eliminate once installed, I wanted to be sure I could live with their permanence wherever I placed them. The first place I planted them was beneath the plum thicket described in "The Plum Patch" on page 129.

Since planting comfrey beneath the plums, I have scattered additional cuttings throughout the garden: at the semi-shaded base of north-facing hügelkultur mounds, between and behind pome and stone fruit trees in hillside orchard-type rows, and in low-lying water-soaked beds. This versatile nutrient accumulator has established and bloomed in each location.

Sea Kale

Crambe maritima

Native Range: Coastal Europe
Height: Up to 2½ feet (75 cm)
Width: Up to 2 feet (60 cm)
Soil Conditions: Fertile, well-drained soil, salt- and drought-tolerant
Sunlight: Prefers full sun
USDA Hardiness Zones: 4 to 8

Sea kale is a perennial vegetable in the brassica family, a relative of cabbage, broccoli, and "regular" kale. I have grown a number of kale varieties in my annual vegetable garden, but I hadn't encountered sea kale until I began the edible forest. Sea kale

Figure 10.32. The edible blooms of sea kale have a sweet, aromatic flavor.

provides edible roots, shoots, collard-like leaves, and flower buds and blooms, each with a distinct flavor. It's best to wait until the plant is well established before harvesting shoots, leaves, or roots, and all of these need to be cooked. Once the plant begins to bloom, the unopened flower heads can be eaten raw or cooked like broccoli. The sweet white flowers are tasty when included in salads.

I planted sea kale seeds in a flat in early spring and kept them indoors. I set out the young seedlings in a well-drained, south-facing bed once the soil warmed in late spring. Whether propagated from seed, division, or root cutting, this plant can take some time to establish. The plants I started from seed have taken more than four years to reach full size. Even when grown from root cuttings, called *thongs*, sea kale plants take at least two years to fully develop.

Sorrel

Rumex species

Native Range: Europe and Asia
Height: 12 to 18 inches (30–45 cm)
Width: Up to 2 feet (60 cm)
Soil Conditions: Moist soil
Sunlight: Full sun to moderate shade
USDA Hardiness Zones: 4 to 8

Sorrel is a clumping perennial plant that begins its growth cycle in early spring, producing new, tender, lemon-flavored leaves that make a delicious addition to salads. As the leaves mature and thicken, use them to add flavor to stews and soups. Sorrels contain oxalic acid, so should not be consumed in large quantities. Garden sorrel's (*Rumex acetosa*) leaves are solid bright green; the leaves of red-veined sorrel (*R. sanguineus*) are more decorative. When planted in full sun, sorrel plants go dormant during hot, dry spells in summer but revive and send out fresh leaves when precipitation resumes. A second growth spurt is common as the weather cools in fall.

Figure 10.33. This garden sorrel, situated in a bed with a slight northeast-facing slope, provides me with tasty leaves every year.

I have grown both garden sorrel and red-veined sorrel from seed. If you start seeds indoors in mid-spring, sorrel seedlings are ready to plant out by early summer. The mature plants send up flower heads in June and may self-seed if not deadheaded. They are also easy to propagate by division.

Sunchoke

Helianthus tuberosus

Native Range: Central North America
Height: 6 to 12 feet (2–3.5 m)
Width: Up to 4 feet (1.25 m)
Soil Conditions: Drought-tolerant
Sunlight: Full sun or partial shade
USDA Hardiness Zones: 3 to 9

Also known as Jerusalem artichokes, sunchokes are tall herbaceous perennials that produce yellow, sunflower-like blossoms in late summer. Once established, these large plants shade out everything growing below. Sunchokes are prolific producers of edible underground tubers, for which Native Americans cultivated them extensively. These tubers are similar in texture to potatoes, but have a sweeter, nuttier flavor. They can be eaten raw or cooked like

Figure 10.34. These tall sunchoke plants flower in September.

Figure 10.35. The fern-like leaves and tiny white flowers of sweet cicely.

potatoes. I like to fry thin slices until they turn light brown, resulting in crisp, sweet-tasting chips. Sunchokes are nutritionally dense in carbohydrates—one of very few perennial vegetables that can be grown in temperate climates supplying this essential nutrient. Moreover, the type of carbohydrate they contain, a form of fructose, does not cause a glycemic spike like that of potatoes but can cause abundant intestinal gas in some people.

I have had success with sunchokes planted in partly shaded earth above a culvert and interplanted between shrubs in a windbreak hedge. I simply planted the tubers 4 to 6 inches (10–15 cm) deep. Three caveats with these plants:

- Place them only in a spot where their outward spread can be controlled.
- Site them to the north of plants you do not want shaded.
- If you intend to harvest roots to eat, choose a site where digging will not disturb the roots of nearby plants.

The tubers are best harvested when the plants go dormant in the fall, or in early spring. To do so, dig underneath and around the stems to find the lumpy growths. It is nearly impossible to dig out every tuber, so the plants will regrow from those left in the ground. And there will always be enough tubers to start a new patch elsewhere in your garden if you wish.

Sweet Cicely

Myrrhis odorata

Native Range: Central Europe
Height: 2 to 3 feet (60–100 cm)
Width: 2 to 3 feet
Soil Conditions: Moist soil
Sunlight: Partial to full shade
USDA Hardiness Zones: 3 to 7

I was unfamiliar with sweet cicely before I learned of it while visiting Martin Crawford at his edible forest garden in Darlington, England.* I am now a huge fan. Not only is every part of this perennial plant—including the root—edible, but its graceful, fern-like leaves and clusters of white flowers make it attractive to behold from spring to late fall. The

* Martin Crawford (author and director of the Agroforestry Research Trust), in conversation with the author, October 2014.

leaves spread out from upright or horizontal stems to form a dense, light-blocking ground cover.

The culinary uses of this herb are as broad as your imagination: You can use the leaves as a feathery garnish or as a sweetener with stewed fruit such as rhubarb. Add the leaves and seeds to any dish for a mild anise flavoring. The thick roots can be peeled and eaten raw like a carrot stick or cooked and eaten like any other root vegetable. I introduced a chef to sweet cicely's oblong green seeds, and he used them in a sauce for salmon that became a huge hit with his clientele.

Unlike most other herbs, sweet cicely flourishes in moist, shaded habitats like those found under mature fruit trees and shrubs. I plant sweet cicely on north-facing slopes of beds and mounds, where it grows lush and tall. When placed in drier, sunnier spots, it struggles to survive. During a recent drought year, the shaded plants became dormant during the dry stretch but greened up and resumed growth after a good rain.

In my garden, sweet cicely germinates wherever its seeds find a compatible habitat. After observing it self-seed in this fashion, I mimicked nature by harvesting the ripe seeds in fall and dropping them in moist, shaded spots, including on my hügelkultur mounds in the woods. The next spring, they germinated and grew well with no additional help from me.

PV EF GC

Turkish Rocket

Bunias orientalis

Native Range: Black Sea steppe
Height: Up to 4 feet (1.25 m)
Width: 1 to 2 feet (30 to 60 cm)
Soil Conditions: Tolerates a variety of soil types; drought-tolerant
Sunlight: Full sun or partial shade
USDA Hardiness Zones: 4 to 9

If you crave cruciferous vegetables such as kale and broccoli, you might want to grow Turkish rocket, a perennial that supplies similar taste and nutritional value. In early spring, its long, pointed leaves form a ground-shading rosette like a many-pointed star. From the center, thin, tall stems emerge covered with green buds. These stems and young leaves can be harvested and prepared like broccoli rabe. The strong flavor mellows with cooking. I like to braise these shoots with oyster sauce to add a savory sweetness to the sharp mustardy taste. If they aren't harvested early, the green buds become yellow flowers that are also edible and a tasty addition to salads.

When permitted to self-seed, Turkish rocket's florets and dense foliage merge together to form an effective ground cover. Its lengthy taproots scavenge for nutrients held deep in the soil, making the plant

Figure 10.36. The edible blooms of Turkish rocket attract bees, butterflies, and other beneficial insects.

oblivious to drought, but also resistant to removal once established. To prevent it growing in unwanted places, deadhead the flowers before seeds form. To save seed for planting in the future, I deadhead all the spent flower heads but one, over which I secure a paper bag so the seeds do not fall to the ground. Once the seeds are mature and have dried, I harvest and freeze them, then plant them in flats or scatter them on the ground in the desired spot the following spring. In my garden I planted sets started from saved seed in a partly shaded area at the top of a small southwest-facing slope, where I allow them to self-seed for ground cover and abundant edible leaves and flowers.

Valerian

Valeriana officinalis

Native Range: Eurasia
Height: 1 to 5 feet (30–150 cm)
Width: 18 to 24 inches (45–60 cm)
Soil Conditions: Adaptable
Sunlight: Full sun
USDA Hardiness Zones: 3 to 9

The tall, narrow flower heads of these clumping perennial plants arise from a tight floret of deeply lobed leaves. The prominent white to pink flowers have a strong, sweet scent that permeates the air. Valerian roots are used in traditional medicine as a sedative and tranquilizer.

Figure 10.37. I mistakenly allowed one valerian plant to go to seed, and the prolific self-sown seedlings produced this impressive mass of plants.

Because of its tendency to self-seed aggressively, valerian has naturalized in many parts of the United States and is considered invasive in several midwestern states as well as Connecticut. I was not aware of valerian's propensity to self-seed until it was too late. I quickly learned to deadhead the flowers before they became seed heads and to use the cut flowers as aromatic additions to bouquets.

Water Celery

Oenanthe javanica

Native Range: Eastern Asia
Height: Up to 2 feet (60 cm)
Width: Spreads widely
Soil Conditions: Prefers moist to wet soil; tolerates drier soil
Sunlight: Full sun or partial shade
USDA Hardiness Zones: 3 to 9

If you like the taste of celery but want to avoid the labor of growing the annual vegetable, herbaceous perennial water celery is an excellent homegrown replacement. The leaves look, smell, and taste like those of celery, with a more intense flavor. I substitute the leaves for celery in all of my cooking: in cold salads, as a garnish, and in soups and stews. Its clusters of small white flowers attract pollinators and other beneficials in summer.

This plant is adapted to a moist habitat: I have even observed it growing from the bottom of a shallow pond. It will grow in drier sites, though at a slower pace. Water celery is also cold-hardy—one of the first plants to leaf out in the spring and the last to die down in the fall. Spreading rapidly by suckering, this plant quickly forms an effective ground cover. It is easily propagated by digging up a clump and transplanting it elsewhere. I planted

Figure 10.38. The dense cover formed by the roots and leaves of water celery outcompetes most other herbaceous plants.

Figure 10.39. Wild fennel flowers attract parasitic wasps, are long lasting in edible bouquets, and produce tasty, nutritious seeds.

small clumps 3 feet (1 m) apart in a semi-shaded saturated area where, over the course of three years, they spread to completely cover the ground. If you need a ground cover for a similar habitat, water celery could be your edible answer.

H EF CF

Wild Fennel

Foeniculum vulgare

Native Range: Mediterranean region
Height: 3 to 5 feet (1–1.5 m)
Width: Up to 2 feet (60 cm)
Soil Conditions: Tolerates dry soil
Sunlight: Full sun
USDA Hardiness Zones: 4 to 9

The leaves, flowers, and seeds of wild fennel are all anise-flavored with multiple culinary applications. The plant comprises slender branching stalks topped with umbels of abundant yellow flowers that mature into the familiar "fennel" seeds. Wild fennel is tall, but it casts minimal shade on other plants. I like to add the leaves and flowers to salads, soups, and fish dishes, and include the seeds in baked goods. The plant extract is sold for its nutritional and possible medicinal value.

Wild fennel can behave as a perennial, biennial, or annual, depending on the climate. The seeds I plant are imported from Italy (I got them from Seeds from Italy, www.growItalian.com), and this strain does not produce a thick bulb at its base. In my garden, one plant lived for three seasons before it died out. Self-sown seedlings sprang up in its place.

Yarrow

Achillea millefolium

Native Range: North America and other temperate regions in the Northern Hemisphere
Height: 2 to 3 feet (60–100 cm)
Width: 2 to 3 feet
Soil Conditions: Adaptable
Sunlight: Full sun
USDA Hardiness Zones: 3 to 9

Yarrow has fine, feather-like leaves and its flower heads, topped with multiple tiny blooms, attract parasitic wasps, ladybugs, and hoverflies. Yarrow's clumping foliage provides habitat for ground beetles and spiders and makes for a serviceable ground cover. This hardy plant is drought-tolerant; its deep taproot scavenges the soil for nitrogen, phosphorous, potassium, and copper, making it a nutritious forage for ruminants and a nutrient accumulator for neighboring plants. The flowers have applications in traditional medicine.

Yarrow is a low-maintenance multipurpose plant that, in addition to its many beneficial qualities, is attractive in the garden landscape and in cut flower arrangements. Hybrids are available sporting pink, yellow, rose, and salmon-colored blooms. In my garden, I interplanted yarrow in the herbaceous layer of an edible windbreak and in a well-drained, south-facing bed.

Figure 10.40. Yarrow's bunched florets and feathery leaves are pretty garden accents.

Whether you are interested in growing leafy greens, perennial vegetables, cut flowers, culinary or medicinal herbs, or all of the above, this chapter offered numerous choices to suit your taste and garden habitats. In the next chapter, I describe an abundance of ground cover plants I have grown that offer even more possibilities to embellish the lower layers of your edible landscape.

CHAPTER 11

GROUND COVERS

A basic principle of perennial gardening is to keep the ground covered with something, preferably something living, everywhere and always. Ground cover helps modulate soil temperature, absorb and preserve moisture, and nurture soil life. Many of the plants described in this chapter can perform these functions for you admirably while simultaneously supplying nutritious edibles, attracting beneficial insects, and enhancing the aesthetics of your landscape. As you learn about my experience with each, I am confident you will find some that are suited to your garden vision and habitats.

I am fond of strawberries as a ground cover, particularly the everbearing varieties. These low-growing plants flower and fruit throughout the growing season, beginning in June. Some produce berries in prodigious amounts while others provide a delightful nibble as I wander through the garden.

While the majority of the ground covers discussed here are ones you will cultivate intentionally, a few are those that have a tendency to move in of their own accord. At first, I fastidiously weeded and weeded and weeded the native intruders to my garden. It took me a while to realize this was a losing battle. I was "doing the same thing over and over expecting a different result"—a sort of insanity. I exerted massive amounts of time and energy, yet the plants continued to encroach. As my frustration mounted, something had to give. That something was me. Gradually, I am adopting a "live and let live" philosophy, making peace with these invaders one by one. In this chapter, I describe three wild ground covers that I have come to accept: ground ivy, horsetail, and wild strawberries. I have come to realize that each saves me the work of installing an intentional ground cover, while at the same time providing its own range of benefits.

Alpine Strawberry

Fragaria vesca

Native Range: Throughout the Northern Hemisphere
Height: 6 to 12 inches (15–30 cm)
Width: 6 to 12 inches, or greater if the running type
Soil Conditions: Adaptable
Sunlight: Full sun or partial shade
USDA Hardiness Zones: 2 to 9

Alpine strawberries are cousins of common garden strawberries (see "Garden Strawberry" on page 221). They are day-neutral, producing fruit over a long period of the growing season, and adapted to a wide range of habitats. Some varieties have a clumping habit; others send out runners. The dainty plants yield small, narrow, cone-shaped berries ranging in color from red to yellow to white, all with intense flavor.

More shade-tolerant than other strawberry types, these plants have done well for me on steep

Figure 11.1. The alpine strawberry, with intensely flavored fruits, provides a tasty nibble in a wander through the edible garden.

north-facing slopes where the sun's rays are indirect and blocked for part of the day. They also work well as borders for partly shaded paths and as ground covers beneath trees and shrubs. The running varieties form a dense ground cover but, in my experience, fruit unreliably. And I've found that after a few years, the clumping varieties tend to die if not dug up, separated, and replanted. However, much to my delight, I have noticed them self-seed either directly in the vicinity of the mother plant or in other parts of the garden, via transportation by rodents or birds. In this way, new plants become established without any effort on my part.

Bellflower

Campanula species

Native Range: Northern Hemisphere
Height: 1 to 2 feet (30–60 cm)
Spread: 1 to 2 feet and farther
Soil Conditions: Well-drained but moist soil
Sunlight: Full sun or partial shade
USDA Hardiness Zones: 3 to 10

There are hundreds of bellflower species and varieties. I am most familiar with clustered bellflower, *Campanula glomerata*, which is native to Europe and Asia. This herbaceous perennial flower has forest-green, heart-shaped leaves. Stalks laden with deep purple, bell-shaped flowers grow from the bellflower's florets. Spreading through rhizomes underground, these plants form a dense mat. The July-blooming flowers, leaves, and roots are all edible. Though the blossoms have a bland taste, their intense color makes for a beautiful embellishment in any dish. You can cut back the first flower stalks to stimulate successive blooms. I have only had success with a single plant, but I am still trying because bellflowers have great potential as an attractive and effective edible ground cover.

Bird's-Foot Trefoil

Lotus corniculatus

Native Range: Grasslands Eurasia and northern Africa
Height: 6 to 12 inches (15–30 cm)
Spread: 12 to 24 inches (30–60 cm)
Soil Conditions: Tolerates dry soil and drought
Sunlight: Full sun
USDA Hardiness Zones: 4 to 8

This legume grows wild on my farm and appeared in my garden as a volunteer. Just a few inches (8 cm) tall, it spreads over the ground from a deep, central root and sports trefoil-shaped leaves and bright yellow flowers that bloom along its stems throughout the summer. Its name is derived from its multi-pronged seed head, which resembles a bird's foot. It is often seen growing along sandy roadsides and is one of the few plants that continues to bloom during periods of drought. Bird's-foot trefoil flowers provide food for insects and nutritious forage for ruminants. This clover is a good

Figure 11.2. The deep purple blooms of bellflower stand out in any landscape.

Acid Lovers

Many species of berry-producing ground covers thrive in acid soil. I have tried to grow them all but have been unsuccessful because my soil has a nearly neutral pH. If you have acidic soil, any of the following diminutive creepers could serve you well as an all-season edible ground cover. All are cold-hardy to at least Zone 3 and some can thrive in winters as mild as those in Zone 7. All are native to the North American continent.

Bearberry (*Arctostaphylos uva-ursi*), also known as kinnikinnick, is an evergreen shrub that grows only 1 foot (30 cm) tall. It has beautiful glossy leaves, pretty clusters of white to pale pink summer flowers, and round red berries that birds and bears love. They are edible for humans, too, but are best cooked to improve palatability. Bearberry leaves turn deep red in fall and remain on the bush throughout winter. This shrub can tolerate tough conditions, including salt spray from roadways and ocean water, but needs good drainage.

Bunchberry (*Cornus canadensis*) is an easy plant to grow. Each stem sports a whorl of green leaves that turn shades of burgundy in fall, and four-petaled white flowers that give way to deep red berries in late spring. The berries are edible raw or cooked. Bunchberry grows 4 to 6 inches (10–15 cm) tall, spreads rapidly, and prefers dappled shade.

Cranberry (*Vaccinium macrocarpon*), best known as the main ingredient of the essential Thanksgiving dinner condiment, grows in full sun on woody evergreen vines, forming dense mats. In June delicate, pink, bell-shaped flowers appear along each stem between small, elliptical leaves, becoming the familiar berries that ripen to deep red in mid-fall. In the wild, cranberries are found in boggy conditions where few other plants can grow. If you have a wet spot, it might be ideal for this plant, but you don't need a bog to grow them in your garden; you just need to keep them well watered. In industrial cranberry production, the fields are flooded so the berries rise to the surface of the water for easier harvest. In a home-scale planting, no flooding is necessary—just pick the berries off by hand and enjoy making a cranberry accompaniment from your very own berries.

Lingonberry (*V. vitis-idaea*), a close relative of cranberry and blueberry, grow naturally in boreal forests and tundra regions of the Northern Hemisphere. Adapted to full sun or partial shade, these evergreen, 6- to 12-inch-tall (15–30 cm) creeping plants sport small, pink, bell-shaped flowers in summer that become clusters of edible, red berries in early to mid-fall. The berries have a delectable sweet/tart flavor.

Wintergreen (*Gaultheria procumbens*) is a cold-hardy, highly shade-tolerant evergreen ground cover native to northeastern North America. Bright green, leathery, spoon-shaped leaves grow from stalks no taller than 6 inches that arise from roots spreading underground as well as from horizontal aboveground stems. White, bell-shaped flowers bloom in summer and yield small, red, edible berries. An extract from the leaves is used for flavoring as well as several medicinal applications.

Figure 11.3. Volunteer bird's-foot trefoil fills a niche between patio stones.

choice if you need a ground cover legume that can take foot traffic and mowing, as well as dry growing conditions. If this plant does not grow wild in your yard or nearby, you can purchase seeds and plant them yourself. The best time to plant is early spring; just scatter the seeds over smoothed bare ground. They may take a while to germinate, during which time you need to keep the ground moist. Covering the area with a light mulch of seedless grass clippings can help retain moisture.

H GC

Creeping Thyme

Thymus praecox

Native Range: Southern, Western, and Central Europe
Height: Up to 6 inches (15 cm), depending on type
Width: Up to 3 feet (1 m)
Soil Conditions: Dry soil, but some types can tolerate more moisture
Sunlight: Full sun or partial shade, depending on type
USDA Hardiness Zones: Range depends on type

There are numerous species and varieties of thyme (Richter's catalog boasts 35 listings of this perennial herb), and the use of common and botanical names is not always consistent, as is true for many garden plants. Most thymes are hardy evergreens with tiny leaves and woody stems. The most popular culinary species is English or German thyme (*Thymus vulgaris*). However, this species has an upright growth habit that only makes a suitable ground cover when the plants are packed closely together. Another group of thymes grow low, creeping over the ground, providing a spreading cover. I like to grow two forms of creeping thyme (also called mother-of-thyme) offered by Richter's: wild thyme and Purple Carpet lemon thyme.

In contrast with most other thymes, in my experience, the wild thyme I've purchased from Richter's does well in partially shaded, moist habitats as well as sunny, dry ones. Subtle, pale purple flowers bloom in late spring and summer. This thyme reaches up to 6 inches tall and has larger, rounder leaves than English, French, and Purple Carpet lemon thyme. It is a favorite of chefs because of its flavor and ease of preparation. I use it fresh and dry it for future use. It is listed as hardy in Zones 4 through 8.

Wild thyme creeps over the soil surface to form a decent ground cover. As it creeps, it sets down roots so it is easy to propagate by layering.

Purple Carpet lemon thyme, which is listed as hardy in Zones 2 through 9, thrives in hot, dry habitats. It sets down a taproot that can reach moisture deep in the soil. Its woody stems grow outward, forming a short, dense mat. In early summer, it is covered with tiny, bright, rose-purple blooms. It grows at most 2 inches (5 cm) tall, but it can take a good deal of foot traffic, and it emits a pleasant lemony-thyme scent when trod upon. Note that there is also a low-growing hybrid species of thyme called lemon thyme (*Thymus × citriodorus*), but I have not tried growing it in my garden.

I planted Purple Carpet between patio stones and as an understory for taller herbs without a dense growth habit. I notice it self-seeding where the habitat is to its liking. Although drought-tolerant, these plants need regular watering until their roots get established. I learned this lesson the hard way one droughty summer when I neglected to water newly installed sets and most of them died.

Figure 11.4. Wild thyme is becoming established as a ground cover under red currants and daylilies in a partially shaded hedgerow. It grows densely enough to snuff out most weeds.

Figure 11.5. Planted between patio stones, Purple Carpet lemon thyme eventually grows to completely cover the slabs in a low, dense mat. With its bright blooms, it is attractive to people, bees, and butterflies.

English Daisy

Bellis perennis

Native Range: Western, Central, and Northern Europe
Height: 5 to 6 inches (12–15 cm)
Width: 5 to 6 inches
Soil Conditions: Loamy, well-drained, and evenly moist soil
Sunlight: Full sun or partial shade
USDA Hardiness Zones: 4 to 8

I first came across the tiny English daisy in a lawn in the Finger Lakes region of New York State. It's no surprise, then, that they are also known as lawn daisies. Similar to but much smaller than the familiar wild daisy, these perennial plants form clumps of oblong leaves from which their many-petaled white blooms arise on short stems. They spread by rhizomes to form a low-growing ground cover. Both the leaves and flowers are edible, and are tasty in salads, sandwiches, soups, and teas.

I found these petite edible flowers so intriguing that I decided to try growing them. In late spring, I planted some young English daisy plants, which I had grown from seed indoors and then hardened off outside, along with Tarpan F1 strawberries to provide ground cover in a row of honeyberries. They did fine there in full sun during a summer with normal precipitation. The few that survived the harsh winter that followed perished during the next summer's drought.

Undeterred, I tried again a few years later. During the late summer I planted daisies alternating with milk vetch in a better-hydrated, deeply mulched, and partly shaded bed. I watered them often as they became established. That fall, they appeared to adjust well, and even sported a few flowers toward the end of the growing season. However, during a dry summer the following year, those with the least shade wilted from the drought, and some even perished. It is my observation that these daisies will do best where the soil remains evenly moist and they can have some shade, especially in the afternoon.

Figure 11.6. The adorable 1-inch-wide (2.5 cm) flowers of English daisy. Hybrids in multiple colors are also available.

Garden Strawberry

Fragaria × ananassa

Native Range: Temperate regions of the Northern Hemisphere
Height: 6 to 10 inches (15–25 cm)
Spread: Varies by type
Soil Conditions: Moist but well-drained soil
Sunlight: Prefers full sun
USDA Hardiness Zones: 3 to 9, depending on type

Strawberries can fill a range of ground cover niches in a forest garden. There are two types of garden strawberry plants: Junebearers and day-neutrals (also called everbearers). Junebearers produce fruit only in late spring; day-neutrals bloom and fruit throughout the growing season, beginning in June. When grown commercially, strawberries have a two-year life span. In my garden, they have persisted

Figure 11.7. The large pink flowers of the day-neutral Gasana F1 strawberry adorn the groundcover layer.

Figure 11.8. These Mara des Bois strawberries are spreading as runners root into the gaps between patio stones.

much longer, and their output waxes and wanes over the years. Whether they produce a bountiful crop or just a few berries, they usually provide a decent ground cover to protect the soil and deter weeds.

Strawberries also have two distinct growth habits: running and clumping. Runner types spread through horizontal stems or stolons that extend out over the ground surface and periodically send out roots to form new plantlets. Clumpers form a mound as they spread slowly outward from the center, producing very few or no runners.

In addition to their sumptuous fruits, strawberry flowers are edible, providing a lovely garnish or ingredient in salads. The leaves can also be used for flavoring teas and medicine. These are some garden strawberry varieties that I am particularly fond of.

Gasana F1. This clumping, day-neutral strawberry has showy, bright pink flowers and large, tasty berries. I grew this variety from seed, and when the young plants were ready for transplanting, I planted them out as a border next to the limestone access route that runs through a portion of my garden. In this sunny location, the extra heat radiating from the stone helps them bloom and fruit from early June to late November.

Mara des Bois. This day-neutral strawberry was developed by a breeder in France. It bears white flowers and small-to-medium-sized, aromatic, and flavorful fruits. It spreads via runners, so it provides a decent ground cover if placed in ample light and evenly moist soil.

I planted this cultivar in a sunny spot where the soil moisture fluctuates naturally with precipitation. When it rains regularly, the plants flower and produce luscious berries from mid-June well into November! In times of drought, they become dormant, then come back to life and flower again with rain. In order to beat the competition (birds, slugs, and chipmunks) to

Figure 11.9. A harvest of Junebearing pineberries.

Figure 11.10. The Tarpan F1 day-neutral strawberry sports deep rose-colored blooms.

the harvest, I often pick the berries before they turn completely red but after they have already developed their high fragrance and flavor. In some years, I have netted the plants to protect them from critters but found that despite my efforts, clever chipmunks figure out how to crawl underneath the nets.

Early and late in the season, I cover these strawberries with a floating row cover to protect the flowers from frost. If frostbitten, the yellow centers of the blooms turn black and will not produce fruit. Strawberry plants also benefit from a winter blanket of leaves applied after the ground begins to freeze, usually sometime in December. As your garden develops, the layers of deciduous plants will apply this mulch for you as their leaves fall in autumn.

Pineberry. A Junebearing cultivar, its unusual berries are white with small red seeds and taste like pineapple. When perfectly ripe, pineberry's aromatic fruit acquires a pale pink blush and is best consumed immediately. These berries are so perishable that once, when I attempted to make a jam with them, they dissolved into a thick syrup instead. It might be possible to preserve pineberries by freezing them, though I haven't tried this method. Pineberries can cease to produce palatable fruit after three or four years, but they persist as a dense ground cover. These hardy plants often remain green all winter underneath the snow, resuming their growth in early spring.

Tarpan F1. This attractive day-neutral strawberry has deep rose-colored blooms and tasty, small- to medium-sized fruit. It spreads slowly via runners. My Tarpan grows well as a ground cover under honeyberries. I have also recently interplanted Tarpan with Gasana, expecting that this combination of clumping and running strawberries will provide beautiful contrasting blooms and suitable ground cover for years to come.

B GC

Ground Cover Raspberry

Rubus hybrid

Native Range: Cultivated hybrid of two plants from the Arctic regions
Height: Up to 5 inches (12 cm)
Spread: Up to 1 foot (30 cm)
Soil Conditions: Well-drained but moist soil
Sunlight: Full sun to light shade
USDA Hardiness Zones: 1 to 7

Ground cover raspberries, also known as all-field berries, fascinated me when I first came across them in a catalog; I had not previously heard of a low-growing, thornless plant that yields fruit that looks and tastes like a raspberry. These dainty plants spread through rhizomes—horizontal underground stems that sprout roots and new plants—until they form a dense mat. I have come across four cultivars: Anna, Beta, Sophia, and Valentina, of which at least two different types are needed for pollination. In late spring, this ground cover is speckled with petite pink blooms that become decorative deep red raspberries as they ripen later in summer. The small leaves—which resemble those of raspberries—turn to shades of red and burgundy in fall.

I first interplanted ground cover raspberries with daylilies and chives. They died out within a year. I ordered more and planted them in a sunny sheet-mulched bed, where they went dormant during dry times but came back to life when moistened by rain. This cycle continued through several seasons until they were outcompeted by quack grass that invaded the bed from below. My most successful planting was underneath honeyberries in small, partially shaded, mounded raised beds where they flowered and produced fruit their second season. By the end of the following season, however, they were overrun by encroaching wild strawberries.

Figure 11.11. The delightful late-spring blooms of diminutive ground cover raspberries morph into slow-ripening edible berries in August.

Based on my experience, I recommend these charming plants if you have a moist, sunny, or partially shaded spot that you can keep free of competition. They take some time to establish, so be patient. If they are happy in your location, you will be rewarded with three seasons of delight.

Ground Ivy

Glechoma hederacea

Native Range: Europe
Height: 2 to 4 inches (5–10 cm)
Spread: Indefinite
Soil Conditions: Moist soil
Sunlight: Prefers shade, but can survive in full sun
USDA Hardiness Zones: 4 to 9

Ground ivy is an evergreen herbaceous perennial in the mint family. Originating in Europe, where it was traditionally eaten as a food and used medicinally, it is considered to be an invasive ground cover in parts of North America where it outcompetes native woodland plants. Nevertheless, ground ivy's florets of scalloped green leaves and bell-shaped lavender blossoms create landscape appeal, and both parts are edible. The leaves and flowers add a strong spicy accent to salads and herb butters.

Figure 11.12. The lovely bell-shaped flowers of ground ivy are very attractive to pollinators.

Ground ivy, which grows wild on our farm, asserted itself in the moister, shadier sections of my garden. It expands aggressively through stolons and by seed. The roots form a thick mat that can be difficult to remove. Once I gave up attempting to stop it, ground ivy rapidly covered shaded paths, the areas under native trees, and parts of one of the raised beds in my garden. Reluctantly, I came to admit it has several positive attributes. Because it can tolerate mowing and foot traffic, it is an effective ground cover for paths, and it even emits a distinctive aroma when walked on. Being evergreen, it also maintains its vitality through the winter, capturing moisture and preventing erosion from snowmelt.

Hairy Vetch

Vicia villosa

Native Range: Europe and western Asia
Height: 3 to 6 feet (1–2 m) if climbing
Spread: 3 to 6 feet
Soil Conditions: Adaptable
Sunlight: Full sun
USDA Hardiness Zones: 1 to 7

Although it's not native to North America, this vining ground cover legume is found throughout the United States and Canada and has naturalized on my farm. Its stems form a dense mat over the ground, and its roots grow densely below. Hairy vetch can also use taller plants as trellises to climb upward. Along the thin, vining stems of this plant are fronds that hold a dozen tiny oval leaves. Each frond is tipped with clinging tendrils. Flower shoots emerge where each stem meets the vine; these

Figure 11.13. In my annual vegetable garden, a blooming hairy vetch vine uses a leek for a trellis.

Figure 11.14. Here, horsetails emerge among violets, helping to both cover and beautify the ground.

shoots hold multiple purple blooms packed closely together, resembling an elongated toothbrush. The small flowers feed pollinators and beneficial insects.

I let hairy vetch grow throughout my edible forest and vegetable gardens, where it adds a splash of color and fixes nitrogen for surrounding plants. Some people find its vining habit a nuisance, but I simply rip it down when its upward spread is robbing the light from a berry bush or another plant. If you mow or cut down vetch when it is in flower, you can usually kill the plant, because its energy stores are concentrated in the blossoms at this stage.

Horsetail

Equisetum arvense

Native Range: Northern Hemisphere
Height: Up to 1 foot (30 cm)
Spread: Indefinite
Soil Conditions: Adaptable
Sunlight: Full sun or partial shade
USDA Hardiness Zones: 3 to 11

Horsetail has existed since the time of the dinosaurs, when its family, Equisetaceae, dominated the understory flora. Several species are native to most parts of the Northern Hemisphere, where they are traditionally used by Indigenous peoples as both a food source and medicine. I first noticed these primitive plants in my garden in early spring, when their rigid fertile shoots shot up from the soil of some recently built hügelkultur mounds. By mid-spring, the plants had also produced sterile shoots, which are covered with the characteristic feathery leaves that give this plant its name. My attempts to eradicate this invader were futile due to its deep, dense, matted root structure.

As time passed and I made my peace with the ineradicable horsetail, I came to appreciate several attributes. I noticed that other plants I transplanted or sowed into the horsetail-infested ground tended to do well despite its presence. The horsetail rarely grew more than a foot tall, so it didn't shade other plants, but rather formed an attractive light green shawl around them. Then my research led me to

Nitrogen Fixers as Trap Crops

A trap crop is any plant that is more attractive to a specific pest than the food crop it is intended to protect. Several nitrogen-fixing legumes can fulfill this function. I first noticed this phenomenon in my annual vegetable garden where I allow hairy vetch to naturalize among the food-producing plants. On several occasions, I noticed that the vetch stems develop a "hairy" surface. In my naïveté, I thought, *That must be why they call it hairy vetch.*

Sometime later, I pointed out this observation to a friend who is a much more experienced gardener. She gently informed me that the "hair" was in fact hundreds of tiny aphids sucking the nutritious sap from the vetch. It was remarkable to me that these pests were not present on the neighboring kale, lettuce, or pepper plants. The more nutritious vetch was fortuitously serving as a trap crop, attracting these bugs away from the more valuable food crops. Enlightened, I later observed aphids clustered on the green tips of blue lupine stems (see "Blue Lupine," page 186) after the flowers below had ceased blooming. *Ah ha,* I thought. *Like vetch, lupine is a trap crop for aphids!* Although I have not observed it, I have since read that perennial sweet pea (see "Perennial Sweet Pea," page 247) can also provide this service. Including one or more of these legumes in your garden plan could provide built-in protection from a future onslaught of aphids.

the discovery that *Equisetum* stems and leaves accumulate silica (silicon dioxide), scavenged by its extensive roots from deep within the soil. This compound gives horsetails an abrasive quality, making them useful as scouring and polishing agents. Moreover, silicon is a beneficial micronutrient that becomes available to surrounding plants once the stems of the horsetail die down and decay, forming a straw-like mulch.

H Saffron Crocus

Crocus sativus

Native Range: Mediterranean region of western Asia
Height: Up to 5 inches (12 cm)
Width: Up to 5 inches
Soil Conditions: Loose, well-drained soil
Sunlight: Full sun
USDA Hardiness Zones: 4 and warmer

Like most people, I was familiar with saffron as an expensive spice originating in the hot, arid climate of the Middle East. Until I attended a workshop about the saffron crocus at the University of Vermont, I had no idea that this plant is acclimated to temperate climates like ours in northern New York State, but it is!

Saffron spice is the dried stigmas of the saffron crocus, a dramatic, deep purple flower containing bright yellow pistons and three vibrant orange stigmas. The flowers are ensconced in a bed of thin, dark green leaves. Saffron is used as a seasoning and a coloring agent in Middle Eastern cuisine, Spanish paella, and French bouillabaisse.

If you are interested in cultivating saffron yourself, beware of your worst enemy: rodents! Unlike spring-blooming crocuses, saffron crocuses bloom in the fall, but like spring crocuses they grow from small

(*continued on page 231*)

Creating a Rodent-Proof Planting Bed

Here are instructions for making a rodent-proof planting bed like the one I made for my saffron crocus crop. This size bed will accommodate 300 corms. You can adjust the length and width of the bed to accommodate the number and spacing of the bulbs you intend to plant, as well as the space you have available. I constructed mine on the soil surface because of drainage issues at the site I chose. An alternative, as long as your soil is well drained and does not have a high water table, would be to dig out an area equal in dimensions to the hardware cloth "cage" and set the cage into the excavated area. In that case, the cage does not need to be reinforced with an outer wood frame.

MATERIALS

Two 14-foot (4.25 m) lengths of 5-foot-wide (1.5 m) hardware cloth (½-inch [13 mm] mesh)

Two 13-foot (4 m) 2 × 6 wooden planks, or one for no wood frame

Two 2 × 6 wooden planks, each 4 feet, 4 inches (132 cm) long, or one for no wood frame

Stapling gun and staples

Two 13-foot 1 × 3 wooden furring strips

Two wooden furring strips, each 4 feet, 4 inches long

3-inch (8 mm) and 1½-inch (4 mm) galvanized deck screws

Wire-cutting tool

PROCESS

Step 1. Cut the corners. Using a wire-cutting tool, make a 6-inch (15 cm) cut at each of the four corners of one of the lengths of hardware cloth. To the extent possible, cut through the tines as close as possible to a crossing wire, which leaves flexible ends that can later be folded over to secure the corners of the cage.

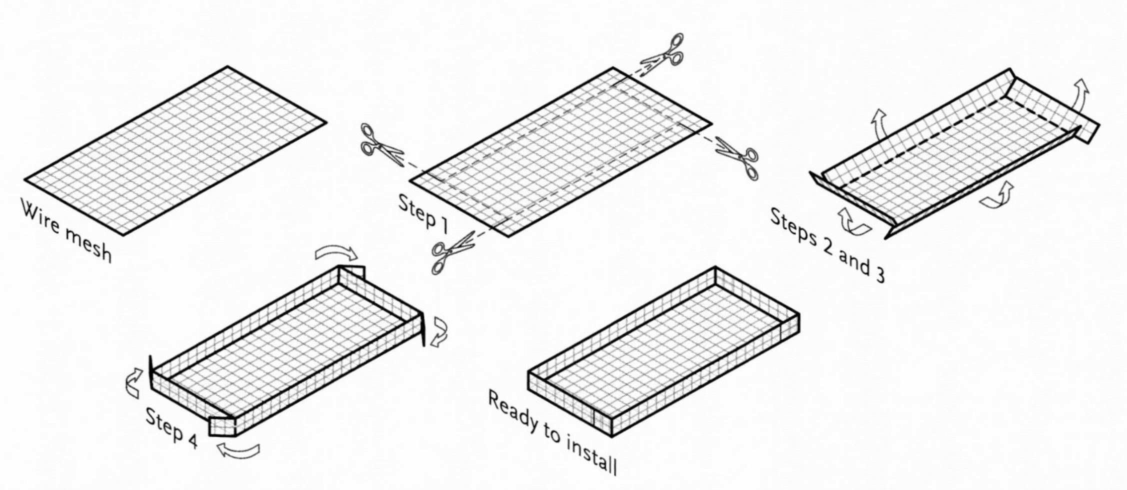

Figure 11.15. Constructing a hardware cloth cage, steps 1 through 4. Illustration by Turner Andrasz.

Step 2. Fold the mesh along one long edge. Place a 13-foot-long plank on edge on top of the hardware cloth, parallel to one of the long sides and 6 inches from the edge. Use a hammer to bend that side of the hardware cloth around the plank so that a 6-inch portion of the hardware cloth juts straight up.

Step 3. Fold along the other edges. Repeat this for the other three sides of the hardware cloth. This creates a "bed" of hardware cloth that is 4 feet (1.25 m) wide and 13 feet (4 m) long with 6-inch-tall sides.

Step 4. Wrap the "flaps" of hardware cloth at each of the four corners around the adjacent upright edge. Secure them in place by folding the tines at the end of each flap through and around the adjacent upright edge of hardware cloth.

Step 5. To install the cage, clear and level a site of the proper dimensions. Place the cage on (or in) the ground.

Step 6. To reinforce the cage if above ground, construct a wood frame around it by lining up the 13-foot planks and 4-foot 4-inch planks. Using the 3-inch screws, screw the planks together to secure them in place.

Step 7. Staple the wire cage to the inside of the wood frame.

Step 8. Sprinkle a thin layer of compost over the bottom of the wire cage and nestle crocus corms (or whatever you are planting) into it at the proper spacing.

Step 9. Cover the corms, bulbs, or crowns with more compost and top that with soil, from excavating the site if available.

Step 10. Top off the bed with a layer of semi-composted wood chips.

Step 11. Lay the second piece of hardware cloth over the bed. To secure it in place and thus rodent-proof the bed, use the 1½-inch screws to attach the 1 × 3 furring strips (placed over the hardware cloth) to the tops of the border planks on all four sides. If your cage is embedded in the ground, you can use heavy rocks to weigh down the hardware cloth cover so that rodents can't slip underneath it.

Figure 11.16. Saffron corms planted in a reinforced cage.

Figure 11.17. A saffron crocus emerging through a hardware cloth cover.

Figure 11.18. A single day's harvest of saffron threads from my garden.

bulbs called *corms*. And rodents of all stripes relish crocus corms of all kinds. In planning a bed of saffron crocuses, the first order of business is to protect against these pilferers. For small quantities of bulbs—around a dozen or so—I have deterred rodents by interplanting the corms with daffodil bulbs, which are poisonous and avoided by rodents. The daffodils emerge in early spring, flower, and die down and decay before the saffron crocuses emerge in autumn. For a larger quantity of saffron, or a pure stand of another plant whose bulbs rodents adore (such as tulips), you may wish to create a rodent-proof planting bed. To do so, I chose a technique I learned about at the workshop I attended—planting in an enclosure of hardware cloth to keep the intruders at bay. I describe how to make such an enclosure in "Creating a Rodent-Proof Planting Bed" on page 228.

Saffron crocus corms send out leaves in late September and flowers in October and continue their growing cycle through winter into spring. During much of October, new flowers emerge daily. The deep orange stigmas, or saffron threads, are best harvested each day in the early morning. When the threads are fully dried, they can be stored in airtight containers. The plants are dormant from June through mid-September. If growing conditions are adequate, they produce new corms each season. Lest they crowd each other out, after four to six years, the corms need to be dug up, separated, and planted in a new bed.

Maintenance tasks for saffron include keeping the planting weed-free and topping the bed with additional compost or wood chips during the dormant period. A 6-inch-thick (15 cm) leaf mulch over the bed will ensure the bulbs stay viable through the winter months.

Sweet William

Dianthus barbatus

EF

Native Range: Southern Europe and eastern Asia
Height: 12 to 24 inches (30–60 cm)
Spread: 6 to 12 inches (15–30 cm)
Soil Conditions: Well-drained soil
Sunlight: Full sun or partial shade
USDA Hardiness Zones: 3 to 9

When a gardener friend of mine was giving away a few flats of perennial flowers, including sweet William, I decided to snap them up once I learned that they are edible. Sweet William leaves form a dense mound that provides ground cover. The flowers, each about ½ inch (13 mm) in diameter, appear in dense clusters atop the stems. Hybrid varieties provide a range of bloom color: shades of white, pink, red, and variegated combinations. These decorative blooms can be added to salads and sandwiches and used as cake decorations. They also make long-lasting cut flowers.

I planted my sets along the lower edge of a well-drained, sloping, south-facing bed. They flowered in early summer, became dormant in mid-summer, then bloomed again later in the season

Figure 11.19. Showy pink sweet William flowers are edible and make great cut flowers.

when there was more precipitation and less intense sunlight. This same pattern occurred over the next few years. These plants are biennial or short-lived perennials and may live only two years, but can increase their longevity through self-seeding.

EF CF

Tulip

Tulipa species and hybrids

Native Range: Netherlands
Height: 9 to 24 inches (23–60 cm)
Spread: Up to 10 inches (25 cm)
Soil Conditions: Well-drained soil
Sunlight: Full sun or partial shade
USDA Hardiness Zones: 3 to 8

Who knew that tulips could be eaten? I was delighted when I discovered this fact while

Figure 11.20. The pastel colors of tulips brighten the spring landscape, and you can eat them!

researching edible flower choices for my garden. Tulip petals have a succulent texture and floral flavor. There are hundreds of shapes and colors to choose from. I prefer the peony-like blooms with multiple layers of petals because they are showier and they make longer-lasting cut flowers.

Tulips grow from bulbs planted in the fall and will regrow for a few years until the bulbs, which multiply underground, start to crowd each other out and need to be dug up, separated, and transplanted for renewed growth and vigor. Rather than dig them up, I just plant fresh bulbs in different locations. (To economize, I wait until mid-fall when stores discount their remaining bulbs by 50 percent or more.) While not particularly palatable to people, the bulbs are a favorite among rodents, who have devoured hundreds of those I planted over the years. (To protect tulips from rodents, see "Creating a Rodent-Proof Planting Bed," page 228). The blooms are favored by deer and rabbits, but I find that placing a floating row cover over the buds keeps these poachers at bay.

Violet and Pansy

EF

GC

Viola species and hybrids

Native Range: Temperate Northern Hemisphere
Height: 4 to 8 inches (10–20 cm)
Spread: 4 to 8 inches
Soil Conditions: Dry to moist soil, depending on species
Sunlight: Full sun to shade, depending on species
USDA Hardiness Zones: 3 to 9 for violets, 4 to 8 for pansies

Garden pansies (*Viola × wittrockiana*), Johnny jump-ups (*V. tricolor*), and other violets are all part of the genus *Viola*. Some are annuals and some are perennials, but all have edible leaves and flowers, can self-seed, and form protective ground covers. Violets are found in temperate climates all over the world, and I have spotted one kind of violet growing wild in the woods and pastures of our farm. The appearance of the flowers also varies, but all flowers in the genus have five petals that form a circular "face." In fact, the multicolored petals of pansies

Figure 11.21. Violets are a particular delight when sited in unexpected places where their seeds have spread, like the crevices between patio stones.

and Johnny jump-ups are configured in such a way that they resemble the eyes and whiskers of a cat. The flowers of many types appear in the spring, with a second blooming later in fall. Some types of pansies are more heat-tolerant and can bloom from spring well into the summer. All of the flowers are edible and can be used in desserts, beverages, syrups, liqueurs, and as garnishes. The leaves make a pleasant addition to salads. Violet leaves and flowers have medicinal uses as well.

Some violet varieties self-seed profusely, forming dense ground covers. In this regard, I have had good luck with small, lavender-flowered violets (*V. sororia*) on semi-shaded, north-facing slopes. The violets I planted early on in this habitat have returned every year since to completely cover the ground and produce lovely blooms. Whether direct-seeded or transplanted, these versatile, low-growing plants are an edible ground cover worthy of inclusion in your edible landscape, large or small.

White Clover

NF EF GC

Trifolium repens

Native Range: Europe and Central Asia
Height: Up to 6 inches (15 cm)
Spread: 18 inches (45 cm) and up
Soil Conditions: Tolerates moisture and clay soil
Sunlight: Tolerates shade
USDA Hardiness Zones: 3 to 10

White clover is planted by farmers and gardeners throughout the temperate world. It forms a dense mat, spreading rapidly by stolons. Its sweet white flowers attract bees—especially honeybees and bumblebees—and can be eaten by livestock and humans. Each leaf is composed of three round leaflets (a trefoil). If you are lucky, you might find the rare four-leaved clover. As a ground cover in your garden, this clover can supply sufficient usable nitrogen to meet the needs of surrounding plants.

Figure 11.22. White clover seeded over a wood-chipped path creates a low-maintenance walkway.

I seeded white clover in wide garden paths where it tolerates heavy foot traffic, some shade, and repeated mowing. A downside can be its tendency to spread aggressively. I had to hand-weed the edges of one stand several times to prevent it from outcompeting the immature plants in the bordering beds.

Wild Strawberry

GC

Fragaria virginiana

Native Range: US and southern Canada
Height: Up to 6 inches (15 cm)
Spread: Indefinite
Soil Conditions: Moist or dry soil
Sunlight: Full sun to shade
USDA Hardiness Zones: 3 to 10

As a child, I enjoyed foraging for wild strawberries while I wandered through the woods and fields near

Figure 11.23. Wild strawberries have occupied the surface of this sheet-mulched bed.

my home. Each of their upright stems holds several white flowers, which begin to produce small, flavorful strawberries over a two- to four-week period in June.

On our farm now, I find wild strawberries growing in pastures, in the understory of brush and woods, and in clearings. In my edible garden, I noticed these wild berries moving into the "clearings" I created by sheet mulching. My first reaction was to rip them out from the edge of a bed they were invading. This routine became tiresome as I realized that my efforts were being outcompeted by the numerous and rapidly spreading stolons of these wild plants. Eventually, I came to accept and even value these strawberries' expanding presence in my garden.

They quickly establish a dense, low-growing carpet of leaves, leaving most intentional plantings unharmed. Their flowers attract pollinators, and their fruits attract everyone who visits my garden during the month of June. In fall, their leaves turn vibrant tones of orange, red, and burgundy, covering the ground with festive color. On paths, they withstand a fair amount of foot traffic. Now, wherever I observe these wild plants entering a newly sheet-mulched bed, I feel grateful to them for performing all these useful functions with no effort on my part.

Instead of covering your beds with mulch alone, installing edible ground covers can bring you visual and culinary delight all season long. Whether as single specimens, massed plantings, or volunteer "invaders," ground cover plants can be a diverse, food-bearing addition to your garden landscape.

Figure 12.1. Climbing skyward, an akebia vine in bloom decorates a black locust trunk.

CHAPTER 12

VINES AND FUNGI

Vines occupy an important niche in the vertical landscape. Because they can climb up built structures and run over soilless surfaces without needing to root, they can occupy space that other plants can't and thus capture solar radiation that would otherwise be lost. They can be trained to grow up trees; ramble over sturdy fences, walls, and rooftops; and even hide eyesores like air-conditioning units. However, use caution when siting vines because many can grow to overtake other plants and even become invasive, especially in warmer hardiness zones. As you will see, some of the vines I describe below are easy to control, and some can be more problematic.

Vines provide nutritious food including edible fruits, shoots, leaves, roots, and flowers. Besides their food-producing value, vines can fix nitrogen, attract beneficials, and provide unique landscape appeal.

Packed with vitamins and minerals, low in calories, and relatively easy to grow, mushrooms are another desirable addition to a perennial garden. It took me a few years to get around to incorporating mushrooms in my edible forest because most require shade, and I didn't have any at first. It wasn't until I expanded the garden into the second ½ acre (0.2 ha), with its patches of natural shade, that I began experimenting with edible fungi. Then oyster, shiitake, and wine cap mushrooms all made their way into my garden. In this chapter, I provide instructions for cultivating these three species.

Akebia Vine

FR

Akebia quinata

Native Range: Japan
Height/Spread: Up to 40 feet (12 m)
Width: Up to 40 feet
Soil Conditions: Loamy, fertile soil
Sunlight: Prefers shade
USDA Hardiness Zones: 4 to 9

Akebia vines are also known as chocolate vines because of the chocolatey scent of their purple or white flowers. Rapid growers, they require a sturdy trellis and can reach 40 feet in height if not checked by annual pruning. Absent a trellis, they will rapidly form a partial ground cover. Akebia vines flower in spring and may bear a sausage-shaped edible fruit if

Figure 12.2. An immature akebia fruit.

a second variety is planted close by for pollination. Although I have not yet tasted one, I have seen the fruit's flesh described as both "insipid" and slightly sweet, with a subtle combination of pear, coconut, and melon flavors, while the seeds provide a hint of bitterness. The outer shell of the fruit is edible, too, and can be stuffed with meat or other ingredients and fried. The vine stems are used in traditional basketmaking. The plant's oval leaves, clustered in groups of five, are decorative in themselves growing up a trunk or trailing over the ground. The leaves remain green all winter if the season is a mild one. The woody stems, withstanding even bitter cold, sprout new leaves along their length each spring.

I planted two different varieties of akebia vines to grow up neighboring black locust trees. In their sixth year, when they both flowered, four fruits appeared on one of the vines.

Arctic Kiwi

Actinidia kolomikta

Native Range: Northeast Asia
Height/Spread: Up to 20 feet (6 m)
Soil Conditions: Moist soil
Sunlight: Partial shade
USDA Hardiness Zones: 3 to 8

Picture a kiwifruit, and the first thing to come to mind is the brown, furry-skinned, green-fleshed fruit sold in grocery stores. That kind of kiwi comes from a vine that grows well in warm, temperate climates but would never survive in my cold climate. There are, however, two cold-tolerant kiwi species that bear smaller, smooth-skinned fruit—hardy kiwis (*Actinidia arguta*) and arctic kiwis. Although hardy kiwis are supposedly hardy to Zone 4, I failed in my attempts to grow several varieties. My only success has been with arctic kiwi. Arctic kiwi is hardy to −40°F (−40°C). A vining deciduous perennial, it has heart-shaped leaves that can be variegated in shades of pink, white, and pale green. The vines bear female flowers and male flowers on separate vines. At least one male plant is needed per up to eight female vines for pollination. The inconspicuous flowers bloom in early summer, maturing to form 1-inch-long (2.5 cm), ½-inch-wide (13 mm) cylindrical fruits that ripen in August. These fruits can be eaten like grapes and have an intense sweet-tart punch. They are also delicious in fruit salads, or processed into jams, salsas, or desserts.

I train arctic kiwi vines on a northeast-facing fence, on wire stays supporting a telephone pole, and along the branches of a native tree. They benefit from frequent watering until their deep roots become established. I neglected to provide this care in my first attempts to grow these fruitful vines, with discouraging results. Another precaution I learned from ill-fated experience is that the first growth in spring is vulnerable to subfreezing temperatures, so it's best to avoid frost pockets when siting these vines. A final safeguard I neglected at first was to protect these plants from deer, rabbits, and goats, which can't resist gorging on the tasty leaves and stems. Once established, the vines require annual pruning for optimal fruit production.

Chinese Yam

Dioscorea polystachya

Native Range: China
Height/Spread: Up to 9 feet (2.75 m)
Width: Up to 4 feet (1.25 m)
Soil Conditions: Deep and moist soil
Sunlight: Full sun to light shade
USDA Hardiness Zones: 4 to 8

Chinese yam, also known as cinnamon vine and air potato, is a vining herbaceous perennial grown from an edible tuber. The first leaves exhibit a shiny reddish hue before turning green with maturity. Tiny, cinnamon-scented flowers bloom at each leaf node to become pea-sized edible bulbils (yam berries) along each vining stem. When ripe, these tiny tubers can

Figure 12.3. The heart-shaped leaves of the arctic kiwi vine shade several ripening fruits.

Figure 12.4. A Chinese yam winds itself along a banister, its bulbils almost ready to harvest.

be harvested by shaking the vine, or else they drop to the ground. They also can be used to propagate new plants. I like to add these to soups and stews, where they impart a mild, earthy, sometimes slightly bitter flavor. Underground, larger cylindrical tubers develop that you can dig up and eat raw or cooked.

I planted one tuber at the base of a dead tree stump in my edible forest's first year. In fall, all of the top growth dies back to the root. The plant has resumed growth every year but one, when a wet season prevented the soil from warming. I thought the plant was dead until it burst forth the next year during a hot, dry spring. Beware that in climates warmer than my Zone 4 garden, Chinese yam will not die back each year and can become hard to control and very invasive.

If you wish to harvest the underground tubers, be sure to plant these vines where you can access the root without disturbing other plants.

FR

Grape

Vitis species and cultivars

Native Range: Temperate Northern Hemisphere
Height/Spread: Dependent on trellis
Width: Best pruned back to 3 to 5 feet (1–1.5 m) in each direction on a horizontal trellis
Soil Conditions: Well-drained soil
Sunlight: Full sun
USDA Hardiness Zones: 3 to 9

Grapes have been cultivated throughout Europe, Asia, and North Africa for millennia. Species native to North America were eaten by Indigenous peoples and bred into what became Concord grapes, the staple ingredient of bottled purple grape juice and jelly in the United States. Until the 21st century, few grape varieties were hardy enough to produce reliable crops in the coldest hardiness zones in North America. Now, thanks to recent breeding programs (especially at the University of Minnesota), there are numerous cold-hardy cultivars available. In my garden, I focus on cultivating table grapes, which I consider a must-have in any edible landscape. I have also eaten the leaves of my grapevines stuffed with rice and herbs—a traditional Greek dish.

Grapes come in white, yellow, pink, green, red, and purplish blue colors, and can be seeded or seedless. There are varieties with intact skins and there are slip-skinned varieties: When you squeeze their skin, the flesh pops out! Some are bred for fresh eating and some for winemaking. Once you determine your preference among these characteristics, be sure to select plants that are hardy to your zone. I planted several fresh-eating varieties in the first year of my garden, all of which have survived over time. It typically takes four years for a grapevine to come into good production, and longer if problems such as winter-kill, girdling by rodents, or browsing by ruminants occur along the way. I have experienced all of these problems with my grapes, and it is too soon to tell which of the varieties I grow will produce the best in the long run. The good news is that once established, grapes are so resilient that they almost always bounce back from an assault.

At the time of this writing, King of the North is my favorite grape. A deep blue, seeded, slip-skinned grape related to the Concord, it has so far shown itself to be the most versatile, vigorous, reliable, productive, cold-hardy, disease-resistant, easy to propagate, and delicious among the cultivars I planted. A friend says the jam I make from these grapes is like "crack cocaine" because it's so addictive. The juice made from them is so rich it needs to be diluted five-fold with water to approximate a commercially produced product.

Choosing a site. Placement is an important consideration with grapevines. They do best in full sun, which promotes vigorous growth and full ripening of the fruit. If possible, give grapevines a southeast exposure so that they receive direct sun first thing in the morning. With this exposure,

Figure 12.5. Ripening King of the North grapes will be ready to harvest in September.

the vines warm up fast and morning dew evaporates readily, helping to protect against fungal diseases, to which they are prone. Ample air circulation is also a must to discourage fungi. I planted my first grapevines on a southeast-facing fence that happens to be parallel to our prevailing southwest winds, which provide excellent ventilation. There is an open area on each side of the fence—a driveway on one side and a large pond on the other—so there are no nearby trees or obstacles casting shade or blocking airflow. Despite several wet summers and falls, I have not observed any fungal problems in this location. In an unintended comparison, I planted one of the same grape varieties on a north-facing fence behind our farmhouse, where early-morning sunlight and wind are both partially blocked by surrounding trees and structures. During wet years, the grapes sited in this unfavorable habitat became inedible due to a late-season fungal infection, while the ones in the favorable environment remained disease-free.

Providing support. Grapes can live from 50 to over 100 years, developing thick trunks and vines over time that become very heavy when laden with a season's growth of canes, leaves, and a ripening crop of grapes. One mature vine can produce 40 pounds (18 kg) of fruit a season. To provide support for them in the long run, you need a strong, sturdy trellising system. This can be a fence, a solid arbor or trellis, or an established tree with robust lower branches. If you have an existing structure that you think will work, go for it. Research conducted by Cornell University in a vineyard near my farm demonstrated that in regions with cold winters and prone to hard frosts in late spring and early fall, grapes that are trellised higher above the ground incur less damage to their vines, leaves, and fruit.* Keep this in mind as you plan your trellis if your habitat is subject to these cold stressors.

Planting. Grapevines can be sourced as bare-root seedlings or in pots, or you can propagate them yourself from dormant cuttings. Bare-root seedling stock is best planted as soon as the ground can be worked in early spring. Potted grapes, if leafed out, can be planted as soon as the ground warms and danger of frost has passed, or anytime thereafter. Once their roots are developed, cuttings are best planted in early spring while still dormant; alternatively, once leaves have emerged, wait until the weather has warmed in late spring to plant them outdoors. As with a newly planted tree, place a trunk guard around the stem of each grape plant to deter rodents from girdling it.

Spacing. Allow plenty of space for a vine to spread, planting no less than 6 feet (2 m) from another grapevine. I planted several different varieties spaced 8 feet (2.5 m) apart along a sturdy fence topped by a high tensile wire. In another location, I installed a single vine to scramble over the top of a tall, 4-foot-wide (1.25 m) trellis. Another runs along a banister above a pond. Three vines, spaced 6 feet apart, are planted at the base of a standing native Seckel pear tree whose lower branches they can traverse. All of these locations face southeast, providing that important morning sunlight.

Pruning. For optimal fruit production and control of the vines, grapes require annual pruning. Before I planted grapevines in my forest garden, I had no experience growing them. Pruning instructions written by experts seemed complex and perplexing. Luckily, I am acquainted with two individuals who were willing to help: a nursery owner who specializes in propagating grapevines and a local vintner with a large vineyard. I sought guidance from both of them, observing their pruning techniques and asking questions. And every year, I still come up with

* Phillip Randazzo (owner of Coyote Moon Vineyards), in conversation with the author, June 2018.

Training and Pruning Pointers

As I've said, I am *not* a pruning expert, but here are a few pointers specific to training grapes on a horizontal trellis in my Zone 4 region.

- Train not just one, but two main vertical stems, which will become permanent trunks, up to the trellising height. With two trunks, you increase the odds that one will survive a brutal winter or the chomping of hungry rodents.
- Once a stem grows above the height of the trellis, prune it just above a bud at that height. This top bud will produce a cane that will become one of the cordons—or permanent arms—that you will train to grow horizontally on top of the trellis. Train one cordon from each trunk to grow in opposite directions. Allow side shoots (canes) to sprout and grow from these two cordons.
- Grapevine buds break their dormancy beginning with the bud farthest from the cordon, then each successive bud gradually begins to develop. If a late frost kills buds that have already produced tender growth, the less advanced buds closer to the cordon may still be viable. Because of this, during the next dormant season, trim each side shoot back so it has four or five buds. When all danger of frost is past, then it is safe to trim each cane back to two buds, which is the right number to allow for optimal fruit production.

new questions! If you are as inexperienced as I was and still consider myself to be, I recommend you find an excellent pruning reference book, a comprehensive YouTube video, or (best choice) an experienced mentor to guide you.

Protecting the harvest. Be prepared to net your grapes before they ripen to protect them from birds. You can tell if they are ripe enough to pick by tasting them, or by purchasing an inexpensive tool called a brix meter that measures the brix, or sugar content, of the grape juice. Table grapes are best harvested when their brix reading is between 17 and 19; wine grapes are best between 24 and 26. Use a clipper or scissors to snip each bunch from the vine, or just remove each bunch by hand. Sometimes the grape clusters on a single vine do not ripen at the same time. In this case, you can use the ripest for fresh eating and the rest for processing into jam, juice, and pies, or wait until each bunch is perfectly ripe to pick it. Grapes can be kept fresh in the fridge for up to two weeks.

Groundnut

Apios americana

Native Range: Southern Canada to Florida and west to Colorado
Height/Spread: 10 feet (3 m) or more
Width: 3 to 4 feet (1–1.25 m)
Soil Conditions: Moist soil
Sunlight: Full sun to considerable shade
USDA Hardiness Zones: 3 to 10

Groundnut is a nitrogen-fixing herbaceous perennial vine that dies back to the ground each fall to emerge again in spring in my region. Oval leaves with pointed tips branch out from its vining stem in arrangements of six. In summer, lovely pinkish

Figure 12.6. A blooming groundnut vine (*top*) and an American hog peanut vine (*bottom*) spreading in a shady spot.

purple flowers appear at each juncture where leaves meet the stem. Groundnut gets its name from small, 1- to 2-inch (2.5–5 cm) round or oval tubers that form along rhizomes, which can run for up to 3 feet underground. The tubers look like dark brown beads arranged along a string and are best harvested once the vine goes dormant in the fall. These starchy "spuds" can be prepared like potatoes, but contain three times as much protein on a per-dry-weight basis. They are a traditional staple food of most Native American tribes that resided within the plant's natural range. The vine is easy to propagate by simply digging up the tubers and planting them elsewhere. If you plan to harvest tubers, plant these vines in a location where digging around their base will not disturb other plants' roots.

Another native tuber-producing vine called American hog peanut (*Amphicarpaea bracteata*) is distinguished from the groundnut by its less-showy flowers and spade-shaped leaves that grow in groups of three. These grow wild in moist, shaded locations on my farm and have appeared as volunteers in my garden at the base of the north-facing slopes of some of my hügelkultur mounds.

Figure 12.7. This Nugget hops vine, loaded with flowers, grows happily up a wire stay that anchors a telephone pole.

Hops

Humulus lupulus

Native Range: North America and Europe
Height/Spread: Up to 20 feet (6 m)
Soil Conditions: Adaptable
Sunlight: Full sun
USDA Hardiness Zones: 3 to 8

Hops are herbaceous perennial vines that regrow from their roots each season. The light green, cone-shaped flowers are used in beer brewing to add flavor and serve as a preservative. Hops have been cultivated for this purpose for hundreds of years, beginning in Germany. The flowers can also be used in cooking and have a history of use as a sedative to reduce anxiety, restlessness, and insomnia. They are ripe when a yellow powder develops under the partially dried petals. Young shoots are edible, too: They have a slightly bitter flavor, but they taste good to me lightly steamed with oyster sauce.

Hops plants can be sourced in multiple varieties, each with different properties of interest to beer connoisseurs. I planted one vine of a cultivar called Nugget. At first, I placed it at the base of a telephone pole, expecting it to wind around and up the pole. It was soon apparent that this vine was not happy with the thick, vertical pole as a trellis, refusing to scale it. Early the next season, I transplanted it to the base of a wire stay that extended up at an angle, connecting to the pole several yards above the ground. The thinner support was much more to the hops's liking. Each year since, it has climbed 20 feet (6 m) up the wire, producing a bountiful crop of flowers.

Unfortunately, this is not the only place my hops vine is growing. It suckers aggressively in all directions through the loosely composted soil in the hügelkultur mound just adjacent to the stay. In order to keep the numerous shoots from overtaking the mound, I have to pull them out by hand three or four times a season. If I had known how aggressively this plant suckers, I would have sited it in undisturbed heavy clay where the dense soil would inhibit its spread, or I would have built in a barrier to stop its movement underground. The good news is that once harvested, these invasive shoots are food for me and for many of the animals in our farm's menagerie.

NF EF CF

Hyacinth Bean

Lablab purpureus

Native Range: Africa
Height/Spread: 10 to 15 feet (3–4.5 m)
Width: 3 to 6 feet (1–2 m)
Soil Conditions: Adaptable, but requires well-drained soil
Sunlight: Full sun to light shade
USDA Hardiness Zones: 10 to 11

Although it is a tender annual rather than a perennial, the nitrogen-fixing hyacinth bean is so useful and beautiful to behold that I couldn't resist including it. Its stems and the undersides of its heart-shaped leaves are a deep purple. Pea-like magenta flowers bloom throughout the summer. The resulting upright flat, purple pods are decorative in themselves. All parts of the plant, including its leaves and roots, are edible when properly prepared. Hyacinth bean pods and mature seeds, which must be boiled repeatedly to remove toxins, are staple foods in many parts of the world. I eat the pea-flavored flowers as well as the deep purple pods, which have a flavor stronger than but similar to the snow peas sold in stores. I have also included the flowers and purple pods in bouquets. Easily propagated from seed once the soil warms, this plant even self-seeded in my garden one season, siting itself near thick limestone slabs that created a favorable microclimate. If you reside in a warmer hardiness zone, this plant may reliably self-seed for you.

PV B

Magnolia Vine

Schisandra chinensis

Native Range: Eastern Russia and China
Height/Spread: Up to 20 feet (6 m)
Soil Conditions: Adaptable
Sunlight: Shade-tolerant
USDA Hardiness Zones: 4 to 9

Magnolia vine, also known as schisandra, is a woody vining perennial with deep-veined, oval leaves. This plant interested me when I first read

Figure 12.8. The decorative hyacinth bean vine is a beautiful addition to any landscape.

about it because of its ability to grow in shade, its attractive, sweet-smelling, magnolia-like flowers, and its 6-inch-long (15 cm) grape-like clusters of bright red nutritious edible berries, which also have medicinal applications. The leaves can be prepared like tea, or lightly steamed along with the stems and eaten as a vegetable.

I can't attest to any of these characteristics personally because these plants have failed to survive under my watch until very recently. I first planted two vines (to ensure pollination) in a naturally shaded spot with very moist, often saturated soil. They struggled to grow and sooner or later succumbed to the inhospitable soil conditions. A couple of years later I tried again, planting two in a better-drained shaded spot along a border hedge with a native tree intended as a trellis. One failed to leaf out. The survivor didn't grow much that season but was still alive the following spring, at which point I planted a replacement companion magnolia vine close by. I am hopeful that both will survive, thrive, and eventually flower and bear fruit.

Perennial Sweet Pea

Lathyrus latifolius

Native Range: Southern Europe and northern Africa
Height/Spread: Up to 10 feet (3 m)
Width: Up to 3 feet (1 m)
Soil Conditions: Adaptable
Sunlight: Full sun or partial shade
USDA Hardiness Zones: 3 to 8

This perennial nitrogen-fixing vine, also known as the everlasting pea, is similar in appearance to annual sweet peas (*Lathyrus odoratus*) and vegetable

Figure 12.9. A recently planted magnolia vine seedling.

Figure 12.10. Perennial sweet pea flowers make lovely additions to bouquets.

garden peas (*Pisum sativum*). It climbs by wrapping its tendrils around another plant or structure and sports flowers in colors ranging from white to pink to lavender. Where these vines find nothing to climb up, they form a dense ground cover instead. The flowers, pods, and seeds are edible but can be toxic to humans in large quantities so should be consumed with caution. This plant is not frost-sensitive; thus, the bright green foliage and pastel flowers bloom from midsummer through fall, adding a delightful and enduring accent to the garden.

I planted these decorative vines at the base of the same fence where my grapes are trellised along its highest wire. When the vigorous pea vines begin to infringe on the grapes, I cut them down to use as nitrogen-rich mulch or a nutritious treat for our goats. Easily propagated from seed, these vines have also self-sown on their own in my garden. In several locations, sweet pea "volunteers" provide a serviceable ground cover, saving me the trouble of planting one. When they choose a location where they could infringe on other plants, I've learned it's best to weed them out at the seedling stage. If you wait until the plants are mature, their deep and extensive roots resist efforts to remove them.

Figure 12.11. Wisteria trained to climb a seckel pear tree.

Wisteria

Wisteria species

Native Range: China, Japan, Korea, and eastern US, depending on species
Height/Spread: 80 feet (25 m) and taller
Width: 20 feet (6 m) and wider
Soil Conditions: Adaptable
Sunlight: Full sun or partial shade, but flowers best in full sun
USDA Hardiness Zones: 4 to 8

While the seeds and possibly all other parts of this woody climbing vine are toxic to humans, wisteria's pendulous bunches of lavender flowers and nitrogen-fixing ability make it an attractive and beneficial addition to an edible garden. There are several species of wisteria, one of which is native to the eastern United States. I was unaware of this when I purchased my Chinese wisteria plants (*Wisteria sinensis*). Non-native wisterias grow more aggressively than the native species and have become invasive in warmer zones. If I had known this, I would have sought out the native species, and I recommend you do so as well if you plan to install this vine.

All species of wisteria can become massive, heavy vines that require strong support. I planted some along a heavy-duty garden fence and trained several to grow up native trees. These vines can grow so rampantly that they will overtake small trees and strangle large ones. As mine grow, I am pruning them back to control this tendency. None of my vines have flowered to date, and some of the woody stems have died back during cold winters. However, all of the plants survived, and I am hopeful that I'll see them flower sometime soon.

Oyster and Shiitake Mushrooms

Pleurotus ostreatus and *Lentinula edodes*

Oyster and shiitake mushrooms are relatively easy to grow on hardwood totems or stumps. Oysters have a smooth, medium brown surface and tend to grow in multilayered clumps that can weigh up to 5 pounds (2.3 kg). Shiitake have a cratered crown with white cracks cross-hatching a darker brown surface and tend to fruit as individual stemmed mushrooms protruding from the sides of the wood substrate.

I came across descriptions of two relatively simple methods for growing these mushrooms in *Farming the Woods* by Ken Mudge and Steve Gabriel. They are called the stump method and the totem method, and the best time to undertake either is in early spring.

Sawdust spawn is a mixture of mushroom spores with moistened sawdust, sold in bricks of varying weights. I procured spawn from Field & Forest Products, which offers a number of certified organic strains of both shiitake and oyster mushrooms. I bought two 5½-pound (2.5 kg) bricks of each variety of mushroom I wanted to grow. One of these bricks is sufficient to seed about seven log segments, as described below.

A limited number of tree species provide a suitable substrate for growing mushrooms, and not every suitable tree may be accessible. My research showed that among the species on my property, oak and ironwood are both suitable for shiitake. Elm and ironwood can host oysters.

Caution: If you use a chain saw for this project, be sure you understand how to do so safely and wear proper protective equipment.

The Totem Method

In this method, fresh logs are cut in 2-foot (60 cm) lengths and prepared as described below.

MATERIALS

Logs 6 to 12 inches (15–30 cm) in diameter
Sawdust spawn
Hand saw or chain saw
Bucket/bin
Flashing tape
Nails or screws
Hammer or screw gun
Shovel
Large paper bags
Twine

PROCESS

1. Cut down one or more live trees or cut off suitably thick live branches to serve as totems. Using a hand saw or chain saw, cut the logs into 2-foot (60 cm) sections, making sure to note the direction in which the tree was growing. After preparing the sections, leave them to season for a minimum of three but no more than six weeks before you inoculate them with spawn. If you leave them for longer, you run the risk of native fungi taking up residence and competing with your intended mushrooms.
2. Cut each log segment into three unequal sections: one 14-inch-tall (35 cm) base (the

bottom section), one 8-inch (20 cm) middle section, and a 2-inch (5 cm) cap.

3. Put a brick of spawn in a bin or bucket and crumble it between your fingers until it is broken into a fine mash.
4. Stand the base segment on the ground, oriented in the same direction it was growing as part of the tree. Correct orientation is important so that the base will naturally siphon water up from the ground, keeping the spawn moist without any maintenance on your part. Spread a ¼- to ⅜-inch (6–10 mm) layer of sawdust spawn over the top cut surface of the base, pressing it down to smooth and compact it.

Figure 12.12. Here, the first spawn layer is complete and enclosed with tape. The second layer of spawn has just been applied to the top of the midsection.

5. Place the 8-inch midsection piece, in its original orientation, on top of the layer of spawn and nail or screw it in place. To accomplish this, hold the nail or screw about an inch (2.5 cm) above the bottom edge of the midsection piece, angling it downward. Use the hammer or screw gun to drive the nail or screw into the wood. Repeat this step three more times, spacing the nails or screws equally around the circumference of the log segments. This step will take practice to master, especially if the wood is dense, so don't be discouraged if it takes a few tries to get it right.
6. Seal the seam between the two pieces of wood with flashing tape to hold in the spawn, preserve its moisture, and protect it from predation by slugs.
7. Apply a layer of spawn to the new exposed top cut surface in the same manner as above.
8. Hammer or screw on the 2-inch-thick "cap," angling the nail or screw at about 45 degrees to the surface to avoid splitting the wood.
9. Tape around the second seam. This is the completed "totem."
10. Using a shovel, dig a hole in a wet area until you reach below the water table and the hole begins to fill with water. Unfortunately, the first time I tried this method I didn't dig deep enough because there was standing water aboveground at the site, and I didn't think I needed to dig deeper. During a subsequent drought year, I had no harvest because the surface water dried up. Be sure to position your totems deep enough

Figure 12.13. This curved row of mushroom totems, placed on the shaded edge of a "swamp," provides nutritious and tasty food while adding function and aesthetic interest as the border of an access route.

so they can access moisture throughout the year, no matter the conditions.

11. Insert the bottom of the totem into the hole and fill in around it with the soil you removed.
12. Cover the totem with a brown paper bag to provide shade. Secure the bag with twine.
13. Make a record of the kind of wood you used and the variety of spawn for future reference.
14. Repeat the above process until all the sawdust spawn is used.
15. Check under the paper bag periodically. When the outside of the log or top of the cap begin to turn white, that's a sign that fungus is growing into the wood. Once this happens, remove the tape and the paper bags, as long as the totems are in natural shade. Then wait. It can take as little as three months for oysters or as long as two years for shiitake to begin to "fruit," or grow mushrooms from the totems. When mushrooms appear, wait until an edge of one mushroom begins to turn upward. This is the ideal time to harvest, before the fruits begin to lose quality. Using a knife, slice them off flush with the log.
16. Check your totems frequently, especially after a rainy period, to see if they are beginning to fruit. The frequency and timing of fruiting depends on the variety of spawn and the weather conditions. On several occasions when I didn't make frequent checks, I discovered mushrooms that had already gone beyond their prime to harvest.

The Stump Method

I think the stump method is even easier than the totem method, and I've had the most success with it. You can use this method if you have a mature tree of a suitable variety that you can spare or want to eliminate. I used an elm tree that I could spare.

Gather the same materials and tools as listed for "The Totem Method" and proceed as follows.

1. Using a chain saw, cut down the tree, leaving a stump 4 to 6 feet (1.25–2 m) tall. This living stump, with its roots intact, will naturally pull sufficient moisture from the ground to keep the spawn well hydrated. You may need to trim new branches as they grow from the base.
2. Use a hand saw or chain saw to cut wedges in the side of the stump and to fashion a cap. First, cut a 2-inch-thick (5 cm) section off the top of the trunk to serve as the cap. Next, 8 inches (20 cm) below that cut, make one horizontal cut a third of the way into the trunk. Then make an angled cut starting 3 to 5 inches (8–13 cm) above the first one. Cut on a downward angle so that the two cuts meet one-third of the way into the trunk, forming a wedge shaped like an orange slice. You should be able to remove this wedge by sliding it out of the trunk.

Figure 12.14. A flush of oyster mushroom emerges from the wedge cut from this live elm stump. This stump yields a reliable harvest of 5 pounds (2.3 kg) or more two times each year.

3. Cut one, two, or three additional similar wedges, spaced equally from top to bottom and spiraling around the trunk. Note which wedge is from which location.
4. Apply spawn on the bottom surface of each wedge opening, to the top surface of each removed wedge, and to the top surface of the stump. All of the spawn layers should be ¼ to ⅜ inch (6–10 mm). Press the spawn to firm it in place.
5. Insert the wedges back into the trunk and replace the cap. Hammer or screw in the wedges and the cap to the stump. One nail or screw angled downward at about 45 degrees in the middle of each wedge and one at a 45-degree angle in the middle of the cap should do it.
6. As you nail or screw each piece back on the stump, tape up all of the cracks along the edges with the flashing tape.
7. Secure a paper lawn bag over the stump so it covers all the wedges, and secure the bag with twine.
8. When you see the top of the stump turn white, you can remove the tape. Begin checking for oyster mushrooms around three months after inoculation; for shiitake, wait a year or longer. When mushrooms are ready to harvest, slice them off the trunk with a knife. Be sure to check for mushrooms along the area where the stump meets the ground, as well as along each crack. One drought year I discovered a huge clump of oyster mushrooms, partially hidden by grassy sod, protruding out of the stump at its base.

My stump, inoculated with oyster spawn, began fruiting that first growing season. It has continued to fruit each spring and fall for the past five years. The stump method is the simplest and most productive I've found for oysters or shiitake. I regret that I had only one tree to try it on.

Wine Cap Mushrooms

Stropharia rugosoannulata

In contrast with oyster and shiitake mushrooms, wine cap mushrooms grow vertically out of their substrate, appearing like multisized reddish brown toadstools popping up from the ground. Harvest them by pulling them out by the stem, which is narrow at the top and bulbous at the bottom.

Wine cap mushrooms are probably the easiest to grow in a garden setting, and unlike most others, they can even be grown in full sun. You can obtain sawdust spawn and crumble it as described for oyster and shiitake mushrooms above. You can use clean straw (not hay), wood chips, or a combination of the two as a substrate. Coniferous and oak chips need to be aged several months before use. Other types can be used fresh. These materials can be applied to the surface of an established garden bed or underlaid by cardboard over sod. Here are the basic steps for cultivating wine caps:

Step 1. Choose a sunny or semi-shaded location. This could be underneath established trees or alongside a garden path.

Step 2. Apply a layer of cardboard if the area is sod rather than garden soil.

Step 3. Apply the substrate atop the cardboard or bare soil: straw about 6 to 8 inches (15–20 cm) deep, wood chips about 3 to 4 inches (8–10 cm) deep.

Step 4. Water the substrate until it is thoroughly moistened. As you work, occasionally, stick your fingers or hand through to the bottom of the substrate to be sure the moisture has reached it.

Step 5. Sprinkle the crumbled sawdust spawn evenly over the substrate and mix it in using a garden fork for straw or a rake for wood chips.

Step 6. Keep the area evenly moist until the mushrooms emerge. You may need to water daily in a sunny spot in midsummer absent natural precipitation, but less frequently in shade or during cooler or rainier times of the year.

Figure 12.15. The reddish brown caps of the wine cap mushrooms can be hard to spot against underlying woodchip mulch.

Wine caps should begin emerging within 8 weeks from a straw substrate, or within 12 weeks if you're using wood chips. While the wood chip substrate takes longer, it also produces mushrooms over a longer period of time. Inspect your bed regularly, looking carefully so you don't miss any mushrooms before they are nibbled away by wild creatures.

The trick with this method is proper maintenance. The first time I tried cultivating wine caps, frequent rains followed the installation. The moisture was maintained naturally, and I had a good flush of mushrooms several weeks later. My second attempt was just before a dry stretch of weather. I became busy with other chores and neglected to water the patch. I never saw a harvest from this inoculation, probably because it died of thirst.

I hope this chapter has inspired you to use an existing tree or structure in your yard to fill some of the vertical space with an edible vine. Perhaps you are also considering devoting a shady corner to the cultivation of some tasty mushrooms. In either case, you will be increasing the diversity of your planting (and your diet) while enhancing the aesthetics of your landscape.

PART 3

CREATING PLANT GROUPINGS

Figure 13.1. An example of herbaceous succession: Daylilies planted below a row of grapevines were joined over time by self-seeded borage, perennial sweet pea, anise hyssop, and wild Queen Anne's lace.

CHAPTER 13

PLANT GROUPING BASICS

Deciding what plants to group together in a particular spot is one of the most engaging, satisfying, and occasionally frustrating enterprises for forest gardeners—both novices and those with experience. Those of you who have cultivated perennial flower gardens will appreciate the challenges of arranging long-lived plants in a group in such a way that they work well together over time. In a forest garden, your goal is to cover the ground completely while also including plants for each layer to fully occupy the vertical space available. Placing plants with diverse functions in close proximity to one another fosters sustainability. For example, in addition to food producers, a selection of plants that attract beneficials, concentrate nutrients, and deter pests can form a mutually supportive group in which the needs of each plant are met by the others. In designing your arrangements, plant characteristics such as growth habit, type of roots, root depth, and light and moisture needs are all important considerations.

Making groupings of edible plants is an undertaking that has few tried-and-true models, so it's good to take an experimental attitude when planning groupings for a forest garden. Every situation is different, and what works for one gardener may not work for another. And because the plants are perennial, changing in height and width over the years, the results may take time to reveal themselves. It is likely that some plants will do well (or too well) in their assigned placements while others will falter. In this chapter and the one that follows, I describe some of the plant groupings I have installed in my garden in sunny, shaded, wet, and dry habitats and how they evolved over time. It's my hope that the successes and failures I share will spark creative solutions for plant groupings in your own garden design.

Succession Is Inevitable

In nature, ecological succession is the way the types of plants occupying an area evolve over time following a disturbance, such as fire, flood, or blowdown, that clears the ground. For example, a common succession sequence in temperate climates is annual ground cover to perennial herbaceous plants to woody shrubs to short-lived trees to a more permanent hardwood forest.

In the early 1800s when our farm was established, it was a significant human-caused disturbance when settlers clear-cut the existing hardwood forest to make way for pasture to graze dairy cattle. Farmers maintained a pasture of perennial herbaceous plants for over two centuries. In the mid-20th century, the dairy went out of business. The fields on the flatter land continued to be hayed, maintaining the pasture, while the hillsides were left uncut.

When we bought the land in 2005, the farming operation had been on pause for a decade or two. The pastures were occupied by large, bushy thickets with elm and ash trees growing up among the scrub. Where the hillsides were left fallow for 50 years, young stands of hardwood trees, including birch, ironwood, and ash, had developed. My decision to turn some of this land into an edible forest garden created a new kind of disturbance, as I cleared brush and cut down some trees.

Succession will take place in a forest garden as it develops, too. As gardeners, we can try to anticipate how successional changes will unfold, but in my experience some of the changes can't be predicted; we have to observe and learn as the garden evolves. For example, it is reasonable to expect that as a multilayered garden develops and grows, more shade will be cast underneath and to the west, north, and east of trees and shrubs as they enlarge. If you place plants that require full sun in these spots that will later be shaded, the plants will struggle, if not die and disappear entirely. Instead, you can anticipate the change and place shade-tolerant plants in these places right from the start. Or, if the sun lovers don't survive, you can repopulate these shaded spaces with new, shade-tolerant specimens. Keep in mind that nature abhors a vacuum. If you don't install new plants in the spots where the sun-loving plants are dying out, nature will step in and "plant" some species suited to the habitat, filling the gap. That might be a desirable self-seeding edible you've planted elsewhere in the garden like borage, a lovely native wildflower like black-eyed Susan, or a noxious invader like Virginia creeper or poison ivy!

Sunlight and shade patterns are predictable phenomena, based on the known trajectory of the sun and the foreseeable ways in which objects blocking its rays will cast shade. Predicting how different varieties of plants grouped together will perform over time is more complex, especially if you have no previous experience growing the plants in the group.

This was my case. I had spent most of my life growing annual vegetables and some annual flowers. Thus, my gardens were perpetually in a beginning state of succession—they didn't evolve. I had little experience with long-lived herbaceous or woody perennials where succession is inevitable. If you have grown perennial flowers in your garden, you already have more experience than I did at this stage. My learning curve in this regard continues to be steep. Some plant groupings I designed are stable and have endured as intended. Others evolved in unanticipated ways. I find it fascinating to observe how plants interact over time, adding to my knowledge about combinations that persist and those that do not. Adapting to the patterns of succession that plants demonstrate is a creative and joyful aspect of managing an edible forest garden. As you embark on this endeavor, I urge you to make your best guess about what combinations will work together. Then watch what happens and use the acquired knowledge to improve your future choices.

How to Make a Plant Grouping

Each type of tree, shrub, or herb, and even each individual species, has a certain habitat preference: the conditions of soil, light, moisture, temperature, climate, and airflow most conducive to its healthy growth. Determining where to place each plant in a forest garden so that it will be most happy can be a challenge. As you develop planting plans, the work you have done identifying the microclimates in your plot will be most helpful.

How do I decide what to include in a particular plant grouping? It varies. As a general rule it is important to plan the tallest layers first because the location of these plants will determine the placement of others below. Sometimes I begin with a single fruit tree and build in shrubs and ground cover plants around it. At other times I have made a list of plants suitable for a specific habitat, such as shady and moist or sunny and dry, and designed an arrangement of plants selected from that list. I have installed a row of trees or bushes, embellishing them a year or two later with supportive plants. In many situations, my vision for a grouping evolves with time, and I enlarge a bed to incorporate the plants in my expanded vision. In all of these cases I concentrate on matching the individual plants with the best possible habitat while considering how each will coordinate with other plants in the group in terms of spacing vertically, horizontally, and over time. In all of these approaches, there is a good deal of trial and error as I observe how the grouping develops and make adjustments in response over time.

Undisturbed soil, covered with living plants at all times, is the cornerstone of an edible perennial garden. At first, the task of covering the 40,000 square feet (3,700 m²) of ground in my garden with edible ground cover plants seemed daunting, if not insurmountable. And as things turned out, for the first few years I was preoccupied with completing the woody layers of my initial forest garden and then planning and planting the garden expansion. For the most part, I did not include edible ground covers in my original plans. Rather, I kept the ground covered with frequently replenished quantities of mulch until I was ready to install herbaceous plants and ground covers.

There are three ways to address the ground cover layer in a forest garden planting. If you have a small plot, you can plant all the layers, including the ground cover, at once. In a larger plot, you can do the same thing in sections: Sheet-mulch or otherwise prepare one manageable area at a time, plant that area completely, and then proceed to the next section. Depending on the size of your plot, it may take you several weeks, months, or even years to execute your entire garden plan.

I chose a third approach. I first planned and planted the entire overstory and understory layers of my forest garden. Next, I incorporated the shrub layer, and finally I added in the herbaceous and ground cover layers. As I write this manuscript, I am still completing the final step by creating planting beds around existing trees and shrubs. By sheet mulching one or more new sections each year, and incorporating ground covers there during the following growing seasons, I am covering the ground with an ever-expanding blanket of intended plants. What once seemed like a Herculean task is made possible by breaking the work down into manageable parts.

Grouping Plants for Sustainability

I want to emphasize yet again how important it is to maximize diversity as you decide what to plant, in terms of both types of plants and the vertical layers they occupy. Mixing plants of different heights and types makes it more difficult for pests and diseases to locate their hosts and spread widely. Greenery in all vertical layers offers beneficial animals and insects the maximum diversity of habitats in which to live, breed, and forage for food. Diversity also ensures a better balance in the soil ecosystem, as different species of plants take up nutrients in varying amounts and proportions. Mixing together plants with different types of root structures maximizes the presence of living roots from the topsoil down into the subsoil, while minimizing root competition among plants. As you plan your groupings, consider the root structure of the plants you want to incorporate and try to avoid clustering together plants whose roots might compete for space, water, or minerals in the same layer of soil.

As mentioned in earlier chapters, there are several categories of plants that contribute to a garden's diversity while providing essential self-sustaining services, including nitrogen fixation and accumulation of other nutrients, attracting beneficials, and deterring pests. Even though some of

Figure 13.2. Multiple layers and types of plants increase the garden's resilience and self-sustainability while creating a landscape that is lush and inviting for humans.

these plants aren't food plants for humans, they are important components of an edible forest garden that you will want to include in your design.

NUTRIENT ACCUMULATORS AND NITROGEN FIXERS

When you integrate a diversity of nutrient accumulators and nitrogen-fixing plants throughout your garden, your planting will contain all the elements to sustain itself with no further effort. If your garden plot is small, one comfrey or sorrel plant may do the job. If it's larger, a scattering of other nutrient scavengers can meet your needs. Even "weeds" like dandelions contribute multiple minerals.

The nutrient accumulator par excellence is comfrey (see "Russian Comfrey" on page 206).

Many ground cover plants, including chives, horsetails, sorrel, strawberries, and violets, are nutrient accumulators. In addition to silicon, horsetails accumulate calcium, copper, iron, and magnesium. Sorrel accumulates potassium, phosphorus, calcium, iron, and sodium. Basswood trees concentrate phosphorus, calcium, and magnesium. When you consume their leaves, your body benefits from these minerals, and as the foliage decays so do the surrounding plants. Appendix 1 provides a comprehensive list of nutrient accumulators.

Regarding nitrogen needs, how do you determine how much nitrogen a non-nitrogen-fixing, food-producing plant requires? That depends on the demand. Nut trees such as walnuts and fruit trees such as apples or peaches require more nitrogen than plants that produce smaller fruit, such as cherry trees. Berry bushes require a moderate amount. The need for nitrogen is lowest in leafy crops or crops grown to supply wood products like garden stakes or firewood.

The goal with an edible forest garden is to build in enough nitrogen-fixing plants to fulfill the needs of the adjacent plants as they grow to maturity. In *Creating a Forest Garden*, Martin Crawford suggests that if the mature canopy area of nitrogen fixers is between 25 and 40 percent of the total canopy area of a forest garden, the need for nitrogen will be met. If most of the plants in your design are fruit- and nut-producing trees, aim for the higher percentage.

For example, let's say your plan calls for one pear tree and two plum trees, each with a mature canopy diameter of 20 feet (6 m). The formula to calculate the area (A) of a circle (assume a tree's canopy is roughly circular) is the old familiar $A = \pi r^2$, where π (pi) = 3.14 and r = radius (half the diameter). Using this formula, the mature canopy of each fruit tree in your plan would be:

$$3.14 \times 10 \text{ ft} \times 10 \text{ ft} = 314 \text{ ft}^2 \ (29 \text{ m}^2)$$

The total fruit tree canopy for the three trees would be 314 ft² multiplied by 3, or 942 ft² (88 m²).

A mature black locust could have a 30-foot-wide (9 m) canopy. Thus, its total canopy would be:

$$3.14 \times 15 \text{ ft} \times 15 \text{ ft} = 706 \text{ ft}^2 \ (66 \text{ m}^2)$$

The total canopy coverage of all four trees would be:

$$942 \text{ ft}^2 + 706 \text{ ft}^2 = 1648 \text{ ft}^2 \ (153 \text{ m}^2)$$

To figure out what 40 percent of the total canopy area would be, multiply 1648 ft² by 0.4. The result is 659 ft² (61 m²). Thus the black locust, with a 706 ft² canopy, makes up more than 40 percent of the total canopy, and it would supply more than enough nitrogen to meet the fruit trees' needs.

In another example, let's say your design calls for five berry bushes, each with a diameter of 4 feet (1.25 m). One bush would have a total canopy of 12.6 ft² (1.17 m²). The canopy of the five would total 63 ft² (5.8 m²). One goumi bush with a 6-foot (2 m) diameter would have a total canopy of 28.3 ft² (2.5 m²).

The total canopy cover of the five berry bushes and the goumi would be:

$$63 \text{ ft}^2 + 28.3 \text{ ft}^2 = 91.3 \text{ ft}^2 \ (8.4 \text{ m}^2)$$

In this case, calculating 30 percent of the total canopy yields 27.3 ft² (2.5 m²). The goumi bush covers a little more than 30 percent of the total combined canopy and therefore should be capable of producing enough nitrogen to meet the needs of the five berry bushes.

At Miracle Farms in southern Québec, Stefan Sobkowiak provides for nitrogen needs by alternating one nitrogen-fixing tree (N) for every two fruit-bearing trees (F).* Thus, his orchard rows follow this pattern:

N - F - F - N - F - F - N - F - F

When planning your plant groupings, incorporate nitrogen fixers in as many layers as possible,

* Stefan Sobkowiak (Miracle Farms proprietor), in conversation with the author, May 2015.

attempting to work in a sufficient total nitrogen fixer canopy equivalent to meet the needs of the other plants. Some nitrogen fixers, like low-growing clovers, may be shaded out as the canopies of the fruit and nut trees mature. At that point, the taller nitrogen fixers will be mature enough to take up the job. Descriptions of many species of cold-hardy nitrogen-fixing plants are included in part 2.

BENEFICIAL ATTRACTORS

Your goal can be to have pollinator-attracting plants flowering in your garden continuously, from as early in the season as possible to as late in the season as possible. If your focus is on diversity, you needn't be concerned with planning out what will bloom every week of the growing season. Just plant an abundant array, and then observe when your plants bloom. If you discover a significant gap in time when nothing is blooming, you can then incorporate a flowering plant to fill that niche. Appendix 2 provides a listing of the flowering times I have observed for plants described in this book.

Shelter and Food for Predators

Pest-devouring insects help keep your garden in balance, eliminating threats to your plants before they become overly destructive. Ground beetles, dragonflies, ladybugs, lacewings, hoverflies, and tiny parasitic wasps are among the insects that provide these services. Spiders also help in insect pest control. The greater the diversity of plants and habitat features you include in your garden, the more predatory insects you can attract.

Ground beetles are generalists that like to burrow in thick mulch, consuming a wide array of insect pests. Spiders will eat anything they can

Elderberries to the Rescue

When I installed a row of elderberry bushes along one side of my vegetable garden to serve as a windbreak, I didn't realize that it would also provide the vegetable crops with other benefits. It so happened that my potato crop, six 100-foot-long (30 m) hills, was located downwind of the elderberries that same year.

Potato beetles are a major pest of potatoes—hard-shelled, yellow-and-black-striped insects that fly in from elsewhere and lay their clutches of yellow eggs under leaves of potato plants. These eggs soon hatch into bulbous, red-orange, black-spotted, soft-skinned larvae. The larvae have insatiable appetites for potato leaves and can strip plants clean virtually overnight. Early on in my market garden experience, I learned the importance of scouting for these pests beginning in June. I would walk up and down each row checking for eggs and larvae, which I dispensed with by squishing them between my fingers. I patrolled twice a day for three weeks. It took a lot of time and was not a pleasant chore.

The year I planted the elderberries, the first time I went on potato beetle patrol, half the larvae I found were already dead! Something had gotten to them before me. I suspect it was parasitic wasps that, attracted by the blooming elderberries, were in the area just in time to lay their eggs inside the potato beetle babies. Within a couple of days, no live potato bugs were to be found. Thanks to the elderberries, the beneficial insects did the work for me. They have done so again every year since.

Figure 13.3. In late summer, bumblebees love the anise hyssop in my garden.

Plants as Pest Barriers

Some plants can even be useful as living barriers to protect vulnerable plants from animal pests. A dense, thorny hedge made up of sea buckthorn, autumn olive, goumi, buffaloberry, Siberian peashrub, rugosa rose, gooseberry, or Maximilian sunflower (which deer abhor) can block or divert deer. At the same time, such a dense hedge provides protective nesting and foraging habitat for beneficial birds and can also supply food for humans. Another example is daffodil bulbs planted around a young tree, which can keep rodents at a distance because they are poisonous and rodents burrowing underground will avoid them. Daffodils and some (but not all) members of the onion family, such as chives and garlic chives, when planted close together, form a dense mat of roots that block weeds as well. Placing these plants along the perimeter of a planting bed can prevent plants with underground runners like quack grass from intruding (see figure 14.11 on page 283 for an example of using daffodils for these two kinds of barriers).

catch and they gravitate toward mulch plants with abundant foliage that provide shelter, including comfrey, borage, and rhubarb. Even a small pond will attract dragonflies that consume mosquitoes and other flying insects. Hoverflies consume aphids, thrips, and other plant-sucking insects and especially like to feed on valerian.

Mites and aphids are favorite foods of ladybugs that winter in leaf mulch. Ladybugs eat the pests' eggs as well. The female ladybug deposits eggs on leaves near aphids or another food source. When the eggs hatch, the ladybug larvae, which resemble tiny alligators with long, bumpy bodies and prominent legs, can consume up to 50 aphids a day.

Trap Crops

Trap cropping is based on the concept of attracting insects, too, but in this case, the goal is to attract pests to a plant *other than* your desired food crops. In chapter 6, I described how certain wild plants and ground covers serendipitously turned out to be trap crops, but you can also intentionally install plants to provide this function. Nasturtiums are an effective trap crop for aphids. Zinnia, a self-seeding annual with attractive flowers, can be a trap crop for Japanese beetles, as can rugosa roses. More suggestions of other plants to "trap" pests of concern in your garden can be found on gardening websites.

In *Sepp Holzer's Permaculture*, Holzer writes about installing perennial root crops to entice rodents away from feeding on the bark of fruit tree trunks, which can girdle the plants and weaken or kill them. Following his model, I interplanted sunchokes among fruit trees and berry bushes in an edible hedge, hoping that the underground tubers would provide enough food so resident voles would leave the stems of the other hedge plants alone. Depending on the size of your plot, use caution if you plan to use sunchokes, though, because they can be aggressive spreaders, as described in "Sunchokes" on page 209.

First Attempts at Plant Groupings

My first attempts at plant groupings yielded mixed results. I was just getting started on planting the forest garden, and I was preoccupied with installing the large fruit trees and some shrubs. In most cases, wood chip mulch served as temporary ground cover. But I did make some attempts at ground cover and herbaceous plant groupings, including planting a mixture of plants from these layers along the length of a 200-foot-long (61 m) fence, in addition to a combination of low growers beneath a shrub layer of beach plums. I decided to view each unanticipated outcome as a learning opportunity, converting disappointments into determination to improve future designs.

FULL SUN: FENCE-MOUND GROUPING

When we constructed the Enchanted Edible Forest, we acquired several dump-truck loads of partially decomposed yard waste from the summer colony on our island. I asked the backhoe operator to dump a portion of this mulch just inside the lower garden fence, forming a raised mound about 2½ feet (75 cm) wide and 200 feet (61 m) long. The site was in full sun, and soil moisture decreased somewhat as the slope rose from a low-lying area up a small hill.

I designed a plant grouping comprising ground cover, herbaceous, vine, and overstory plants. I chose day-neutral strawberries and hardy kiwi vines for fruit, herbaceous plants for aesthetic appeal and cut flowers, and an overstory of locust trees for edible blooms and eventually a wood harvest. The locusts would also provide a windbreak for the entire forest garden, supply nectar for pollinators, and fix nitrogen to nourish neighboring plants. When installing long-lived, large trees, it is important to consider the eventual size of their trunks and allow ample space to expand. I positioned the locusts 4 feet (1.25 m) inside the fence to allow room for their trunks to expand without impinging on the mound or the fence. I expected that the kiwis would benefit from the eventual dappled shade cast by the locusts.

The newly formed raised mound rested for the winter, and the following spring, I planted the locusts, branchless whips no more than a foot tall. I planted 15 hardy kiwis interplanted with 4 wisteria vines (another nitrogen fixer), all spaced 10 feet (3 m) apart along the fence, which would eventually serve as trellis for the vines. As a ground cover, 500 hybrid day-neutral strawberry sets filled three rows across the width and down the entire length of the raised bed. I also scattered three kinds of seed over the bed surface: blue lupine, a third nitrogen fixer, for its prominent flower spikes; perennial blue flax (*Linum perenne*) for its edible seeds and ability to accumulate copper, magnesium, and phosphorus; and coneflowers to complement the planting with their showy, long-lasting blossoms and for their potential medicinal applications. All of these flowers would attract pollinators.

Note that in this grouping, I applied the permaculture principle of providing at least three sources of important functions—three independent sources of nitrogen are included in the plan, as well as nutrient accumulators and beneficial insect-attracting plants.

By late June of the first year, the day-neutral strawberries produced fruit, and the lupine, coneflowers, and flax had germinated, with the lupine most prominent. The locusts, kiwis, and wisteria leafed out. My first foray into plant grouping looked promising.

Figure 13.4. In early summer, flower seedlings emerged around the strawberry sets planted all along the fence-side bed.

By the second year, I was rewarded with bodacious blooms of lupine and blue flax in early summer, followed by coneflowers blooming till October. I reaped a small harvest of strawberries, but the plants in the interior of the bed, deprived of light by the taller flowers, did not produce much fruit. The locusts grew by leaps and bounds. Although the kiwis were rated hardy to Zone 4, almost all of them had perished during the harsh intervening winter.

By the third year, the strawberry plants had all but disappeared with the exception of the few that sent out runners to the sunny, southwest-facing edge of the row. The flax faded as well, outcompeted by the bushier lupines and coneflowers. I replaced the dead "hardy" kiwis with 10 new female plants of two different varieties, along with two new males to ensure pollination. The locusts grew even taller. The wisteria survived but grew slowly, sometimes set back by partial winter-kill.

By the time two more years passed, all the female kiwis again met their demise. I replaced them with grapes, though I knew the northeast exposure was not ideal for these fungus-prone vines. I trimmed the lower branches of the locust trees so that enough sunlight would reach the young grapevines as they climbed up and along the fence.

As the years progressed, coneflowers took over most of the raised row with a few clusters of lupine remaining. The lupine self-seeded elsewhere in the garden, providing blooms in unexpected places. The dense stand of coneflowers looks magnificent in summer and feeds birds in winter. The locusts grow taller each year. I continue to prune off lower

Figure 13.5. In the second year of this planting, lupine and blue flax are blooming in tandem in early summer.

limbs to allow full light to reach the grapes. It is too soon to tell how the grapes will fare over time. I am gradually sheet mulching over part of the coneflower row, replacing the flowers with a low-growing perennial shrub layer that does not need full sun. I started with aronia planted under the trellised grapevines.

This Fence-Mound Grouping evolved over time, as some members of the original group thrived and some met their demise. For me, responding to such changes as a creative endeavor only adds to the pleasure of stewarding a developing edible forest garden.

Figure 13.6. In this late-spring view, violet blossoms along with several kinds of volunteers form a lush, attractive group beneath the flowering beach plum. The volunteers (which aren't yet blooming) include black-eyed Susan, lupine, and hairy vetch.

FULL SUN TO PARTIAL SHADE: BEACH PLUM GROUPING

My second foray into plant groupings proved to be more durable. This group includes ground cover, herbaceous, and shrub layers.

I prepared a 10-foot by 40-foot (3 × 12 m) bed by applying some semi-decomposed yard waste that I transported downhill from a pile deposited at the high point of my garden, as I described in chapter 2. The intensity of sun varies across the bed, with full sun at the top. The rest of the bed receives indirect rays due to its northeastern slant. The ground stays moist due to its northward slant and proximity to an intermittent aboveground water flow fed by spring snowmelt and heavy rains.

I planted five bare-root beach plum bushes for their attractive flowers and yummy fruit, placing them 6 to 7 feet (2–2.25 m) apart in a staggered row near the top of the slope. To provide nitrogen, I inserted three licorice seedlings between the bushes. I planted two varieties of day-neutral strawberries throughout the bed, and I interplanted 25 blue wild violets spanning the entire bed. I expected the violets to provide edible leaves and flowers, attract beneficials, concentrate phosphorous, and self-seed to cover the ground. The reduction in light intensity because of the northward slope combined with the shade of the beach plums would benefit the violets, which do best in partial shade. The intermittent flow of surface water kept the ground moist, benefiting them as well.

The grouping was successful with just one outlier. The beach plums took root and flowered the next year. The violets self-seeded profusely, covering the bed with low, dense foliage and bright flowers in spring and again in fall. The licorice suckered underground, sending up spikes a few feet away from the original plants. Feathery shoots of indigenous horsetails arose throughout the bed. Native hairy vetch vines appeared providing additional nitrogen for the soil, flowers that attracted beneficials, and stems that trapped aphids. The strawberries, however, fared poorly. They tried to move to the sunny top edge of the bed, but only a few succeeded. Based on this outcome and my experience with the Fence-Mound Grouping, I concluded that hybrid strawberries (Junebearers and day-neutrals) need full sun and don't like competition, a lesson I applied in later installations.

The third year, I noticed some volunteer herbaceous seedlings that looked like coneflowers growing

Figure 13.7. In mid-summer, black-eyed Susan's ostentatious blooms in the Beach Plum Grouping command the attention of the garden visitor.

in the bed. I assumed that the coneflowers, blooming not far away in the Fence-Mound Grouping, had self-seeded under the beach plums. I was happy to let them remain there. When they began to flower, I was surprised to see black-eyed Susans (a close relative) emerge from the buds instead. Seeds of this native perennial must have found the spot habitable. The adage I apply to volunteers is this, "If I like it, I leave it. If I don't like it, I weed it." In this case, I liked the showy blooms and left them. They have bloomed each year since. When they appear to be impinging on light needed by ripening beach plum fruits, I cut them back to open up the space. One year, a blue lupine self-seeded on the sunniest side of the bed. I let it remain to provide additional nitrogen, a blue accent, and a possible trap for aphids.

Inspiring Stories

Now that I've shared my first attempts at grouping plants to foster self-sustainability and mutual support, I want to share stories of four other gardeners I know who have applied some of these grouping strategies in their own edible gardens with gratifying results.

LORI'S STORY

Lori, a young mother, had done some gardening but was unacquainted with permaculture principles. After visiting my garden and learning about the permaculture principles incorporated in it, she populated her small front yard with perennial edibles. She underplanted her roses with garlic and her blueberries with strawberries. An elderberry bush she underplanted with strawberries grew rapidly, becoming a sight and sound barrier between the house and a busy street. Lori says the planting turned that area of the front yard into a "kind of oasis." She installed a wildflower garden that attracted butterflies and birds and added aesthetic appeal. In her backyard vegetable garden, she planted lettuce and basil beneath tomatoes. The tomatoes shaded the lettuce, extending that harvest, while the basil protected the tomatoes. Mulching with leaves kept weeds out and the moisture in.

JAIMIE'S STORY

When I met Jaimie she told me she had a "black thumb." Every plant she tried to grow died. She volunteered to work with me to learn how to successfully grow plants, and apprenticed herself once a week for a year and a half on our farm. Following that, her family moved to Georgia where they bought a newly built house that had no landscaping. Applying the lessons she'd learned, Jaimie went to work landscaping her lot with edibles combined with decorative flowers. First, she tested her soil to gauge its acidity. She did her best to match plants with their preferred habitats: strawberries where there was good sun, lemongrass with ample moisture, tayberries and blueberries near each other where the soil was acidic and drainage was good. She planted spring bulbs with the strawberries "to add beauty" where they wouldn't deprive the berries of sun. She placed blackberries and a peach tree in advantageous habitats where they rewarded her with fruit the second year after planting! Chives, chamomile, oregano, sage, and thyme surrounding rosebushes along a foundation provided tasty herbs and attracted beneficial insects.

After two and a half years in Georgia, Jaimie had to move again. Her house sold before several newer ones in the same development. It was the unique edible landscaping that sealed the deal.

SARAH'S STORY

Sarah, an empty nester, has gardened most of her life. When she toured my edible forest, she was intrigued by my description of microclimates. She said she realized she was "making a huge mistake" by trying to force ill-suited plants to adapt to her landscape rather than match plants with her existing habitats. As a result of this insight, she drastically changed her gardening strategy. "Rather than struggling to put in a plant from South Carolina," she states, she began "choosing plants made for our climate." Reexamining her property with a fresh perspective, she incorporated a diversity of plants suited to her natural landscape—currants, Korean nut pines, sour cherry trees, and a hazelbert hedge to feed wildlife and block strong winds coming from an open field. To reduce the need to water, she chose some plants that could deal with drought. She discovered that comfrey turned "not-so-great soil into beautiful soil." At the end of each season, she chops up the comfrey leaves into "beautiful mulch," which she distributes around other plants.

CAROLYN'S STORY

Carolyn, now retired, describes herself as a "dabbler in gardening" who grew a small patch of vegetables most years of her life. Her soil was heavy clay, shale, and rock with poor drainage. To expand her garden space, she felt her only option was to build a raised bed, which would entail the expense of trucking in soil, purchasing materials, and hiring labor to construct it. She had a lot of trees on her property with dead branches and leaves aplenty. When she learned about hügelkultur mounds, she immediately realized this was a way "to use what I have at hand instead of spending money." She chose the sunniest spot in her yard to construct a 3-by-6-foot (1 × 2 m) mound, oriented so the longer sides faced east and west, minimizing the northern exposure. She arranged the branches as a base, covered with layers of weeds and leaves, topped with some clay mixed with the droppings from her pet rabbit. She built the mound in fall so the seasonal rains and winter snow would moisten the leaves, hastening their decomposition. The next spring, she planted a combination of annual vegetables and berry bushes, looking forward to the improved drainage and growing space the mound would afford.

Each of these gardeners applied what they had learned about forest gardening in their own unique way, taking what they liked and leaving the rest. Their stories show that the principles of forest gardening can be applied successfully even on a small property. You, too, can choose to apply what fits your goals and setting. Don't be discouraged if everything doesn't work out as expected. Each disappointment is an opportunity to improve future choices.

In the next chapter, you'll see that I follow my own advice as I describe a variety of plant groupings I have tried in diverse habitats and comment on how they have evolved over time. I hope all of these examples provide inspiration for you to make your best guess about what combinations will work, and enjoy the process of learning from your plants as they develop together.

CHAPTER 14

GROUPING PLANTS IN DIVERSE HABITATS

This is where I share my experience grouping plants in a diversity of habitats. As I developed my edible forest garden, I combined plants I thought would do well in soil and light conditions that ranged from sunny and dry to shady and saturated. I did my best to incorporate supportive plants so the needs of the group would be met by its members. I created and expanded plant groupings to encompass all vertical layers. In this chapter, I describe a variety of these groupings matched to habitats to help you envision edible combinations that could work in the niches afforded by your plot.

I used a paper model to plan each new grouping. As described in chapter 3, I began with a to-scale map of the garden section, then cut out circles of paper representing the mature canopies of the trees and bushes I wanted to include and adjusted their arrangement on my scale drawing. While completing a design, I frequently traveled back and forth from the map to the bed location, measuring and remeasuring distances and adjusting my plan accordingly to be sure everything would fit. This was important in order to ensure that in subsequent nursery orders, I would specify the right number of plants. Frequently I built out beds in stages, beginning with some key woody plants and later expanding the growing area to include additional bushes and herbaceous edibles.

The planting plans and garden views in this chapter illustrate the factors considered in each grouping. They are derived from my original paper-and-pencil arrangements supplemented by actual measurements and observations of the growing plants in the field. Though some of the drawings may appear complex and a bit overwhelming, it is my hope that they, along with my explanations, will further illustrate the factors to take into account as you plan for your garden, large or small. In my discussion of the illustrations, I also reveal how the plant groupings depicted evolved over time: which ones endured and which developed in unanticipated ways.

Compact Groupings

When space is limited, often a single specimen tree underplanted with shrub and ground cover is the most desirable design. Here are two simple examples from my garden, one in sun and one in partial shade.

A SUNNY SHIPOVA GROUPING

Most standard fruit trees require full sun and occupy a space between 15 and 20 feet (4.5–6 m) in diameter. I designed a shipova tree group to fill such a space that includes four layers in a sunny, well-drained spot, with each plant providing functions supportive of the other members. For ground cover, strawberries provide edible flowers, fruit, and nourishment for pollinators. Herbaceous clumping chives supply edible leaves and flowers, attract beneficial insects, accumulate potassium and other minerals, and possibly deter pests. The contorted flowering quince (*Chaenomeles speciosa* 'Contorta'), a shrub that grows no more than 3 feet (1 m) tall and 6 feet (2 m) wide, contributes attractive flowers and edible fruit that is similar to that of the fruit borne on quince trees (see "Quince" on page 136). Its bare, distorted branches bring visual interest to the winter landscape. The shipova tree is the visual center of the grouping, attractive through three seasons with its spring blooms, dense light green summer foliage, and pear-like fruits that ripen in fall.

Shipova requires a second tree for pollination as well as a nearby nitrogen fixer. In my large garden, I had plenty of space to plant both near this compact grouping. If your space is limited, you could choose a self-fertile fruit tree to fill the center niche in a grouping instead, such as apricot, peach, Asian pear, or sour cherry. A nitrogen-fixing goumi (which also bears tasty fruit) could substitute for the quince.

As it matures, the tree will cast significant shade, and the sun-loving strawberries may have to be replaced with a more shade-tolerant ground cover like violets or sweet cicely. Or you could choose shrubs and herbaceous plants from the start that can tolerate both sun and some shade. Locating the specimen tree to the north of the surrounding plants will minimize the loss of light to the lower layers over time.

Figure 14.1. This late-spring scene features a young shipova fruit tree, Junebearing strawberries, a clump of blooming chives, and a small contorted flowering quince with yellow-green leaves.

A SEMI-SHADED REDBUD GROUPING

I used a redbud as the specimen tree in a compact group located in a spot where full sun alternates with dappled shade throughout the day, and the soil stays evenly moist. The redbud supplies nitrogen, beautiful edible flowers in spring, edible pea-like pods along with a dense canopy of heart-shaped leaves in summer, striking yellow color in fall, and an attractive, vase-shaped form in winter. Beneath it, I grow raspberries that tolerate partial shade and bear a delectable fall crop. Daylilies supply cut flowers, salad ingredients, and a habitat for beneficial spiders. Low-growing, semi-shade-tolerant wild thyme serves as ground cover under the other plants. The thyme is a nutrient accumulator, bears flowers that attract beneficials and has an aroma that may deter pests, plus it is a flavorful culinary herb. A nitrogen-fixing perennial sweet pea vine that came up as a volunteer rounds out the group.

If you have limited space in semi-shade, perhaps on the eastern side of your home or east of existing shade trees, a compact grouping such as this could provide both food and four-season aesthetic appeal.

Figure 14.2. Raspberries and a clump of strappy daylily foliage nestle around the base of a redbud in this late-spring view of a semi-shaded grouping.

In place of the redbud, you might prefer a shade-tolerant fruit tree such as Cornelian cherry, medlar, quince, persimmon, or pawpaw. A shade-tolerant groundnut vine or a ground cover of white clover could supply nitrogen. In lieu of the raspberry, you could try an aronia, gooseberry, currant, or any other type of shade-tolerant berry bush.

A SHADY BLACK LOCUST GROUPING

Finding edible plants that will grow well in shade can be a challenge. At first, there was barely any shade in my forest garden because I had removed preexisting brush and trees, and the first young trees I planted did not cast significant shadow. Incorporating shade-loving plants had to wait.

The first trees to cast shade were two black locusts set beside a fence and about 20 feet (6 m) apart. These trees grew tall and wide enough in three years to shade the ground below. When I created the bed surrounding the locusts, I added extra organic material to form an exaggerated downward slope toward the northwest, which also enhanced the potential for shade there. Not far from this slope and farther to the northwest was a stand of tall trees. In high summer, this stand cast considerable shade over the bed late in the day. The scene was set to incorporate plants that would thrive under these conditions.

The year after I planted the locusts, I planted four gooseberries and three jostaberries on the southeast side of the bed. Here they would receive sun early in the day and, as the trees grew taller over time, increasing shade as the day progressed. I also planted two goji berries just inside the fence, so they could use it for support, and two Chinese chestnut seedlings.

By the third year, the locusts had grown enough to serve as trellises for two shade-tolerant akebia vines. I added five hosta plants, interplanted with 16 houttuynia on the now well-shaded northwest slope of the bed. Some wintergreen that I placed just inside the fence did not survive. (I concluded it was because the soil was not acidic enough.) During the fourth year, I interplanted three giant Solomon's seal between the hostas and added sweet cicely on the north edge of the bed.

This grouping encompasses all seven vertical layers within a 20-foot-long (6 m), 6-foot-wide (2 m) space. The houttuynia, hosta, sweet cicely, and Solomon's seal yield edible leaves, shoots, and roots. Their flowers also provide food and habitat for beneficials. The berry shrubs and vines supply fruit for birds and people, while the chestnuts may someday provide carbohydrate- and mineral-rich nuts. The locusts provide edible flowers and living trellises for vines; they also accumulate calcium, potassium, and nitrogen to benefit nearby plants.

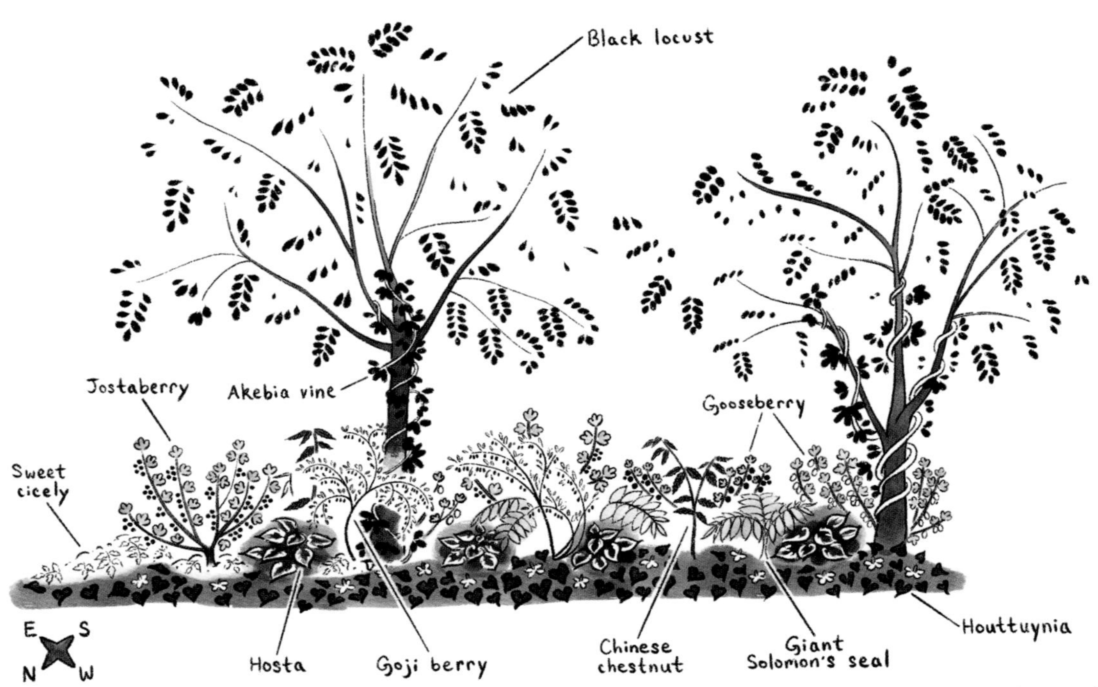

Figure 14.3. In this plant grouping, shade-tolerant ground cover, herbaceous, woody, and vining plants form a dense array under two black locusts. Other trees adjacent to this grouping lend additional shade. Illustration by Zoe Chan.

Overall, this shade-tolerant edible grouping of plants has proved compatible. Within five years, the houttuynia spread to form a dense, low-growing ground cover. Above the houttuynia, hosta, and giant Solomon's seal expanded each year. The sweet cicely self-seeded, filling gaps between the original plants. All the berry bushes flowered and produced fruit each season. In its fifth year, one akebia vine flowered for the first time. The Chinese chestnuts, despite facing winter-kill more than once, continued to survive in the understory. Both locust trees recovered vigorously after strong winds broke their trunks.

Many shade-loving plants could be substituted in this grouping. For a ground cover, you could try wild ginger (*Asarum canadense*), ginseng (*Panax ginseng*), ramps (*Allium tricoccum*), or violets. Ostrich ferns (*Matteuccia struthiopteris*), which yield edible fiddleheads, are a taller option. Groundnut and hog peanut are shade-tolerant vines. Thimbleberries (*Rubus parviflorus*) or red and pink currants could occupy the shrub layer. For the understory, medlar, redbud, basswood, persimmon, and pawpaw could be good choices.

Designing Orchard Rows

As I described in part 1, I decided to include some orchard-like rows of fruit trees and fruiting bushes in my forest garden. Over a period of years, I designed six of them, three each across the southeast- and south-facing slopes, keeping in mind all of the permaculture principles discussed in chapter 2.

These rows are a good example of my design approach of starting with the tallest layers and working downward and outward from there. For me, the tallest layers in a design were usually overstory and fruit trees, but if you are working on a smaller scale, the tallest layer in your orchard row might be the shrubs. Plan the placement of those plants first, and add in the lower layers later.

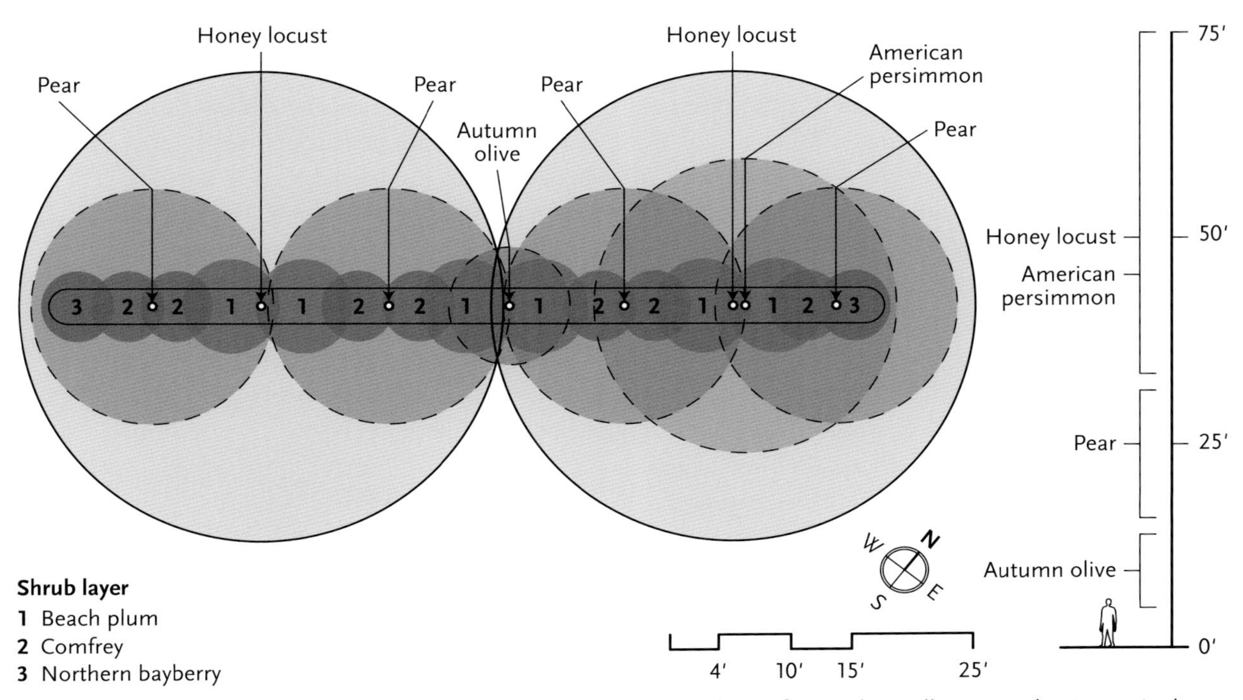

Figure 14.4. Planting plan for one of my orchard-type rows on a southeast-facing slope. Illustration by Turner Andrasz.

Although I designed six different orchard rows, I'll describe just one of them here. (Later in this chapter, I describe how I expanded another of my orchard row plantings.) When I first designed this simple orchard row, shown in figure 14.4, the only plants I planned for and planted were three pear trees and an autumn olive. There are two reasons why I didn't aim for more diversity of tree species. First, I had observed two mature pear trees growing beside our house for more than 10 years and had rarely seen signs of pest or disease infestation. Thus, I figured it would be safe to plant pear trees relatively close together in one orchard row. Second, I had a general idea that in the future, I would plant a shrub layer that included plants from other botanical families, and the row would be surrounded by wide swaths of native grasses and wildflowers. Those layers would supply ample diversity.

Since pear trees can grow up to 30 feet (9 m) tall, I located them high enough to the northwest on the hillside so their shadows would not impinge on other sun-loving species placed farther down the slope. I spaced the pear trees 20 feet (6 m) apart so their mature canopies would not impinge on one another.

In the shrub layer of this row, the beach plums benefit from the improved drainage on the hillside and sun exposure from morning through mid-afternoon, 8 to 10 hours a day. I did not plant them until four years after planting the pears, and when I did I set them as far from the trunks of the pears as I could to limit damage to the pear roots while digging. I also planted comfrey at this time, but because I chose small potted comfrey transplants, it was safe to situate them close to the pear tree trunks.

The honey locusts, autumn olive, and northern bayberries are all nitrogen fixers. I planted the honey locusts in the third year, and over time I plan to trim their branches up to 30 feet aboveground level if necessary. Pears grow in a pyramid shape, wide at the bottom and narrow at the top. Pruning the honey locust limbs will ensure that ample light can reach the pear tree crowns.

As the beach plums widen to their mature girth, I can trim back the nearby nitrogen-fixing shrubs to make room for them. If need be, those shrubs can eventually be eliminated, because the maturing honey locusts will provide ample nitrogen for this row on their own. They should even have capacity to fix nitrogen for the rows adjacent to this one. If necessary I will also trim the lower branches of the pear trees as the beach plums reach their mature height.

The American persimmon was an afterthought. I got the idea to include it so the neighboring locust could serve as a nurse tree. I figure I can trim back the persimmon or even pollard it as it grows so it won't impinge on the pears. If the trees become too crowded, I could simply cut down the persimmon. In the meantime, if it produces fruit, I will harvest it or leave it as food for birds and to add interest to the winter landscape.

This row has developed well, but some aspects haven't turned out as planned. I originally planted a seventh beach plum at the southwest end of the row, which is the lowest and wettest spot overall. That beach plum died within a year. I replaced it with a northern bayberry, and added a second at the other end of the row to match. The pear at the northeast end of the row was a late addition, too, replacing an apricot tree that failed to survive. Other than the comfrey, to date I've yet to complete the herbaceous and ground cover layers. For now, nature has filled those layers with healthy thistle plants.

Hedge and Windbreak Groupings

The most obvious function of hedges and windbreaks is to block wind. This is especially important in the first years of a forest garden, when the plants protected by the windbreak hedge are establishing themselves and approaching their mature height. Once the fruit and nut trees in a garden are full grown, they will form a windbreak themselves as the mature tree canopies guide the wind over their tops.

Windbreak hedges also provide food, shelter, and nesting habitat for beneficial insects, birds, and animals. They can block access to the garden by deer and other pests, or serve to detour those animals around the garden. For the gardener, they provide culinary, medicinal, and other useful products as well as a pleasing vista or visual screen. I have incorporated windbreak hedges in my gardens that fulfill all of these functions, and I describe four of them below.

HAZELBERT AND SEA BUCKTHORN HEDGE

Hazelbert and sea buckthorn are excellent hedge plants because they grow thickly, with no gaps between branches even at the base. I designed a hedge of these two species along a garden fence that faces the northwest, interplanting two or three sea buckthorns in a group with two or three hazelnuts, repeating that pattern for the length of the hedge so that all the plants would have neighbors close enough to ensure good pollination within species. Hazelberts have multiple straight stems extending from the base up to 15 feet (4.5 m) tall. Sea buckthorn has thorn-covered branches, produces suckers, and fixes ample nitrogen to meet the needs of the hazelberts.

I sourced the hazelberts from three different nurseries. Two groups grew quickly as anticipated. The third grew much slower, which left a gap in the windbreak. Luckily, I had also planted some tamaracks and Korean nut pines just inside this hedge, and they helped to screen the gap. (This planting of tamaracks and pines is described in "My Tamarack Travesty" on page 115.)

THREE DIVERSE EDIBLE HEDGEROWS

Over the course of developing my forest garden, I designed three diverse edible hedges to block wind

Figure 14.5. *Left*: A sea buckthorn (*foreground*) and a hazelbert (*rear*) seedling the year after they were planted to form a windbreak hedge that would also keep out deer. *Right*: Years later, a fall view shows the hazelbert festooned with catkins as well as Korean nut pine and tamarack trees planted nearby. The tamaracks will eventually grow much taller than the hazelberts.

and add plants that in some cases I had no space for elsewhere. Below I describe each at inception and as observed over time.

The Pecan Hedgerow

A mature native hedgerow runs along the border of our property on one side of my edible forest garden. I decided to design and install an edible hedgerow on a raised bed that I built adjacent and parallel to the native hedgerow. The existing hedgerow was mostly populated with ash trees, which are projected to succumb to the invasive pest called the emerald ash borer during the next few years. My vision was that as the ash trees died out, the young trees in my hedge would gradually grow to occupy their space. The hedge slopes gently downhill from the northeast. At the upper end, the habitat is well drained and sunny. In the mid-section, sun and shade alternate throughout the day and the ground stays evenly moist. Toward the lower, southwest end, the ground becomes progressively wetter and more shaded by the nearby woods.

I chose a diversity of tree species for the overstory and understory layers, as illustrated in figure 14.6. The food-producing trees in this design are spaced to maximize yield. I spaced the non-food-bearing trees more closely, at distances that I feel will result in aesthetically pleasing arrangements over time. For a hedge designed to break the wind, planting trees and bushes closer than their anticipated mature width makes sense. As in a native forest, this dense arrangement forces the trees to stretch upward to compete for the available light.

As I figured out plant placement, I considered each tree's tolerance for shade and moisture. The peach tree occupies the highest, best-drained, and sunniest spot in accordance with its preferred habitat. Its expected life is only about 10 years, and I figured it would likely die before it is shaded by neighboring trees. The pecans will grow the tallest and be exposed to full sunlight as they mature. (I chose a cold-hardy variety called Michigan, as described in "Pecan" on page 111.) It could take 100 years before their canopies achieve their full widths. Meanwhile, the honey locusts, sassafras, and wild yellow cherry trees will receive enough sunlight to meet their needs. The wild yellow cherry trees intrigued me when I saw them in the Oikos catalog; I thought this dense hedge placement would suit them because in nature, wild cherries are found growing close to other trees in mixed hardwood forests. The slower-growing and most shade-tolerant species—basswood, redbud, and witch hazel—should do fine in the upper reaches of the understory. Basswood and witch hazel can also tolerate wet soil, so I positioned them in the spot that was both shady and wettest.

I considered the sun and moisture needs of the berry bushes and brambles in the shrub layer and placed them accordingly. I did the same as I chose plants for the herbaceous and ground cover layers that I added later on.

Most of the outcomes I envision for this planting may or may not come to pass. It will take many years, perhaps beyond my lifetime, to determine whether some of my choices are correct. However, looking back over the relatively few years of this hedge's existence, I can already see that I made some misjudgments.

The brambles were the first casualties. One by one, the black raspberry canes disappeared, probably due to too little light and too much moisture. Next, the blackberries began moving from their initial locations between and behind the red currants toward the outside of the hedge—they were trying to escape the shade. Even in their new digs, they had insufficient light to produce fruit before the growing season ended. The red currants, on the

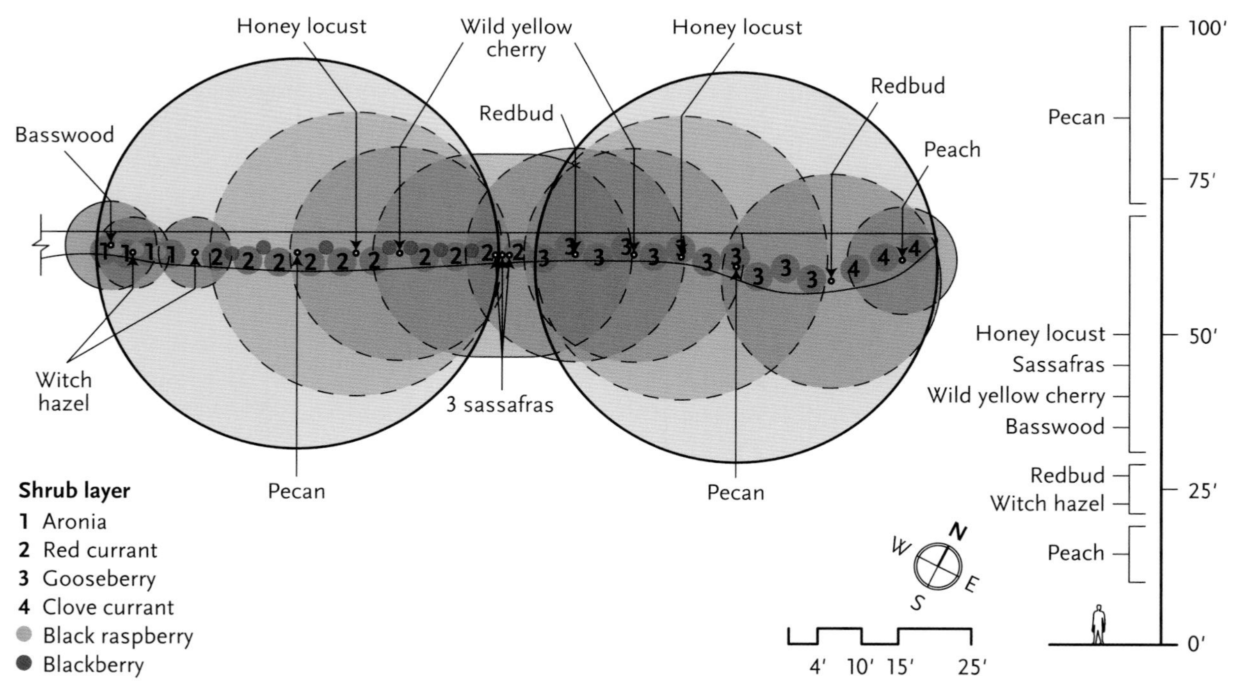

Figure 14.6. This subsection of an edible hedgerow shows the densely planted, many-layered overstory. Over time the trees will fill up to 100 feet (30 m) of vertical space. They will also expand horizontally to become an effective windbreak. Illustration by Turner Andrasz.

other hand, are thriving in this semi-shaded spot where they produce a more bountiful crop of berries each year. Two varieties of fall raspberries, originally planted along the hedge under canopy of a pecan tree, slowly succumbed to repeated infestations of cane borer. I replaced them with the gooseberries and clove currants shown in figure 14.6. One of the redbuds was originally planted where the witch hazel is now. When I noticed that the redbud was failing to thrive in the wet ground, I moved it to its current location on higher ground. Meanwhile, the second redbud has flourished in its better drained location. And lastly, since I am insistent on succeeding with sassafras even though they are borderline Zone 4, I have replanted two of them more than once, sourcing the plants from different nurseries each time in the hope of finding ones that will adapt rather than perish in my setting.

The Juneberry and Cherry Hedgerow

I have also designed and planted an edible hedgerow along the southwest side of the original ½-acre (0.2 ha) portion of my garden. This hedge has gone through changes over time, and it now comprises nitrogen fixers, several types of edible berry plants, and herbaceous plants (including yarrow and sunchokes) that feed pollinators, attract beneficials, accumulate nutrients, and distract pests. At the moister, low point of this hedgerow, I planted a decorative elderberry and aronia bushes, which all tolerate the occasionally wet ground there. I interplanted two juneberry trees, which may reach 20 feet (6 m) in height and create a canopy of foliage above the shrubs. The juneberries flower early, feeding hungry pollinators, and provide berries and perches for birds. The hedge also includes alternating sand cherries, Nanking cherries, and Siberian peashrubs (for nitrogen). The decorative flowers of all three feed pollinators, and they all bear edible fruits.

The grouping has its pros and cons. All the shrubs are fairly fast growers, and they form a dense hedge from the ground up. The sand cherry does not grow as tall as I would like for this windbreak, however. The sun-loving Siberian peashrubs tend to be partially shaded by the adjacent cherry bushes, which inhibit the peashrubs' growth. To keep these nitrogen fixers thriving, I periodically prune back the cherries, but that's somewhat counterproductive because it reduces the density and thus the wind-blocking ability of the hedge. Rodents like to girdle the cherry stems, which also reduces their growth. I have protected the plants with metal screen trunk guards and planted a trap crop, but this hasn't yet eliminated the problem.

Figure 14.7. Juneberries and cherries blooming in a windbreak hedge.

The Curving Hedgerow

In the area where I expanded my edible forest garden, there are native woods on three sides that provide windbreak services. On the exposed

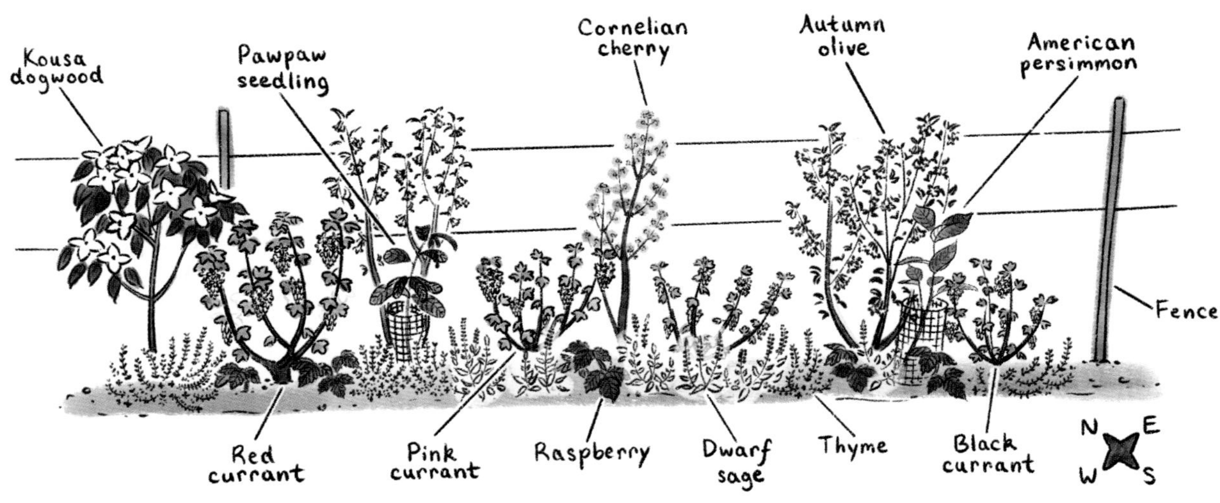

Figure 14.8. This is a section of the Curving Hedgerow, featuring plants that can tolerate some shade. Illustration by Zoe Chan.

northeast side of the garden, I designed a windbreak hedge in an undulating curve to add visual interest and maximize edge habitat. This site is located on a well-drained, northeast-facing slope with bedrock close to the surface in spots. In one place, black snakes created an underground den in a crevice in the rock. When I see these beneficial reptiles slithering in and out of their home, I welcome them because they eat rodents!

I chose hedge plants that could tolerate some shade, because I had planted a row of fruit trees on the slope above, and I expected that those trees would eventually shade the hedge. My design included nitrogen-fixing autumn olive shrubs spaced 10 feet (3 m) apart. Between the autumn olives, I interplanted red, pink, and black currants.

As time went on, I incorporated more plants to increase the height and diversity of the group. For ground cover, I installed a dwarf variety of sage (*Salvia officinalis*) and English thyme as a pest-deterring border on the sunniest edge of the planting. I dug raspberries from a stand in a friend's garden and interplanted them as well. I added two American persimmon seedlings, figuring the autumn olives would make good nurse plants for them. I included some kousa dogwoods (*Cornus kousa*) and Cornelian cherries for their early blooms and late-summer fruits, along with two pawpaw seedlings, the babies of trees from the original part of the forest garden.

Overall, this design performed as expected. I have reaped a reliable yearly harvest of currants. The raspberries produced a small crop each fall until cane borers decimated them. But they are gradually regrowing. The Cornelian cherries have flowered but not fruited as yet. The persimmons are growing slowly. I replaced the pawpaw babies, which did not survive their first winter, with new ones that had grown in pots for a year to allow their roots to develop. It is too soon to evaluate their success.

Revamping a Planting

As I gained experience with forest gardening, I realized that the more sod I eliminated and replaced with ground cover plants, the less mowing would be necessary. In line with my favorite permaculture principle, to "reduce human labor going forward," I decided to replace sod with beneficial plants everywhere except in pathways. Expanding plantings in

A Planting Dilemma

One year while I was in the midst of expanding my forest garden, I ordered over 100 trees, 200 bushes, 250 ground cover plants, and 10 vines. As the plants arrived, I was overwhelmed. How was I going to find the time to install all of them?

One group of plants included 28 trees and 35 shrubs destined to form a white-birch-lined path. My design called for planting clusters of one speckled alder with three white birch surrounded by five shrubs at 25-foot (7.6 m) intervals along the path. The task of digging 63 individual holes to install these groupings seemed insurmountable. I thought about my observations of plant groupings in developing woods. There, trees often grow up among bushy shrubs. I thought, *That's what I'll do! Instead of digging 63 holes, I'll dig 7 and plant each group of nine plants in the same hole.* I had observed that white birch are often found in clumps of three or more in nature. Most of the shrubs had a suckering habit, so I thought, *If they don't like being in the same hole with other plants, they can move!*

In each of the seven holes, I combined three white birch, a speckled alder, an aronia, a northern bayberry, and three winterberry (*Ilex verticillata*) shrubs. Winterberry is a type of holly that provides food for wildlife (but is toxic to humans), and it occurs as separate male and female plants. I put three in each group to increase the odds of having male and female plants close together. What could be closer than planting all three in the same hole?

How did this experiment work out? Most of the 63 plants survived. The major exceptions were all but one of the speckled alder and all but two of the winterberry. I attribute their demise to the soil conditions. The upper hillside is well drained, and I believe they needed more moisture.

Figure 14.9. A cluster of nine tree and shrub seedlings that were planted together in one hole.

Figure 14.10. Three expanded beds: herbaceous plants of the peach tree group in foreground, the recently sheet-mulched pawpaw bed at the upper left, and a lush pear grouping on the right. The wide access paths allow light to enter from all sides.

this way has dramatically reduced the time spent (or the money paid) on mowing. Next, I describe how I expanded three groupings and an orchard row.

EXPANDING A PEACH/PERSIMMON GROUPING

On a south-facing slope of a small hill in my garden, I decided to try creating a "sun trap"—an extra-warm microclimate for a tender Yates persimmon tree. I knew the spot was sunny, well drained, and even dry in mid-summer. The extra heat captured by the southern slope would help the persimmon grow well and produce ripe fruit, while the tree's taproot would reach down for moisture during dry summer months. To block the north wind and concentrate the heat of the sun, I planted three sun-loving, drought-tolerant Siberian peashrubs along the circumference of what would be the persimmon's drip line at maturity. To further protect the tree, I planned an additional windbreak of Korean nut pines, later supplemented by tamaracks, as described in "My Tamarack Travesty" on page 115.

Five years after that initial planting, I decided to expand it. The persimmon was growing slowly and suffered from winter-kill despite my built-in "insurances." I thought, *What if I plant a peach tree just to the south of the persimmon?* In nature, persimmons germinate and grow in the shade under taller trees; perhaps the shade cast by the faster-growing peach might help the persimmon. *Besides,* I thought, *I will be lucky if the peach lives more than 10 years. After that the persimmon can take over the space.* So I planted a Contender peach just a few feet downhill and to the south of the Yates persimmon. The persimmon has grown much taller since then, and I was so impressed with the succession scheme I had come up with that I installed a second peach to shade a Meader persimmon in another location in the garden.

My next step was to sheet-mulch the area around the Yates persimmon and peach extending out to the peashrubs, about 8 feet (2.4 m) out in all directions. I placed two layers of overlapped cardboard over the sod, shaped the edges to form a curve, and covered it with a thick layer of dry tree leaves held down with wood chips. I completed this work by late spring. By fall the sheet mulch had done its job—the dense sod was replaced by crumbly soil. That October, I purchased hundreds of daffodil and tulip bulbs at marked-down prices. I planted the daffodils in two dense circles, one close to the peach and persimmon trees, and one around the circumference of the bed. Then I populated the interior of the bed with patches of several varieties of tulips. The daffodils successfully curtailed weed incursions, and discouraged rodents from girdling the peach and persimmon trunks, but unfortunately did not keep rodents from decimating the tulips, most of which disappeared within two years.

That winter, I designed some of the understory, choosing sun-loving and drought-tolerant plants. I grew some lavender (beneficial attracter and pest confuser), milk vetch (nitrogen fixer with flowers for pollinators and medicinal uses), and sea kale (a nutritious vegetable with flowers that feed parasitic wasps) from seed to supply some herbaceous ground cover. I ordered five low-growing native junipers (*Juniperus communis*) (which supply edible berries and bird habitat) for the shrub layer. These evergreen shrubs grow wild on the barren tops of rocky ridges in our woods. I also looked through my seed packets to find sun-loving, drought-tolerant supportive plants and came up with yarrow and wild fennel: Both attract beneficials, accumulate nutrients, and are used for tea.

That spring, I installed all the plants and shrubs and scattered the yarrow and fennel seeds. I interplanted the lavender amid the kale, hoping it would deter pests. Almost all the plants took root and grew as expected, and a blue lupine was a welcome volunteer. Soon, the clumping yarrow spread, shading and stunting a couple of lavender plants. It was more aggressive than I anticipated. Going forward, I controlled its spread by deadheading its flowers before they went to seed.

Over time, I added more plants. When I discovered that sweet William flowers are edible, I incorporated them along the lower border of the bed. Then I decided to include some edible peonies and interplant them between the sweet Williams. I grew some hollyhocks and Maximilian sunflowers from seed, including them along the upper border.

As is evident from this example, a garden expands as the gardener has time and inspiration to

Figure 14.11. Daffodils in concentric circles surround the dormant peach and persimmon trees as well as the bed, serving as both a rodent deterrent and a weed barrier.

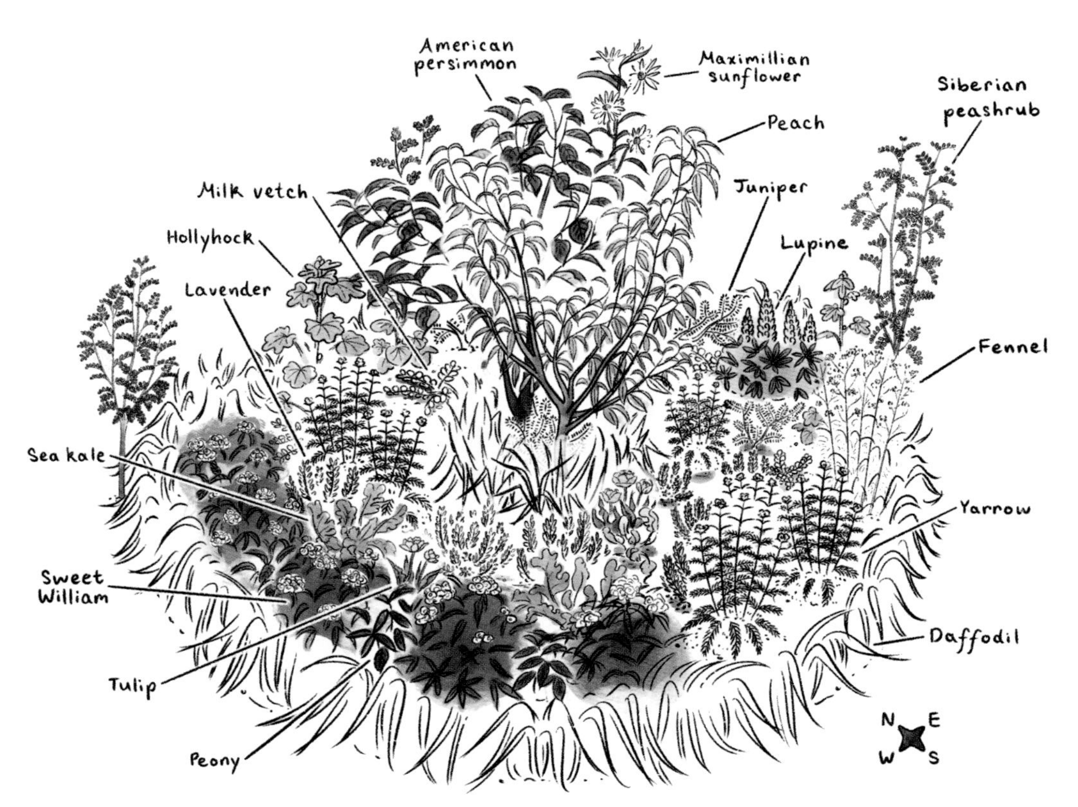

Figure 14.12. This planting plan shows a grouping for a sunny, high-and-dry habitat centered on a persimmon tree and a peach tree. Illustration by Zoe Chan.

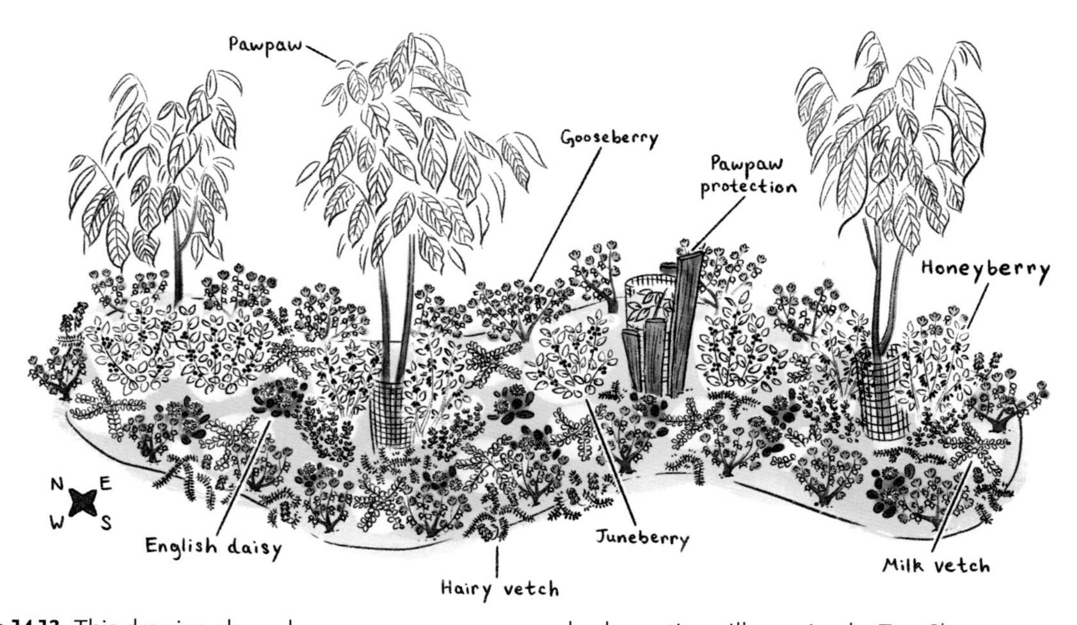

Figure 14.13. This drawing shows how my pawpaw grove evolved over time. Illustration by Zoe Chan.

embellish it. When I planted the persimmon on this slope, I had no specific plan to develop its surroundings. As the years passed and my experience grew, a vision emerged. So it will be with you. Plant one tree. With the principles of permaculture in the forefront of your mind, your self-sustaining garden will evolve with time.

RENOVATING A PAWPAW GROUPING

I started this grouping by planting a staggered row of four pawpaw trees spaced 20 feet (6.1 m) apart into bare ground. (An adjacent row of black locust provided nitrogen for the pawpaws.) Immediately after planting, I mulched each young tree with a 2-foot-wide (61 cm) circle of wood chips. From there, I expanded the planting over the next few years. The next year, I laid mulch paper and covered it with wood chips to create a small bed around each tree. The following spring, I planted two honeyberry bushes apiece in two of the beds, and two juneberry shrubs in each of the other two. In the fall, I planted tulip bulbs. The next spring, I added oregano to cover the ground under the honeyberries along with ground cover raspberries under the juneberries.

The oregano was the best performer, spreading out to shield the ground and providing food for beneficials and a tasty herb for the gardener. The ground cover raspberries fared poorly; periodic droughts caused them to go dormant. They rarely flowered and never bore fruit. Quack grass intruded in the beds and eventually took over. The juneberries never bore edible fruit. The honeyberries did not perform as well as others in my garden have. Rodents decimated the tulip bulbs. I decided it was time to start over.

I began by sheet mulching the entire 60-foot (18.3 m) stretch of the pawpaw row, filling all the areas around the trees and berry bushes with a double layer of cardboard covered by a 10-inch-thick (25.4 cm) layer of leaves topped with wood chips. The result was one long 10-foot-wide (3 m) bed with edges that undulated around the staggered trees.

Then I took some time to think about what might grow well under the growing expanse of dense shade cast by the maturing pawpaw. Gooseberries! The demand for these delectable fruits exceeded the supply from the few plants I had already installed. That winter, I ordered 20 more. The following spring, I planted them 4 feet (1.2 m) apart and set them 2 feet from the edges of the bed around its entire perimeter. When I dug the planting holes, I was encouraged to find moist ground under the sheet mulch. Though the quack grass was still vital under the mulch, the two layers of cardboard were fairly intact. The small holes I punched through the mulch to insert the small root balls of the gooseberry seedlings would not give the aggressive grass much opportunity to emerge.

I grew a quantity of milk vetch and English daisies from seed. I kept them going in pots throughout the summer while I pondered where to place them. Suddenly, a light went on in my brain—the spreading, sun-loving vetch combined with the clumping and self-seeding daisies might provide a decent ground cover, at least in the short run. Over time, the diminishing light would affect the vetch, but the daisies, more tolerant of shade, might expand to fill the space. That September, I planted all the sets in the pawpaw bed. To date, they are performing as expected. Time will tell if the ground cover succession takes place as projected. If not, I could always insert a shade-tolerant plant like alpine strawberries to fill gaps. If the honeyberries and juneberries continue to underperform, I will eliminate them as the gooseberries spread out to fill their space.

As I did with my pawpaw grove, you may find the need to start over in some area of your garden. Observing and learning to improve choices in the future is the name of the game. Whenever you design a grouping of plants, you make your best guess about their compatibility with each other and the intended habitat. Over time, you will discover what works and what doesn't and modify the design accordingly, or even restart from scratch if necessary.

EXTENDING A PEAR TREE GROUPING

My next story of creative expansion concerns a grouping with a sunny exposure to the south, and a moist, partially shaded exposure to the north. My plan for the sunny side included two pear trees spaced 20 feet (6.1 m) apart with an autumn olive between them, surrounded by five Regent juneberry bushes. As with other groupings, I planted the woody plants first and then covered the entire bed with wood chip mulch to keep out weeds until I was ready to install ground cover plants.

The following year, I decided that rhubarb might provide robust ground cover toward the bottom of the north-facing slope of the bed where it was nice and moist. In my vegetable garden I had a well-established rhubarb plant of a variety I liked. I chopped off a chunk of its thick root. Next I separated the severed root into four pieces, and planted the segments low on that slope. I also purchased three rhubarb plants of other varieties (hoping to find a new favorite) and planted them there. I included two garlic chives near the pear trees to attract beneficials and deter pests and, planted some garden sorrel around the rhubarb to accumulate nutrients and provide salad ingredients. In the third year, I planted day-neutral Mara des Bois strawberries to the south of the pears and juneberries where they would receive enough sun and have no nearby competition.

Two years later, I expanded this grouping by taking root chunks from my new favorite rhubarb variety and planting them farther up the slope. I added a sorrel variety called Perpetual, which does not produce seeds, paired with red-veined sorrel

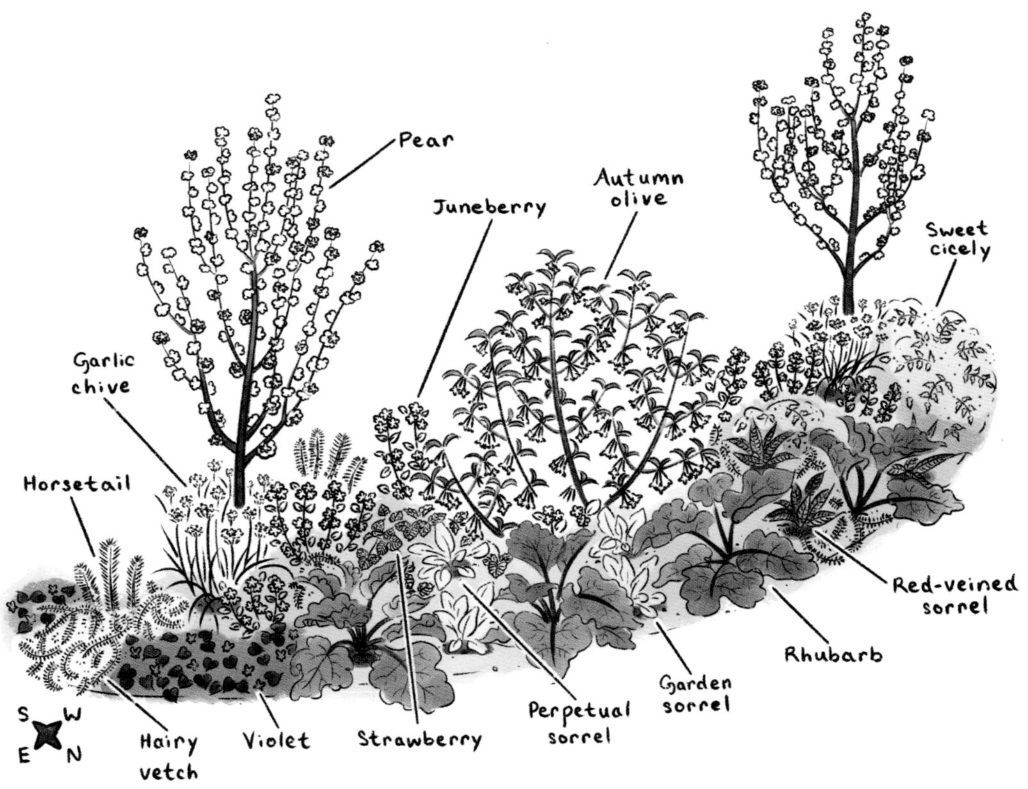

Figure 14.14. My pear tree grouping for a partially shaded, moist site is one of my most successful plant groupings. Illustration by Zoe Chan.

grown from seed. Over time, the former grew well but had tougher, less desirable leaves than regular garden sorrel. The red-veined sorrel, although prettier, was less hardy than either. As a final touch, I added a few sweet cicely plants at a corner of the bed. I had planted this herb a year earlier in a similar location where it did well, so I figured it would be happy in a comparable habitat. In fact, it was so happy, it began to self-seed here and there across the semi-shaded side of the bed.

In the years since, several changes occurred. The juneberries suckered outward to occupy more of the space to the south and north of the pear trees. Some Mara des Bois strawberries sent runners down the northern slope. Violets from a neighboring bed moved in as the sweet cicely spread. Meanwhile, hairy vetch and horsetail emerged from the native seed bed. Both the completeness of ground cover and the bed's diversity increased with no effort on my end! As you observe the changes that occur naturally in your garden over time, I am sure you will be both intrigued and delighted to witness the initiative your plants take on their own to embellish your garden space.

WIDENING AN ORCHARD ROW

Earlier in this chapter, I described one of the orchard rows I planted. A few years after the original planting, I decided to widen some of these rows by sheet mulching on both sides of the fruit trees.

The orchard row depicted in figure 14.15 is the lowest one on the southeast slope. In the original planting, I first arranged the diverse fruit trees far enough apart to allow for their mature canopies, and interplanted two autumn olives to provide nitrogen. When I sheet-mulched the ground below the trees to expand the row, I extended the bed downhill to meet the curve of an adjacent access route and extended the opposite side up the gentle slope of the hill an additional 6 feet (1.8 m) or so. The previous year, I had taken cuttings of black currants and potted them up. I decided to place them just inside the lower border of the new bed in a curved arrangement, interplanting two varieties to provide proper pollination. Here they would benefit from good sun and would not mind the very moist soil typical at the bottom of slopes on my land.

I originally planned a row of asparagus behind the fruit trees, but later reconsidered. Raspberries had proven very popular with U-pick customers. But the two summer-bearing varieties planted elsewhere in my forest garden stopped producing berries in early August when demand for raspberries is high. I decided a row of late-fruiting summer raspberries would be a better choice than asparagus, and the raspberries would not mind the shade cast by the fruit trees and could be accessed for harvest from both sides. That spring I attended a workshop at the University of Vermont on growing saffron crocuses. Intrigued by the prospect of

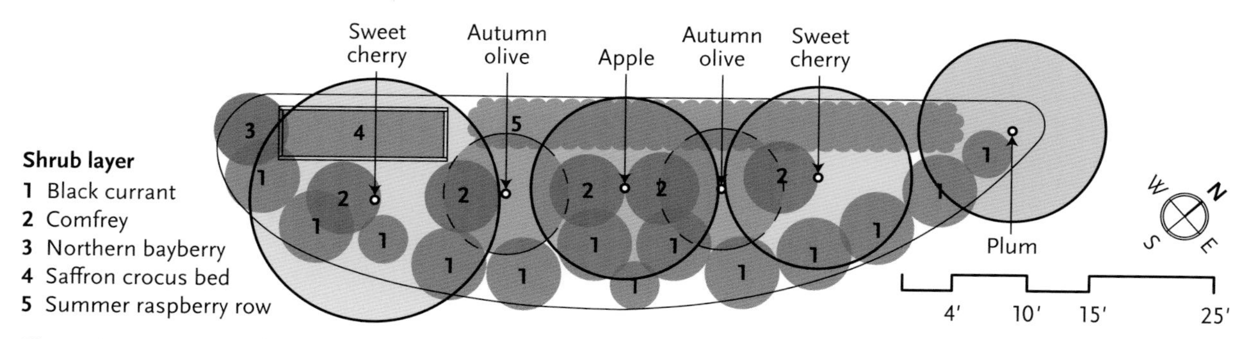

Figure 14.15. An expanded orchard-type row containing diversified understory, shrub, and ground cover plants. Illustration by Turner Andrasz.

harvesting this exotic spice from my own garden, I decided to devote some of the space upslope from the fruit trees to this new venture.

With the expansion of the growing space, I was concerned about the supply of nitrogen. This was another rationale for planting two honey locusts in the row uphill. Judging by the size of their projected mature canopies, I knew their roots would extend into this lower row. I surmised that the growing supply of nitrogen would be ample for the added currants and raspberries. I also decided to interplant comfrey to supply additional nutrients.

Over time, I did make some adjustments. In order to plant the raspberries and install the saffron bed, I transplanted two black currant bushes that were in the way. In turn I moved four thimbleberry plants to another location entirely, and swapped in the black currants there. I added a plum tree as an afterthought when I realized there was enough space available for another small tree. I also added northern bayberry after I became enamored with the plant and sought places for more of them. Wild strawberries moved in to supply ground cover. I installed a densely planted border of chives and garlic chives to deter weed incursion.

In the summer of the third year of this expanded planting, I was delighted to see all of the black currants loaded with ripening berries. When I went out to pick them a few days later, there were few to be found. Since it was a drought year with a scarcity of food in the woods, and I neglected to net these currants, they were almost completely consumed by birds. Another lesson learned.

Groupings for Raised Beds

As you know by now, I have had to learn how to work creatively with the areas in my forest garden that have saturated soil. Here are two examples of bed renovations and plant groupings I have tried in these areas that may spark your interest, especially if you, too, are contending with sodden ground. I have also used hügelkultur mounds to raise plants above a high water table as described in "Mounds Built on Saturated Ground" on page 291.

HIGHBUSH CRANBERRY GROUPING

When I designed the ½-acre (0.2 ha) expansion of the Enchanted Edible Forest, I decided to create a bed in a spot that receives full sun but where the ground is sodden for most of the year. I planted five highbush cranberry bushes spaced at least 10 feet (3 m) apart and a thicket of four American plum trees, each paired with a speckled alder to serve as a nitrogen-fixing nurse plant. I placed all these seedlings directly into the native sod in patches of the highest ground and mulched them with wood chips. My research suggested that these plants would do well in the wet and sunny conditions. What I didn't plan on was deer incursions in the first years of this garden. While the deer ignored the cranberries for the most part, they repeatedly browsed on the plums and alders. When I enclosed these seedlings in protective tomato cages, the night marauders ceased their unwelcome pruning, but the plums took years to recover. I think the repeated assaults plus soil too wet for their liking inhibited their growth. The cranberries and alders, though, grew larger each year.

To eliminate native sod and compensate for the sodden soil condition, I sheet-mulched the area around these plants three consecutive times. Each application of cardboard, leaves, and wood chips raised the level of the ground surface a little. After the second application, I installed additional shrubs: aronia encircling most of the bed to attract pollinators and bear edible berries, and northern bayberry for its aromatic leaves and nitrogen production. For the ground cover layer, I added violets, mountain mint (*Pycnanthemum incanum*) for its flavorful leaves and possible pest deterrence, and Russian comfrey to accumulate nutrients. I needed to sheet-mulch a third time to eliminate quack grass that survived the second application. I sheet-mulched right over

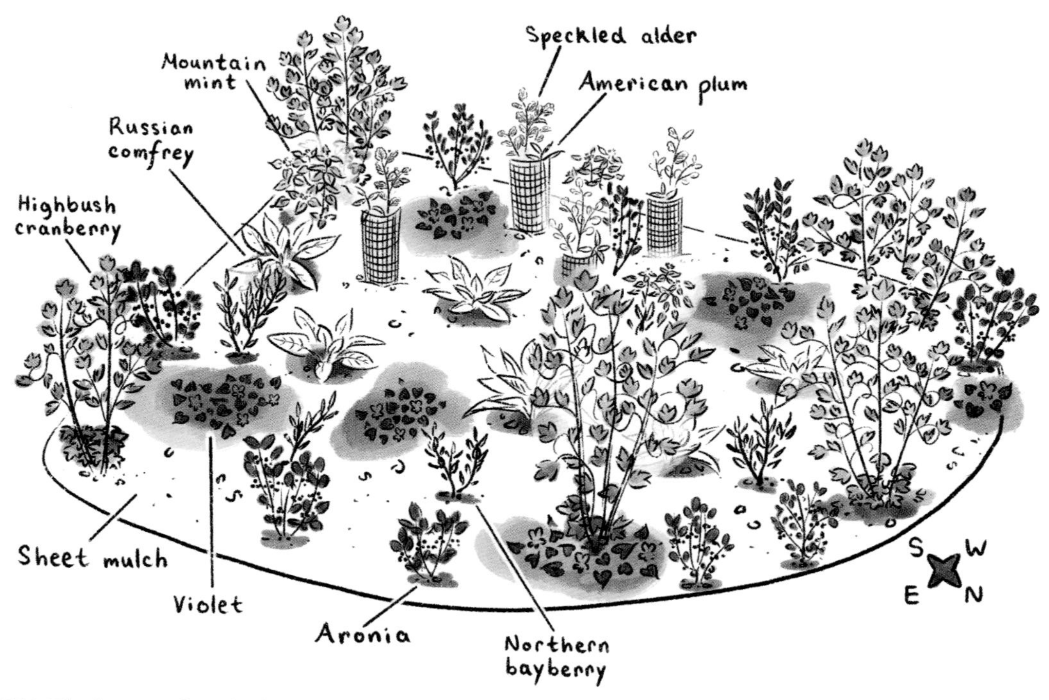

Figure 14.16. The layout of my highbush cranberry bed in a sunny, wet site. Illustration by Zoe Chan.

Figure 14.17. Mature highbush cranberries in a re-sheet-mulched bed.

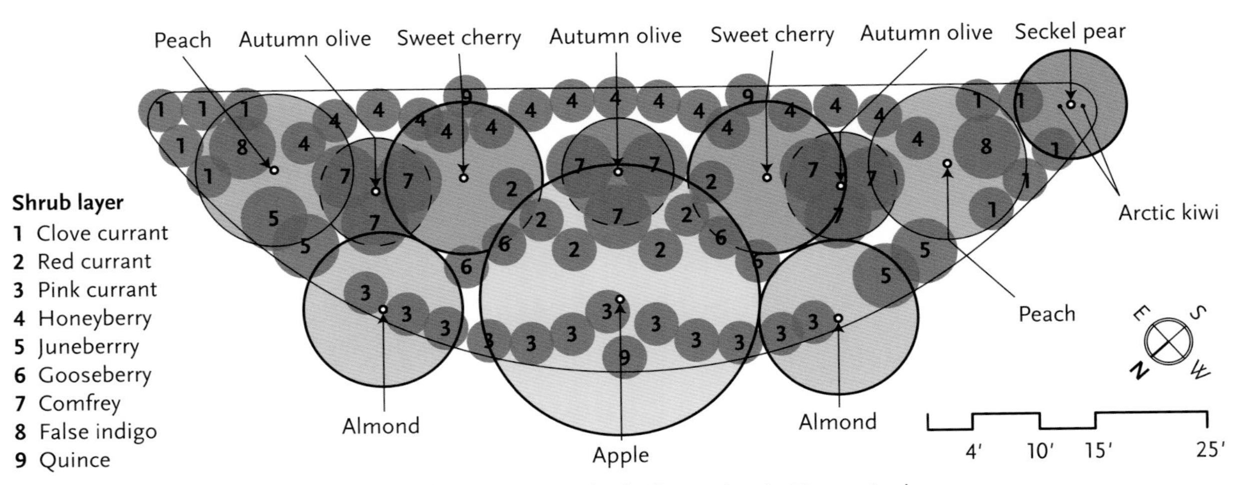

Figure 14.18. The original planting plan for my crescent bed. Illustration by Turner Andrasz.

the comfrey in late fall after it died down, knowing from experience that these indomitable plants would have no trouble emerging through the cardboard the following spring.

If you have a sunny, sodden area in your plot, all the plants here, except possibly the plums, could do well. Other options for ground cover with edible leaves and flowers are water celery, houttuynia, marshmallow, and members of the mint family, including apple mint. You could try elderberry, as well as a honey locust tree as an overstory nitrogen-fixer. Other suitable native plants for the shrub layer with useful or edible parts include Carolina allspice (*Calycanthus floridus*), spicebush (*Lindera benzoin*), and sweetbay magnolia (*Magnolia virginiana*).

THE CRESCENT BED

My crescent bed is an 80-foot-long (24.4 m) half-moon-shaped expanse bordered by two access routes. As part of the overall landscape design for my forest garden, I decided to turn this area into a network of narrow raised beds: scallop-shaped mounds punctuated with access footpaths around the outskirts and two T-shaped mounded structures in the interior. I decided I would build small, rounded mounds for individual trees, as described in chapter 4.

As I constructed the mounds, I planned out the initial planting as detailed in figure 14.18. I first chose locations for the understory fruit and nut trees, making sure their mature canopies had ample spacing. Below and between the peach and cherry trees, I sited three nitrogen-fixing autumn olive shrubs. To increase interest and edge, I designed scalloped rows of berry bushes along each edge of the garden. I planned honeyberries to span the southeast edge (the sunniest exposure), with clove currants surrounding the bed ends. I sited shade-adapted gooseberries and red and pink currants in areas where fruit trees would eventually create considerable shade. A few juneberries, placed between canopies, would receive direct sun early and late in the day. I would use an existing Seckel pear as a trellis for vines that would benefit from its shade. Several narrow footpaths would allow entry into the bed and access to the interior to harvest tree fruits and berries.

Some of the plants in this bed have struggled or died out. One sweet cherry and the two almonds met their demise when an unusually temperate fall abruptly transitioned to a frigid winter with a windchill below –30°F (–34°C). The apple tree continues to struggle in the inhospitably wet, poorly drained clay soil, even though it was planted in a raised mound. In the shrub layer, deer munched repeatedly on the juneberries until we installed an effective barrier. Two separate plantings of akebia vines failed to survive next to the pear tree, which prompted me to build up the soil around the tree and try planting arctic kiwi vines instead. In the herbaceous layer, peonies perished in the saturated ground and rodents devoured spring crocuses. On the bright side, all the other shrubs did fine and the peach trees have grown larger every year.

Hügelkultur Mound Groupings

Hügelkultur mounds both create and intensify habitats, allowing for a greater diversity of plant groupings. In this section, I illustrate the plantings on two mounds, one in shade and one in full sun, built atop saturated soil. I also describe plans for plant groups to occupy several mounds built in existing woods where there is standing water.

MOUNDS BUILT ON SATURATED GROUND

I have built and planted two hügelkultur mounds in areas where the water table is at the surface for much of the year, rather than repeatedly sheet-mulch to attempt to raise the soil level. One is in full sun and one is mostly shaded by a stand of ash. Each of these two oval mounds cover approximately 100 square feet (9.3 m²) of ground and were 5 feet (1.5 m) tall when first constructed. Through the force of gravity and decomposition of the contents, they shrank to about 3½ feet (1.1 m) tall. The mound surface affords at least 150 square feet (13.9 m²) of growing area.

In the mounds' first year, I planted them with annual vegetables. As soon as the shady mound was constructed in the spring, I populated it with some leftover kale sets. I scattered turnip, radish, and bok choy seeds over the sunny mound in the fall. In addition to providing an edible harvest, the roots from the annuals hold the soil in place, encourage soil life, add organic matter, and prepare the ground for perennial plants to be installed later. My plan for both these mounds was to install plants that I propagated myself from seed or cuttings, rather than purchased ones.

The sunny mound receives direct sunlight for at least three-quarters of the day. I planted drought-tolerant honeyberries over the top of the mound where the ground is the driest, with encircling black currants midway down the sides. I topped the mound with a false indigo shrub to fix nitrogen and attract beneficials. Strawberries and pineberries provided herbaceous ground cover and fruit. Some scattered herbs and bunching onions also included could attract beneficial insects while deterring insect pests and yielding flavorful leaves. Licorice along with volunteer hairy vetch supplemented nitrogen and other nutrients.

The shady mound receives some early-morning sun on the southeastern face and varying amounts of shade there for the balance of the day. The other exposures receive minimal, if any, direct light. For the shrub layer, I included red and pink currants and gooseberries, all started from dormant hardwood cuttings. A redbud tree I grew from seed, placed on top of the mound, provided nitrogen,

Figure 14.19. Annual root vegetables growing on the sunny mound in the fall.

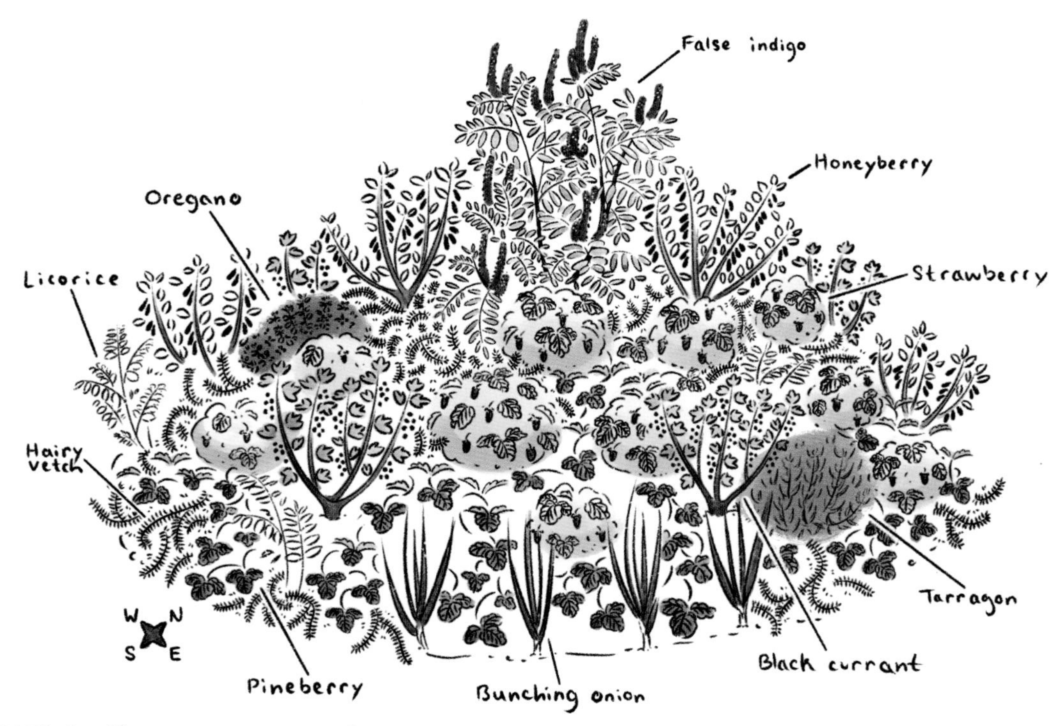

Figure 14.20. A self-sustaining grouping for a sunny hügelkultur mound. Illustration by Zoe Chan.

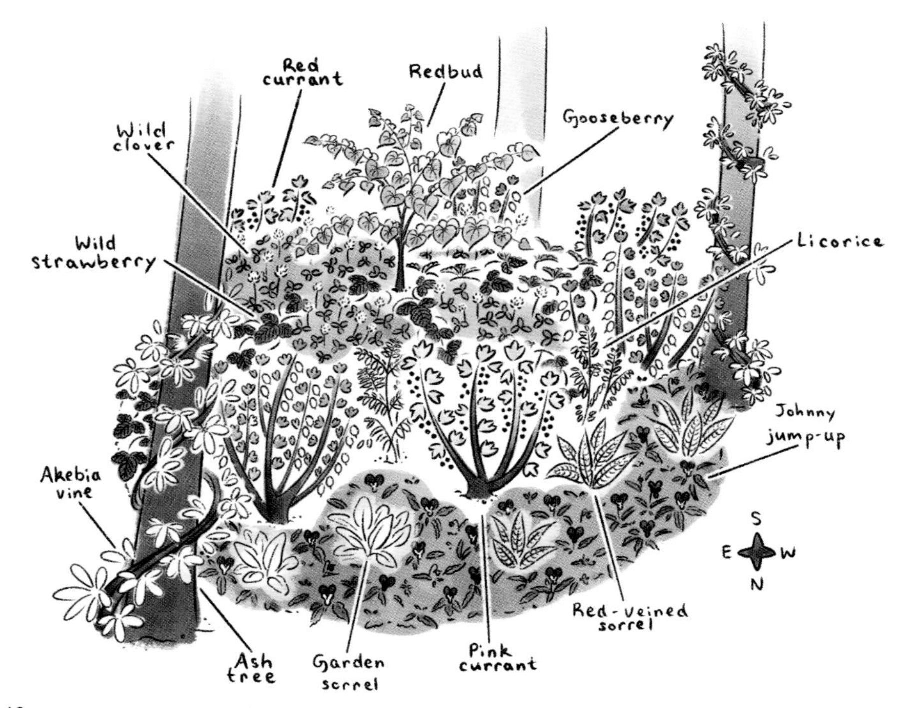

Figure 14.21. A self-sustaining grouping for a shady hügelkultur mound. Illustration by Zoe Chan.

Figure 14.22. A section of the shady mound in its third year. It is amazing how quickly plants fill the space!

added edible flowers, and gave year-round aesthetic interest. Sorrel, licorice, and Johnny jump-ups—all started from seed—provided some ground cover, additional nutrients for the fruiting plants, and a nice harvest for the kitchen. Wild strawberries and clover moved in of their own accord to cover gaps in the ground cover. Two akebia vines from layered cuttings climbed up two of the existing ash trees.

GROUPINGS FOR MOUNDS IN THE WOODS

Nine mounds that were built in late 2020 extend into the woods adjacent to my forest garden. There is usually standing surface water around these mounds nine months out of the year. During the growing season, the mounds closest to the forest garden fence, just inside the wood's edge, receive some direct sunlight, especially during the morning hours. Those located farther inside the stand of trees are in shade for most of the day.

Every garden is a work in progress. At this writing, I have shaped my ideas for planting these mounds, but I haven't figured out specific locations for all the plants I want to grow. It's not critical to plan out shrub and ground cover layers as meticulously as the overstory layer, so I can remain more flexible in playing with the details of those layers.

At the time of this writing, I have made lists of the plants I intend to incorporate on these mounds, but my design ideas are still percolating. Listed next are the shade-tolerant plants by category that I plan to install. The native plants are chosen primarily to

Figure 14.23. In this midwinter view, standing water covered with surface ice is visible at the base of the mounds in the woods. Mounds are a viable option to grow a variety of edible plants under these conditions.

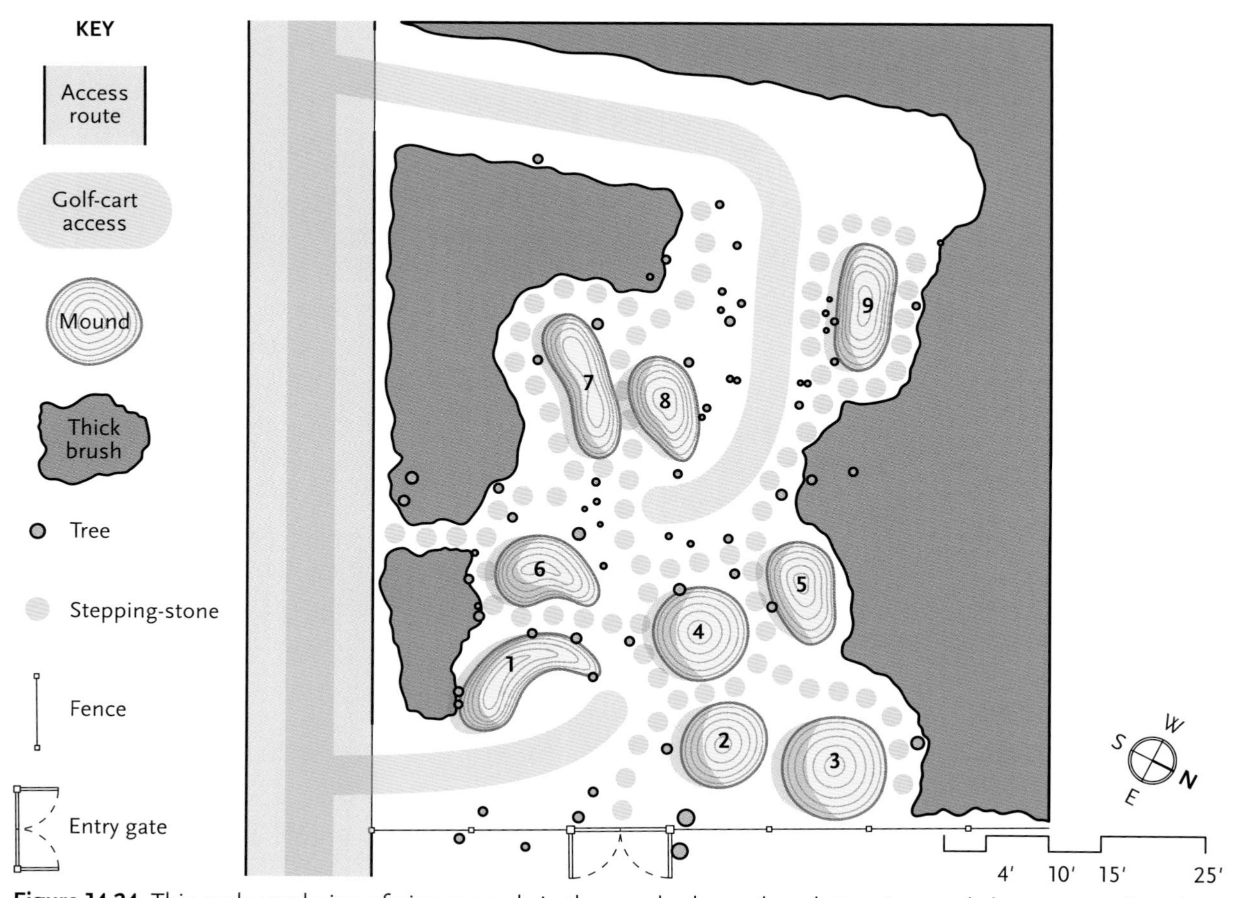

Figure 14.24. This scale rendering of nine mounds in the woods shows the relative sizes and shapes, as well as the location of surrounding trees. Illustration by Turner Andrasz.

attract and provide food for wildlife in addition to visual interest.

Nitrogen fixers: false indigo, goumi, groundnut, licorice, redbud
Fruit trees: American persimmon, pawpaw
Berries: aronia, black currants, clove currant, goji berries, gooseberries, jostaberries, raspberries, red and pink currants
Ground covers: alpine strawberries, chives, daylilies, English daisies, garlic chives, giant Solomon's seal, hosta, sorrel, sweet cicely, violets, wild thyme
Flowers: bee balm, campanula, forget-me-nots
Native plants: buttonbush, Carolina allspice, ground ivy, hog peanut, Indian currant, Oregon grape snowberry, spicebush, thimbleberry, wild strawberry
Mushrooms: totems inoculated with lion's mane sawdust spawn
Ephemeral vegetables: fiddlehead ferns, ramps

Before I install any plants, I will extend the electrified T-post fence to protect the mounds from browsing deer. Over time, as I observe the amount of light that reaches each plant I include and how each is faring, I may decide to relocate some of them and will probably add in others. I might also remove additional trees from the woods to allow more light to enter.

My experience thus far gives me some idea of what plants will work in this new habitat, but the planting will be experimental in the first few years. It is likely that some plants will do well in their assigned placements while others will falter. This is part of the excitement and challenge of creating an edible forest; all the discoveries and adjustments made along the way make the project continually engrossing.

I hope this chapter provides many ideas you can apply to groupings in whatever conditions of sun, shade, and moisture exist in your garden. If you think a group of plants will be compatible with one another and with the intended habitat, go for it! Like I have, you will find that your groupings evolve over time in ways that are fascinating to observe. Some will work out as expected, and others will diverge from your plan. To me (and I hope for you), these latter cases are even more intriguing because they are motivation to experimentally add new plants and over time to discover if they will adapt better or "play nice" with their neighbors.

Developing a planting of edible perennials is not a trivial undertaking. It requires careful observation of your plot and thoughtful planning of your landscape. Matching a diversity of useful plants with your available habitats will maximize their health and your harvest. Using as much of the vertical space as your situation affords will provide a bounty of edibles from the smallest area. When your design includes features that provide nutrients, attract beneficials, and deter pests, there will be little need for you to expend energy addressing these functions.

With the knowledge that there will be some losses along the way, I encourage you to "just do it." Design a plan and plant! Focus on your successes and analyze your losses to improve future outcomes. Your efforts will surely be rewarded with growing abundance.

As you continue to develop your garden applying the principles described here, you will reap growing dividends. With increased diversity, your garden will burst with flowers and fruit throughout the growing season. Water conservation and solar absorption will be enhanced as your plants grow to cover the ground and occupy more of the vertical space. The beneficial plants you incorporate will reduce or eliminate the need for additional inputs and pest and disease management. As your garden matures and reveals a life of its own, you will be filled with a sense of wonder and gratitude as your dining table—and your life—become immeasurably enriched.

ACKNOWLEDGMENTS

First, I wish to acknowledge the people whose work inspired and informed me: Martin Crawford, Steve Gabriel, Sepp Holzer, Toby Hemenway, David Jacke, Lee Reich, Eric Toensmeier, Jonathan Bates, and Stefan Sobkowiak. Without their insights and examples, my garden and this book would not have come to be. I'd like to thank my partner, David Belding, without whose initial and continuing support the garden would not have been created. I am appreciative of a host of volunteers and interns over the years who helped construct and maintain every aspect of the edible forest. And I am beholden to my builder, Dick Edgar, who created the magnificent infrastructure that makes the garden the showpiece it has become.

The help of the following people made the completion of this book possible. My friend Dr. Gretchen Starks-Martin first suggested that I write a book about my garden and encouraged me to begin it. She also named the garden the Enchanted Edible Forest. Emily Levenson was indispensable in reviewing my rough drafts and providing countless helpful suggestions as well as helping me assemble and categorize the photos and illustrations. My illustrator, the talented Zoe Chan, patiently incorporated my sequential suggestions. Turner Andrasz flawlessly executed the CAD drawings. David Veldran helped me navigate various obstacles my computer confronted me with. My sister, BethAnn Friedman, was both encouraging and meticulous in her constructive criticism of each chapter. Helen Vaughan reviewed the draft and helped compile supporting documents. Lauren Miller offered many constructive suggestions to clarify the text, and was key in helping to assemble the complete manuscript. I wish to thank Steve Diehl for his helpful suggestions to improve my photography, Vici Diehl for her knowledgeable insights into plant and insect life, and Michael Davis, PhD, who kindly reviewed the manuscript for scientific validity. Among Chelsea Green personnel, I wish to thank Margo Baldwin for believing in this book and accepting it for publication; Melissa Jacobson for creating a true work of art with her inspired page design; Patricia Stone and Alex Bullett for patiently marshalling me through the numerous phases of production, and especially my editors, Fern Marshall Bradley and Natalie Wallace, who did not let up until the contents of this book became better than I could have imagined.

APPENDIX 1

NUTRIENT ACCUMULATORS

Table 1. Dynamic Nutrient Accumulators

Common Name	Botanical Name	Nutrient Accumulated											
		N	P	K	Ca	S	Mg	Mn	Fe	Cu	Co	Zn	Si
Alfalfa	*Medicago sativa*	X							X				
Apple	*Malus* spp.			X									
Beech	*Fagus* spp.			X									
Alyssum	*Alyssum murale*				X			X				X	
Bentgrass	*Agrostis* spp.					X		X		X		X	
Birch	*Betula* spp.		X										
Borage	*Borago officinalis*			X									X
Bracken, eastern	*Pteridium aquilinum*		X	X				X	X	X	X	X	
Buckwheat	*Fagopyrum esculentum*		X	X									
Burdock	*Arctium minus*							X					
Caraway	*Carum carvi*		X										
Carrot leaves	*Daucus carota*			X			X						
Cattail	*Typha latifolia*	X											
Chamomile, corn	*Anthemis arvensis*			X	X								
Chamomile, German	*Chamomilla recutita*		X	X	X								
Chickweed	*Stellaria media*		X	X				X					
Chicory	*Cichorium intybus*			X	X								
Chives	*Allium schoenoprasum*			X	X								
Cleavers	*Galium aparine*				X								
Clovers	*Trifolium* spp.	X	X										
Clover, hop	*Medicago lupulina*	X	X										
Coltsfoot	*Tussilago farfara*			X	X	X	X		X	X			
Comfrey	*Symphytum officinale*	X		X	X		X		X				X

Table 1 (*continued*)

Common Name	Botanical Name	Nutrient Accumulated											
		N	P	K	Ca	S	Mg	Mn	Fe	Cu	Co	Zn	Si
Creosote bush	*Larrea tridentata*									X			
Dandelion	*Taraxacum vulgare*		X	X	X		X		X	X			X
Dock, broad-leaved	*Rumex obtusifolius*		X	X	X				X				
Dogwood, flowering	*Cornus florida*		X	X	X								
Duckweed	*Lemna minor*	X								X		X	
Dulse	*Palmaria palmata*				X		X		X				
Fat hen	*Atriplex hastata*			X					X				
Fennel	*Foeniculum vulgare*	X	X										
Fescue, red	*Festuca rubra*									X		X	
Flax	*Linum usitatissimum*			X				X	X				
Garlic	*Allium sativum*					X		X					
Geranium, scented	*Pelargonium* spp.							X	X	X	X	X	
Groundsel	*Senecio vulgaris*								X				
Hickory, shagbark	*Carya ovata*		X	X	X								
Horsetail	*Equisetum* spp.			X	X		X		X		X		X
Kelp	(several genera)	X			X		X		X				
Lamb's-quarters	*Chenopodium album*	X	X	X	X			X					
Lemon balm	*Melissa officinalis*		X										
Licorice	*Glycyrrhiza* spp.	X	X										
Locust, black	*Robinia pseudoacacia*	X		X	X								
Lupine	*Lupinus* spp.	X	X										
Marigold	*Tagetes* spp.		X										
Meadowsweet	*Astilbe* spp.		X		X	X	X		X				
Mosquitofern, Pacific	*Azolla filiculoides*							X		X			
Mullein, common	*Verbascum* spp.			X		X	X		X				
Mustard	*Brassica* spp.		X		X	X		X		X		X	
Nettles, stinging	*Urtica urens*	X		X	X	X			X	X			
Oak bark	*Quercus* spp.			X									
Parsley	*Petroselinum crispum*			X	X		X		X				
Pennycress, alpine	*Thlaspi caerulescens*									X		X	
Peppermint	*Mentha piperita*			X			X						
Pigweed, red root	*Amaranthus retroflexus*		X	X	X				X				
Plantain	*Plantago* spp.				X	X	X	X	X				X
Primrose	*Oenothera biennis*	X											
Purslane	*Portulaca oleracea*			X			X	X					
Rapeseed	*Brassica napus*		X		X	X		X		X		X	

Table 1 (*continued*)

Common Name	Botanical Name	Nutrient Accumulated											
		N	P	K	Ca	S	Mg	Mn	Fe	Cu	Co	Zn	Si
Salad burnet	*Poterium sanguisorba*				X	X	X		X				
Savory	*Satureja* spp.			X									
Scarlet pimpernel	*Anagallis arvensis*						X						
Shepherd's purse	*Capsella bursa-pastoris*				X	X							
Silverweed	*Potentilla anserina*			X	X					X			
Skunk cabbage	*Navarretia squanosa*						X						
Sorrel, sheep's	*Rumex acetosella*		X		X								
Sow thistle	*Sonchus arvensis*			X			X			X			
Sunflower	*Helianthus annuus*				X			X		X		X	
Strawberry	*Fragaria* spp.								X				
Tansy	*Tanacetum vulgare*			X									
Thistle, Canada	*Cirsium arvense*								X				
Thistle, creeping	*Sonchus arvensis*		X	X					X				
Thistle, nodding	*Carduus nutans*								X				
Thistle, Russian	*Salsola pestifer*								X				
Toadflax	*Linaria vulgaris*				X		X		X				
Tobacco stems/stalk	*Nicotiana* spp.	X											
Valerian	*Valeriana officinalis*												X
Vetches	*Vicia* spp.	X	X	X						X	X		
Walnut	*Juglans* spp.		X	X	X								
Watercress	*Nasturtium officinale*		X	X	X	X	X		X				
Willow	*Salix* spp.						X					X	
Yarrow	*Achillea millefolium*	X	X	X						X			

Sources: Hemenway, Toby. *Gaia's Garden: A Guide to Home-Scale Permaculture.* 2nd ed. Chelsea Green, 2009; Cocannouer, Joseph. *Weeds: Guardians of the Soil.* Devin-Adair, 1976; Famulari, Stevie. University of New Mexico. Unpublished; Jacke, David, and Eric Toensmeier. *Edible Forest Gardens.* Chelsea Green, 2005; Kourik, Robert. *Designing and Maintaining Your Edible Landscape—Naturally.* Metamorphic, 1984; Ehrenfried Pfeiffer. *Weeds and What They Tell.* Biodynamic Farming and Gardening, 1970.

APPENDIX 2

FLOWERING AND HARVEST TIMES

The following tables, which indicate the range of flowering and harvest times for plants described in this book, reflect my experience over several years in my Zone 4 garden. Times will vary depending on your local climate conditions and variations in the weather from year to year.

Table 2. Flowering Times

	Jan	Feb	Mar	Apr	May	Jun	Jul	Aug	Sept	Oct	Nov	Dec
Akebia vine					X	X						
Almond					X							
Alpine strawberry					X	X	X	X	X	X	X	
American basswood					X	X						
Anise hyssop							X	X	X	X		
Apple					X							
Apple mint							X	X	X	X		
Apricot				X	X							
Arctic kiwi						X						
Aronia					X	X						
Asian pear					X							
Autumn olive					X	X						
Beach plum					X	X						
Bellflower						X	X	X				
Bird's-foot trefoil						X	X	X	X			
Blackberry							X					
Black currant					X							
Black locust					X	X						
Black locust (Purple Robe)					X	X						

Note: Times are based on the author's experience in her Zone 4 garden.

Table 2 (*continued*)

	Jan	Feb	Mar	Apr	May	Jun	Jul	Aug	Sept	Oct	Nov	Dec
Black raspberry						X						
Blue lupine					X	X						
Blue vervain							X	X	X			
Borage						X	X	X	X	X	X	
Buffaloberry				X	X							
Bush cherry					X							
Chamomile					X	X	X	X	X	X		
Chinese yam							X					
Chive					X	X	X	X	X	X		
Clove currant					X							
Coneflower							X	X	X			
Cornelian cherry				X								
Crab apple					X							
Crocus			X	X								
Daffodil				X	X							
Dandelion				X	X	X	X	X	X	X	X	
Daylily						X	X	X	X	X	X	
Elderberry						X	X	X				
English daisy					X	X	X	X	X	X	X	
False indigo						X	X					
Garden strawberry (Gasana F1)					X	X	X	X	X	X	X	
Garden strawberry (Mara des Bois)				X	X	X	X	X	X	X	X	
Garden strawberry (Pineberry)					X							
Garden strawberry (Tarpan F1)					X	X	X	X	X	X	X	
Garlic chive								X	X			
Giant Solomon's seal					X	X						
Gloriosa daisy							X	X	X			
Goji berry					X	X	X	X	X	X	X	
Goldenrod								X	X	X	X	
Gooseberry				X	X							
Goumi					X							
Grape					X							
Ground cover raspberry					X	X						
Ground ivy				X	X	X						

Note: Times are based on the author's experience in her Zone 4 garden.

Table 2 (*continued*)

	Jan	Feb	Mar	Apr	May	Jun	Jul	Aug	Sept	Oct	Nov	Dec
Groundnut									X			
Hairy vetch					X	X	X	X	X			
Hazelbert			X	X								
Highbush cranberry					X	X						
Hollyhock							X	X	X	X		
Honeyberry				X	X							
Honey locust					X	X						
Hops						X	X	X				
Hosta							X					
Houttuynia						X	X	X				
Hyacinth bean							X	X	X	X		
Jostaberry					X							
Juneberry				X	X							
Lavender						X	X	X	X	X		
Lemon balm						X	X	X	X	X		
Lilac					X							
Marshmallow							X	X	X			
Maximilian sunflower								X	X	X		
Milk vetch							X	X	X			
Mountain ash					X	X						
Nanking cherry				X	X							
Nasturtium						X	X	X	X			
New Jersey tea							X					
Oregano						X	X	X	X	X		
Pawpaw					X	X						
Pear					X							
Peony						X	X					
Perennial sweet pea						X	X	X	X	X		
Persimmon						X	X					
Plum					X							
Queen Anne's lace							X	X	X	X	X	
Quince					X	X						
Red and pink currant					X							
Redbud					X							
Red raspberry (Nova)						X			X	X		

Note: Times are based on the author's experience in her Zone 4 garden.

Table 2 *(continued)*

	Jan	Feb	Mar	Apr	May	Jun	Jul	Aug	Sept	Oct	Nov	Dec
Red raspberry (Prelude)					X	X		X	X			
Rhubarb					X	X						
Rugosa rose					X	X	X	X	X	X		
Russian comfrey					X	X	X	X	X	X		
Saffron crocus										X		
Sand cherry					X							
Sea buckthorn					X							
Sea kale						X						
Shipova					X							
Siberian peashrub					X							
Sorrel						X						
Sour cherry					X							
Speckled alder			X	X								
Sunchoke										X		
Sweet cherry					X							
Sweet cicely					X	X						
Sweet William						X	X	X	X	X		
Sweet yellow clover						X	X	X	X	X		
Thyme					X	X	X	X	X			
Tulip					X							
Turkish rocket					X	X	X	X				
Valerian						X	X					
Violet				X	X				X	X		
Water celery							X	X				
White birch			X	X	X							
White clover						X	X	X	X	X		
Wild aster								X	X	X		
Wild strawberry				X	X							
Wisteria					X	X						
Witch hazel	X										X	X
Yarrow						X	X	X	X	X		

Note: Times are based on the author's experience in her Zone 4 garden.

Table 3. Harvest Times

	Jan	Feb	Mar	Apr	May	Jun	Jul	Aug	Sept	Oct	Nov	Dec
Alpine strawberry						X	X	X	X	X	X	
American plum								X	X			
Anise hyssop				X	X	X	X	X	X	X	X	
Apple mint				X	X	X	X	X	X	X	X	
Arctic kiwi								X				
Aronia								X	X	X		
Asian pear								X	X	X		
Asparagus					X	X						
Autumn olive								X	X	X	X	
Beach plum								X	X			
Blackberry								X	X			
Black currant							X	X				
Black raspberry							X					
Bush cherry								X				
Chamomile						X	X	X	X	X		
Cherry						X	X	X				
Chinese yam								X				
Chive				X	X	X	X	X	X	X	X	
Clove currant								X				
Cornelian cherry								X				
Crab apple									X	X		
Egyptian onion				X	X	X		X	X	X	X	
Elderberry								X	X			
Garden strawberry (day-neutral)						X	X	X	X	X	X	
Garden strawberry (Pineberry)						X						
Garlic chive				X	X	X	X	X	X	X	X	X
Giant Solomon's seal					X							
Goji berry									X	X	X	X
Good King Henry				X	X	X	X	X	X	X	X	
Gooseberry						X	X					
Goumi						X	X					
Grape								X	X			
Groundnut			X	X							X	
Hazelbert								X	X			

Note: Times are based on the author's experience in her Zone 4 garden.

Table 3 (*continued*)

	Jan	Feb	Mar	Apr	May	Jun	Jul	Aug	Sept	Oct	Nov	Dec
Honeyberry						X	X					
Hosta					X							
Houttuynia					X	X	X	X	X			
Jostaberry						X	X					
Juneberry						X						
Lavender						X	X	X	X	X	X	X
Mint					X	X	X	X	X	X	X	
Mountain ash								X	X			
Mulberry						X	X	X	X			
Nanking cherry						X	X	X				
Northern bayberry						X	X	X	X	X	X	
Oregano			X	X	X	X	X	X	X	X	X	
Oyster mushroom		X	X	X	X			X	X	X		
Peach								X	X			
Pear								X	X	X		
Pink currant						X	X					
Plum							X	X	X			
Quince										X	X	
Red currant						X	X	X				
Red raspberry						X	X	X	X	X	X	
Rhubarb					X	X						
Saffron crocus										X	X	
Sage					X	X	X	X	X	X	X	X
Sand cherry							X	X				
Sea buckthorn								X				
Shiitake mushroom				X	X	X			X	X		
Sorrel				X	X	X	X	X	X	X	X	
Sunchoke			X	X								X
Sweet cicely				X	X	X	X	X	X	X	X	
Thyme				X	X	X	X	X	X	X	X	
Turkish rocket					X	X	X	X	X	X	X	
Water celery				X	X	X	X	X	X	X	X	
Wild fennel							X	X	X	X		
Wild strawberry						X						
Wine cap mushroom					X				X	X		

Note: Times are based on the author's experience in her Zone 4 garden.

APPENDIX 3

STEWARDSHIP THROUGH THE SEASONS

The following lists reflect tasks that I undertake during various seasons of the year to manage and maintain my Zone 4 garden. Some garden tasks can be completed at any time of year, weather permitting, and I have not listed those below. Such tasks include staking out new beds and paths, sheet mulching, replenishing wood chip mulch, spot weeding, building hügelkultur mounds, checking and repairing fencing, and readjusting trunk guards and protective cages.

Early winter

- Scout for and remove snow from trunk guards.
- Clean out old nests from birdhouses; put fresh wood shavings in bottom of chickadee houses.
- Transplant woody herb seedlings into individual containers.
- Order trees/plants/seeds for spring planting.

Mid-Winter

- Scout for and remove snow from trunk guards.
- Check for deer hoofprints inside garden and repair or rebait fence as necessary.
- Harvest seeds from plants that hold their seeds through winter and may require stratification, such as New Jersey tea, garlic chives, black locusts.
- Plant chive and garlic chive seeds in flats.
- In event of thaw, check for mushrooms fruiting.

Late Winter

- Remove cages around trees that have grown taller than deer can reach.
- Remove screen trunk guards
- Prune berry bushes and nitrogen-fixing shrubs.
- Prune and thin raspberries and blackberries and add new mulch.
- Layer berry bushes to propagate.
- Place dormant wood cuttings in pots or in ground outdoors to start new plants.
- Trim dead tops of woody-stemmed herbs such as oregano, thyme, lavender.
- Stomp down dried stems of herbaceous plants such as coneflowers, anise hyssop.
- Frost-seed Dutch white clover on paths.
- Install new birdhouses for migrating birds.
- Clean out kestrel and owl houses.
- Check for mushrooms fruiting.
- Order mushroom spawn.

- Remove mulch from top of strawberries and saffron crocus bed and cover with row cover.
- Harvest mushrooms, root vegetables, first herbs.

Early Spring

- Transplant dormant young trees from nursery to garden as soon as ground thaws.
- Build and install mushroom totems.
- Check for mushrooms fruiting.
- Sow perennial vegetable and flower seeds in flats.
- Continue pruning trees and shrubs.
- Replenish crusher run over culverts.
- Rebait electric fence with peanut butter.
- Plant bare-root trees and bushes.
- Plant dormant bare-root herbaceous plants such as strawberries, daylilies, and pineberries.
- Put row cover over tulips to deter rabbits and deer.
- Plant root cuttings from herbaceous perennials such as rhubarb, comfrey, groundnuts, sunchokes.
- Plant out cold hardy herbaceous perennial sets such as chives and garlic chives.
- Prune grapes.
- Transplant bushes and trees.
- Dig out thistles.
- Stake trees to straighten as needed.
- Tie up young grapevines to direct growth.
- Harvest herbs, flowers, mushrooms, perennial vegetables.

Mid-Spring

- Mow paths and other grassy areas as needed.
- Complete pruning grapes.
- Scout for mushrooms fruiting.
- Scatter flower seeds in established beds.
- Plant out herbs, flowers, perennial vegetables grown from seed.
- Plant annual vegetable seeds and sets in new hügelkultur mounds.
- Plant herbaceous plants as well as bare-root and potted trees, bushes, and vines received from nurseries.
- Transplant clumps of self-seeded ground covers such as violets.
- Pot up or transplant layered and suckered seedlings marked with stakes in fall.
- Pot up rooted cuttings.
- Dig out more thistles.
- Scout for dead trees using the scratch test.
- Prune out dead wood and stems on trees and bushes.
- Harvest herbs, flowers, perennial vegetables, mushrooms.

Late Spring

- Plant leafed-out nursery stock and herbs and flowers started from seed.
- Clear brush growing under and around fences.
- Scout for mushrooms fruiting.
- Scout for dead trees and plants—order replacements.
- Net honeyberries.
- Tie up peonies.
- Trellis blackberries.
- Cut comfrey, and drop cut tops around base of trees.
- Mow grass.
- Spread bird netting over red currants and gooseberries.
- Harvest herbs, flowers, perennial vegetables, berries, mushrooms.
- Freeze strawberries.

Early Summer

- Net sweet cherries.
- Water newly planted herbaceous plants as needed.
- Deadhead valerian, herbs, weeds.
- Thin tree fruits such as pears, peaches, Asian pears, apples.
- Prune tips of black raspberries and blackberries.
- Mow paths.
- Trim under fence lines.
- Water fruiting plants such as raspberries if needed.
- Harvest herbs, flowers, berries, perennial vegetables, fruits.

- Make gooseberry, raspberry, currant, and cherry preserves.
- Freeze honeyberries, red and pink currants, gooseberries.

Mid-Summer

- Plant cool-season annual vegetable seeds and cover crops on new hügelkultur mounds.
- Cut and drop comfrey.
- Prune out spent summer raspberry canes.
- Clear brush growing under and around fences.
- Plant out remaining herbaceous plants and herbs started from seed in spring.
- Mow between orchard rows and around beds.
- Reseed paths with Dutch white clover.
- Water plants such as blackberries if needed.
- Cut back flower heads on marshmallow and other herbs.
- Protect grapes with netting.
- Harvest herbs, flowers, berries, fruits, perennial vegetables.
- Make black currant jam.

Late Summer

- Plant oat cover crop on new mounds and sheet-mulched beds.
- Mow grass.
- Scout for mushrooms fruiting.
- Plant saffron crocuses, weed any established bed, and cover with row cover to protect from rabbits.
- Cover day-neutral strawberries with row cover before frosts.
- Order replacement trees for next year.
- Harvest herbs, flowers, perennial vegetables, fruits, nuts, berries, grapes, mushrooms.
- Make beach plum, American plum, grape, and aronia jam.
- Freeze raspberries.
- Make grape juice.
- Dry herbs.

Early Fall

- Transplant bushes, trees to new spots.
- Plant spring bulbs such as crocuses, daffodils, tulips.
- Scout for mushrooms fruiting.
- Clear brush growing under and around fence.
- Replenish chips around base of tender trees such as Asian pears.
- Place screen guards around bushes.
- Prune out spent summer raspberry and blackberry canes.
- Layer berry bushes to propagate.
- Order plants for spring.
- Scatter some flower seeds.
- Paint fruit tree trunks white.
- Harvest herbs, flowers, perennial vegetables, fruits, nuts, berries, mushrooms.
- Make grape and raspberry jam.
- Freeze goji berries and autumn olive berries.

Mid-Fall

- Continue painting trunks.
- Continue installing screen guards around bushes.
- Cut down all fall raspberry canes.
- Mark location of new suckers with bamboo stakes.
- Transplant groundnuts and sunchokes.
- Replenish mulch around rhubarb, peonies, blackberries.
- Plan for and order plants for spring.
- Harvest herbs, flowers, perennial vegetables, berries, fruits, nuts.
- Freeze berries.

Late Fall

- Rebait electric fence with peanut butter.
- Scout for and remove snow from over trunk guards.
- Check for deer incursion and repair fence if necessary.
- Apply leaf mulch over strawberries and saffron crocuses.
- Clean out birdhouses.
- Plant woody herb seeds in flats indoors, such as thyme, oregano, lavender, sage.
- Plan for and order plants, seeds for spring.

GLOSSARY

Aromatic pest confuser: A plant with a strong odor that is thought to distract pests.

Bare-root: A plant in a dormant state, with soil removed from the roots.

Beneficial attractor: A plant that attracts pollinators as well as predators and parasites of plant pests, or one that provides habitat for helpful insects, animals, and birds.

Biennial: A plant that dies after flowering in its second year of growth.

Bramble: A genus (*Rubus*) of prickly shrubs, including raspberries, blackberries, and black raspberries.

Bulbil: A tiny bulb-like structure that, if planted, can give rise to a new plant.

Cane: A long, thin, flexible stem or branch such as those produced by brambles and grapevines.

Clumper: A plant that forms a mound as it spreads outward from the center.

Clumping: Having a growth habit wherein multiple trunks or stems arise close together.

Coppice: To cut a tree close to ground level to stimulate production of new shoots.

Cordon: A permanent portion of a grapevine trimmed of all lateral branches and trained to grow horizontally.

Cross-pollinate: To transfer pollen from the flowers of one plant to the flowers of another plant.

Cultivar: A cultivated plant variety.

Culvert: A construct that channels water.

Deciduous: Falling off or shed at a particular season.

Dew point: The temperature at which water begins to condense out of the air.

Dormancy: The state wherein a plant ceases to grow and slows its metabolism as a survival strategy during winter or times of drought.

Drip line: An irrigation hose with small holes along its length that allow water to slowly drip out; the circumference at ground level that corresponds to the farthest extent of a tree's canopy.

Floating row cover: A type of synthetic fabric that is permeable to air and water; when secured over plants, it can protect them from pests and affords a few degrees of frost protection.

Floricane: A bramble cane in its second year; also, a variety of bramble that produces its crop the second year.

Forb: A broad-leaved, herbaceous flowering plant, as distinguished from a grass.

Frost seeding: A seeding technique in which seeds are sown over bare ground during a time of year when frosts occur.

Germinate: To sprout; to begin to grow from seed.

Girdling: The removal of bark from the entire circumference of a woody trunk or stem.

Grafting: A technique for joining together the woody or green tissue of two different varieties or species of plants.

Graft union: The spot at which tissues of two different varieties or species of plants are joined together; in fruit trees, this is usually located near the base of the trunk.

Hardwood: A deciduous tree that yields hardwood, as distinguished from a softwood (evergreen) tree.

Hardwood cutting: A leafless piece of a woody stem or branch used to propagate a new plant.

Hedgerow: A hedge of shrubs and trees typically bordering a field or road.

Herbaceous: Having succulent stems that die back in autumn.

Hügelkultur mound: A mound built up from a base layer of logs or branches and covered with other organic materials.

Hybrid: An offspring of two parents of different species, subspecies, or varieties.

Interplant: To plant one type of plant alternating with or among other types.

Layering: A propagation method in which a plant stem is anchored to the ground to encourage the formation of new roots.

Microclimate: The environmental conditions of a small area that differ from the surrounding area with respect to heat, shade, moisture, wind, and so on.

Monoculture: An area in which only a single species of plant grows.

Mulch plant: An herbaceous plant whose growth habit produces copious amounts of leaves and stems, which enrich the soil as they decay.

Nitrogen fixer: A plant that can convert atmospheric nitrogen to a form usable by plants.

Nutrient accumulator: A plant that concentrates one or more elements in its leaves and stems.

Overstory: The top layer of a forest garden that includes trees at least 30 feet (9.1 m) tall.

Permaculture: An agricultural system mirroring nature, designed to be sustainable and self-sufficient.

Photosynthesis: The process by which plants use the energy of sunlight to fuel the production of simple sugars from carbon dioxide and water.

Pollard: To prune a tree heavily, leaving just the trunk and branch stumps.

Pome fruit: A fruit such as an apple comprising a core containing a few seeds enclosed in a tough membrane.

Primocane: A bramble cane in its first year; also, a variety of bramble that bears fruit in its first year.

Propagation: The process of creating a new plant using seeds, cuttings, or other plant parts.

Rhizome: A horizontal underground root stem that at intervals sprouts roots and new plants.

Root layer: The vertical space below the soil surface occupied by roots.

Sanitation: The practice of removing spent fruit and dead and diseased plant tissue to prevent disease and pest infestations.

Scarification: The process of abrading the hard outer covering on a seed to allow water to penetrate.

Self-fertile: Able to use its own pollen to form fruit.

Self-pollinating: Self-fertile.

Shrub layer: The layer of vegetation in a forest garden that extends from 3 to 12 feet (1–3.7 m) above ground level, comprising woody-stemmed bushes and shrubs.

Specimen: In landscape design, a plant with a unique appearance placed apart from other plants so its features can be appreciated.

Stolon: A horizontal stem that extends just over or under the ground surface and is capable of rooting at nodes as it grows; also called a runner.

Stone fruit: A fruit such as a peach that has flesh surrounding a single pit.

Stratification: The practice of exposing a seed to a period of cold temperature to induce germination.

Subsoil: The less-fertile layer of soil underlying the topsoil layer.

Sucker: A shoot that develops from a root of a plant or from the lower portion of a plant stem.

Tipping: Clipping off the top of a bramble cane or the tip of a shoot to promote growth below.

Tree canopy: The upper layer of leaves of a group of trees; also, the total volume of the leaves of a single tree.

Tuber: An enlarged, fleshy underground stem, such as a potato.

Umbel: A cluster of flowers that branches out from a single stem in a shape like an umbrella.

Understory: The vertical layer of woody plants in a forest garden that extend from 10 to 30 feet (3–9.1 m) above the ground surface.

Vine layer: The plant layer in a forest garden composed of vines.

Weir: A dam installed to raise the level of water uphill from it.

RESOURCES

Arbor Day Foundation

Arborday.org

Plants can be ordered individually or in bulk through this nonprofit organization's website; offers a user-friendly "tree finder" to help you identify plants compatible with your hardiness zone, soil type, and habitats.

Burnt Ridge Nursery & Orchards

Burntridgenursery.com

Located in Washington State; has a large collection of fruit and nut trees and bushes including some hard-to-find cold-hardy cultivars.

Cold Stream Farm

Coldstreamfarm.net

Located in west-central Michigan; offers a large variety of bare-root woody plants, including many natives. Reasonable pricing, both wholesale and retail, with choices of size and quantity.

Cornell Small Farms Program

Smallfarms.cornell.edu

An abundance of resources on farming techniques, including online courses, videos, and guides.

Fedco Seeds

Fedcoseeds.com

Located in central Maine; a source of seeds and bare-root nursery plants. Informative and beautifully illustrated catalogs.

Field & Forest Products

Fieldforest.net

Located in central Wisconsin; offers multiple varieties of certified organic shiitake and oyster mushroom spawn and that of other less common varieties. Knowledgeable and accessible customer service.

The Garden Conservancy

Gardenconservancy.org

A national not-for-profit organization promoting gardens; my garden participated in an "Open Day" event they sponsored.

Hartmann's Plant Company

Hartmannsplantcompany.com

Located in southwest Michigan; offers a large variety of hardy edible shrubs grown in pots and priced very reasonably when purchased in quantity.

Johnny's Selected Seeds

Johnnyseeds.com

Located in central Maine; a high-quality seed source where I sourced the seeds for strawberries with pink and rose-colored flowers.

Jung Seeds & Plants

Jungseed.com

Located in south-central Wisconsin; offers good-quality woody plants with some unusual varieties.

Nourse Farms

Noursefarms.com

Located in western Massachusetts; specializes in high-quality berry plants; also offers perennial vegetables.

Oikos Tree Crops

Oikostreecrops.com

Located in southwest Michigan; a unique selection of woody and herbaceous food plants, many of which are quite cold-hardy and vigorous. In 2021, discontinued operating as a nursery but still sells seeds.

Prairie Moon Nursery

Prairiemoon.com

Located in southeast Minnesota; offers an extensive selection of native seeds and plants, some of which are provided in mixes matched to habitats.

Raintree Nursery

Raintreenursery.com

Located in Washington State; offers an extensive collection of fruit, nut, and berry plants with many rare varieties.

Richters Herbs

Richters.com

Located just north of Toronto; offers a multitude of herb plants. International shipping available. Before ordering, make sure the plants you select are compatible with your hardiness zone.

Roco Saffron

Rocosaffron.com

Located in the Netherlands; a source of quality saffron corms in small and large quantities.

St. Lawrence Nurseries

Stlawrencenurseries.com

Located in northern New York State; offers a selection of cold-hardy fruit trees and berry plants.

Twisted Tree Farm

Twisted-tree.net

Located in south-central New York State; specializes in easy-to-grow food plants grown with no artificial substances.

USDA Natural Resources Conservation Service

plants.usda.gov

Extensive searchable database listing all plants in the United States and its territories, with detailed descriptions of each.

WeedGuardPlus

Weedguardplus.com

A source of OMRI-listed biodegradable paper mulch.

Wild Farm Alliance

Wildfarmalliance.org

Located in California; nonprofit offering useful information on how to create habitat to attract wild birds that help control garden insect pests.

SELECTED BIBLIOGRAPHY

Crawford, Martin. *Creating a Forest Garden: Working with Nature to Grow Edible Crops*. Totnes, U.K.: Green Books, 2010. Martin Crawford created and heads a research institute in southwest England with a 2-acre (0.8 ha) edible forest garden. This user-friendly exhaustive manual on how to design, plant, and care for an edible forest garden is one of the reference "bibles" I used to develop my garden.

Crawford, Martin. *How to Grow Perennial Vegetables: Low-Maintenance, Low-Impact Vegetable Gardening*. Totnes, U.K.: Green Books, 2012. Descriptions of how to cultivate, use, and maintain over 250 varieties of edible perennial plants, many of which can grow in cold climates.

Crawford, Martin. *Trees for Gardens, Orchards, and Permaculture*. Hampshire, U.K.: Permanent Publications, 2015. Extensive descriptions of nitrogen-fixing and food-producing trees, including uses and problems. Also includes a helpful section classifying trees based on climate and hardiness zones.

Crawford, Martin, and Caroline Aitken. *Food from Your Forest Garden: How to Harvest, Cook and Preserve Your Forest Garden Produce*. Cambridge, U.K.: Green Books, 2013. Detailed instructions for harvesting, cooking, and preserving nearly 250 kinds of fruits, nuts, and vegetables that can be grown in a perennial planting, including helpful recipes.

Fukuoka, Masanobu. *The Natural Way of Farming: The Theory and Practice of Green Philosophy*. Translated by Frederic P. Metreaud. Madras, India: Bookventure, 1985. Fukuoka's treatise on working with rather than against nature to produce year-round resilient crops with no tillage and minimal labor.

Fukuoka, Masanobu. *The One Straw Revolution: An Introduction to Natural Farming*. Translated by Larry Korn, Chris Pearce, and Tsune Kurosawa. New York: New York Review of Books, 1978. The classic summary of Fukuoka's philosophy and practice of no-till organic farming in Japan. This book inspires creative ways to solve no-till dilemmas in annual vegetables.

Hemenway, Toby. *Gaia's Garden: A Guide to Home-Scale Permaculture*. 2nd ed. White River Junction, VT: Chelsea Green, 2000. This classic text, another of my "bibles," makes permaculture principles accessible to the home gardener. Detailed examples of the application of these principles in arid and mild temperate climates.

Hill, Lewis. *Secrets of Plant Propagation: Starting Your Own Flowers, Vegetables, Fruits, Berries, Shrubs, Trees, and Houseplants*. Pownal, VT: Storey Communications, 1985. A useful reference for step-by-step propagation techniques, including by seeds, division, layering, cuttings, and grafting.

Holzer, Sepp. *Sepp Holzer's Permaculture: A Practical Guide to Small-Scale, Integrative Farming and Gardening*. White River Junction, VT: Chelsea Green, 2004. Sepp Holzer's large and highly diversified farm is in the Austrian Alps, where he has implemented permaculture principles to great advantage. This description of how he developed his farm includes many widely applicable, practical tips for any size of planting.

Jacke, Dave, and Eric Toensmeier. *Edible Forest Gardens, Vol. 2: Ecological Vision and Theory for Temperate Climate Permaculture*. White River Junction, VT: Chelsea Green, 2005. The second volume of a comprehensive work delineating permaculture principles as they apply in a temperate climate. Dense and technical in some sections; I rely on this book for the numerous tables in the appendix listing characteristics of over 600 plants as well as those of beneficial creatures.

Kelsey, Anni. *Edible Perennial Gardening: Growing Successful Polycultures in Small Spaces*. Hampshire, U.K.: Permanent Publications, 2014. Using a permaculture approach, this slim volume offers examples of herbaceous plant groupings, including herbs and perennial vegetables for a small garden plot.

Kindred, Glennie. *Walking with Trees*. Hampshire, U.K.: Permanent Publications, 2019. Lyrical descriptions of 13 common trees along with their historical significance and present-day uses with emphasis on medicinal applications. Passages describing the author's personal connection with trees and the natural world are transformative.

MacKentley, Diana, and Bill MacKentley. *St. Lawrence Nurseries Planting Guide: Planting and Care of Northern Climate Trees & Shrubs*. Potsdam, NY: St. Lawrence Nurseries. Useful instructions for planting and caring for edible trees and shrubs.

McNair, James K. *All about Herbs*. San Ramon, CA: Ortho Books, 1973. A user-friendly guide to growing, planting, propagating, preserving, and using easy-to-grow herbs.

Mollison, Bill, and Reny Mia Slay. *Introduction to Permaculture, Second Edition*. Tasmania, Australia: Tagari Publications, 2011. This classic work explicates permaculture principles as they apply to arid, tropical, and temperate climates. Many practical examples described in detail are universally relevant.

Mudge, Ken, and Steve Gabriel. *Farming the Woods: An Integrated Permaculture Approach to Growing Food and Medicinals in Temperate Forests*. White River Junction, VT: Chelsea Green, 2014. Steve Gabriel is the teacher who first inspired me to plant an edible forest. This handy reference covers practices for cultivating food, medicine, mushrooms, and wood products inside the woods.

Phillips, Michael. *The Holistic Orchard: Tree Fruits and Berries the Biological Way*. White River Junction, VT: Chelsea Green, 2011. A comprehensive reference work for understanding techniques for planning, planting, propagating, and nourishing tree fruits and berries in an orchard setting. Includes detailed information on preventing, diagnosing, and remedying pest and disease problems.

Phillips, Michael. *Mycorrhizal Planet: How Symbiotic Fungi Work with Roots to Support Plant Health and Build Soil Fertility*. White River Junction, VT: Chelsea Green, 2017. Technical discussion of the indispensable role multiple types of mycorrhizal fungi play in supporting soil fertility and plant health, and how the gardener can support and nourish these invaluable helpers.

Reich, Lee. *Landscaping with Fruit: A Homeowner's Guide*. North Adams, MA: Storey, 2009. Beautifully illustrated; this was my inspiration to grow unfamiliar fruits like the shipova in my edible garden. Includes simple landscape designs plus a catalog of tree fruits and berries and easy-to-follow instructions on pruning these plants for best production.

Silver, Akiva. *Trees of Power: Ten Essential Arboreal Allies*. White River Junction, VT: Chelsea Green, 2019. Silver describes 10 of his favorite trees in detail, cataloging their aesthetic and useful properties. Techniques for planting and propagating trees are also included.

Toensmeier, Eric, and Jonathan Bates. *Paradise Lot: Two Plant Geeks, One-Tenth of an Acre, and the Making of an Edible Garden Oasis in the City*. White River Junction, VT: Chelsea Green, 2007. This description of the integration of permaculture principles in designing and planting an edible forest garden on a small city lot in Holyoke, Massachusetts, helped me conceive the possibilities for my own edible forest garden.

Wohlleben, Peter. *The Hidden Life of Trees: What They Feel, How They Communicate*. Vancouver, Canada: David Suzuki Institute, 2015. Evocative descriptions of the ways in which trees evince a unique intelligence as they communicate with one another for mutual benefit. Wohlleben's admiration and love for trees is contagious.

INDEX

Page numbers in *italics* indicate figures and illustrations. Page numbers followed by *t* indicate tables.

ABOUT THE AUTHOR

Larry Asam

Dani Baker is a retired clinical psychologist and a self-taught gardener and photographer who learned her craft by immersing herself in reading, poring over nursery catalogs, attending workshops on permaculture and gardening, and enthusiastic trial-and-error experimentation. Her Enchanted Edible Forest garden is located at Cross Island Farms on Wellesley Island in the St. Lawrence River between New York and Canada. There, she and her partner, David Belding, grow certified organic produce and raise grass-fed beef and goats. Dani conducts workshops and tours in her edible forest garden as well as giving presentations at organic farming conferences and other venues. She takes particular pleasure in inspiring others to try their hand at incorporating permaculture principles in their edible gardens.